Engineering Tables and Data

Published by Chapman & Hall, 2-6 Boundary Row, London SE1 8HN, UK

Chapman & Hall, 2-6 Boundary Row, London SE1 8HN, UK

Blackie Academic & Professional, Wester Cleddens Road, Bishopbriggs, Glasgow G64 2NZ, UK

Chapman & Hall GmbH, Pappelallee 3, 69469 Weinheim, Germany

Chapman & Hall USA., One Penn Plaza, 41st Floor, New York, NY10119, USA

Chapman & Hall Japan, ITP - Japan, Kyowa Building, 3F, 2-2-1 Hirakawacho, Chiyoda-ku, Tokyo 102, Japan

Chapman & Hall Australia, Thomas Nelson Australia, 102 Dodds Street, South Melbourne, Victoria 3205, Australia

Chapman & Hall India, R. Seshadri, 32 Second Main Road, CIT East, Madras 600 035, India

First edition 1972
Second edition 1991
Reprinted 1994

Typeset in 10/12 Times by Thomson Press (India) Ltd, New Dehli
Printed and bound in Singapore by Kin Keong Co. PTE. Ltd.

ISBN 0 412 38970 3

A Catalogue record for this book is available from the British Library

Howatson, A. M.
Engineering tables and data.-2nd. ed.
I. Title II. Lund, P. G. III. Todd, J. D.
345.645034

ISBN 0-412-38970-3

Library of Congress Cataloging-in-Publication Data

Howatson, A. M.
Engineering tables and data / A. M. Howatson, P. G. Lund, J. D. Todd
-New ed.
p. cm.
Includes bibliographical references.
ISBN 0-442-31368-3
1. Engineering-Tables. I. Lund, P.G. (Peter Gradwell)
II. Todd, J. D. (Joseph Derwent) III. Title.
TA151.H68 1991
620'.00212-dc20 91-6296
CIP

Engineering Tables and Data

A. M. HOWATSON
P. G. LUND
J. D. TODD

*Department of Engineering Science,
University of Oxford*

CHAPMAN & HALL
London · Glasgow · Weinheim · New York · Tokyo · Melbourne · Madras

Preface

In producing this new edition our purpose has been the same as for the first, namely to provide a comprehensive collection of information covering all branches of engineering. We have aimed, as before, to make it helpful to practising engineers, in matters outside their immediate expertise, as well as to students of the various engineering disciplines.

For this edition a number of changes have been made. Some material, made redundant by two decades of developing technology, has been excised. New material has been added not only to incorporate information suggested by hindsight but also to reflect the march of progress.

We gladly acknowledge our gratitude to Drs B. Derby and C. R. M. Grovenor of the Department of Materials at Oxford, who very kindly provided revised and extended data for large parts of the section *Properties of Matter*.

The tables on pages 31–35 are reproduced from *Elementary Statistical Tables*: Lindley and Scott, by permission of Cambridge University Press.

The graphs on page 46 are based on data published in *Handbook of Optical Constants of Solids*: Palik, by permission of Academic Press.

The graph on page 47 was provided by the Cable Products Division of STC Telecommunications, whose permission to publish it is gratefully acknowledged.

The properties of water and steam tabulated on pages 48–65 inclusive are based on *U.K. Steam Tables in S.I. Units* published by Edward Arnold.

The tables on pages 66 and 67, the upper table on page 68 and the thermochemical data on pages 70 and 71 inclusive are reproduced from *Thermodynamic Tables*: Haywood by permission of Cambridge University Press.

The chart on page 77 is reproduced from *Engineering Thermodynamics Work and Heat Transfer*: Rogers and Mayhew, by permission of Longman.

Tables 1–3 on pages 78–87 inclusive and the charts on pages 106 and 107 are reproduced from *Elements of Gas Dynamics*: Liepmann and Roshko, by permission of John Wiley.

Tables 4 and 5 on pages 87–105 inclusive are reproduced from *Introduction to Gas Dynamics*: Rotty, by permission of John Wiley.

The tables on pages 119–137 inclusive are reproduced from the publication *Structural Sections*, by permission of British Steel General Steels.

The graphs on pages 141–143 are reproduced from *Theory of Vibration with Applications*, William T. Thomson, 3rd edition, 1988, pp. 53, 56, 64. Reprinted by permission of Prentice-Hall, Inc., Englewood Cliffs, NJ.

The diagrams on page 152 are reproduced from *Waves*: Connor, by permission of Edward Arnold.

The diagrams on page 154 are reproduced from *Electronic Engineer's Reference Book*: Mazda, by permission of Butterworths.

The graphs on pages 161 and 162 are reproduced from *Electronic Circuits and Systems*: King, by permission of Thomas Nelson.

We are grateful to these companies and authors for their collaboration. It is difficult adequately to express thanks to our colleagues in the Department of Engineering Science at Oxford: many have helped us, with hard information as well as wise advice, and some to a very considerable extent; but these very ones would be the last to wish to be singled out. We therefore offer only this acknowledgement, no less sincere for its anonymity.

As before, we shall be grateful to hear of those mistakes which, inevitably, will have escaped our notice up to this point.

A.M.H.
P.G.L.
J.D.T.

Oxford 1991.

Contents

Preface

General

Mathematics

Properties of matter

Thermodynamics and fluid mechanics

Solid mechanics and structures

General

The Greek alphabet

α	A	alpha
β	B	beta
γ	Γ	gamma
δ	Δ	delta
ϵ	E	epsilon
ζ	Z	zeta
η	H	eta
θ	Θ	theta
ι	I	iota
κ	K	kappa
λ	Λ	lambda
μ	M	mu
ν	N	nu
ξ	Ξ	xi
o	O	omicron
π	Π	pi
ρ	P	rho
σ, ς	Σ	sigma
τ	T	tau
υ	Υ	upsilon
ϕ	Φ	phi
χ	X	chi
ψ	Ψ	psi
ω	Ω	omega

SI units

Quantity	*Unit*	*Symbol*	*Equivalent*
Base-units			
Length	metre	m	–
Mass	kilogramme	kg	–
Time	second	s	–
Electric current	ampere	A	–
Temperature	kelvin	K	–
Luminous intensity	candela	cd	–
Amount of substance	mole	mol	–
Supplementary units			
Plane angle	radian	rad	–
Solid angle	steradian	sr	–
Derived units			
Frequency	hertz	Hz	s^{-1}
Force	newton	N	$kg\,m\,s^{-2}$
Pressure, stress	pascal	Pa	$N\,m^{-2}$
Work, energy, quantity of heat	joule	J	N m
Power	watt	W	$J\,s^{-1}$
Electric charge	coulomb	C	A s
Electric potential, electromotive force	volt	V	$J\,C^{-1}$
Electric capacitance	farad	F	$C\,V^{-1}$
Electric resistance	ohm	Ω	$V\,A^{-1}$
Electric conductance	siemens	S	Ω^{-1}
Magnetic flux	weber	Wb	V s
Magnetic flux density	tesla	T	$Wb\,m^{-2}$
Inductance	henry	H	$Wb\,A^{-1}$
Luminous flux	lumen	lm	cd sr
Illumination	lux	lx	$lm\,m^{-2}$

Other metric units

Quantity	*Unit*	*Symbol*	*Equivalent*
Length	Angstrom	Å	10^{-10} m
	micron	μm	10^{-6} m
Area	are	a	10^{2} m^2
Volume	litre	l	10^{-3} m^3
Mass	tonne	t	10^{3} kg
Force	dyne	–	10^{-5} N
Pressure	bar	bar	10^{5} Pa
Energy	erg	–	10^{-7} J
Viscosity (dynamic)	poise	P	10^{-1} Pa s
Viscosity (kinematic)	stokes	St	10^{-4} $m^2 s^{-1}$
Magnetic field strength	oersted	–	$10^{3}/4\pi$ A m^{-1}
Magnetomotive force	gilbert	–	$10/4\pi$ A
Magnetic flux density	gauss	–	10^{-4} T
Magnetic flux	maxwell (or line)	–	10^{-8} Wb
Luminance	stilb	–	10^{4} cd m^{-2}
*Temperature	degree Celsius	°C	1 K

*The degree Celsius is used only for the temperature scale having zero at the absolute temperature 273·15 K, the ice point of water.

Multiples and sub-multiples

Factor	*Prefix*	*Symbol*
10^{18}	exa	E
10^{15}	peta	P
10^{12}	tera	T
10^{9}	giga	G
10^{6}	mega	M
10^{3}	kilo	k
10^{2}	hecto	h
10	deca	da
10^{-1}	deci	d
10^{-2}	centi	c
10^{-3}	milli	m
10^{-6}	micro	μ
10^{-9}	nano	n
10^{-12}	pico	p
10^{-15}	femto	f
10^{-18}	atto	a

Conversion factors

Length

1 mil = 25·4 μm (1 mil = 10^{-3} inch)
1 foot = 0·3048 m
1 yard = 0·9144 m
1 fathom = 1·829 m (1 fathom = 6 feet)
1 furlong = 0·2012 km (1 furlong = 220 yards)
1 mile = 1·609 km
1 nautical mile = 1·852 km (1 nautical mile = 1·15 miles)

Area

1 square inch = 645·2 mm^2
1 square foot = 929·0 cm^2
1 square yard = 0·8361 m^2
1 acre = 4047 m^2 (1 acre = 4840 sq. yd)
1 square mile = 2·590 km^2

Volume

1 cubic inch = 16·39 cm^3
1 fluid ounce = 28·41 cm^3
1 pint = 568·3 cm^3
1 Imperial gallon = 4·546 l
1 U.S. gallon = 3·785 l
1 cubic foot = 28·32 l (1 cu. ft = 6·24 gallons)
1 cubic yard = 0·7646 m^3

Angle

1 degree = 0·0175 rad
1 minute = $\frac{1}{60}$ degree = 2·909 × 10^{-4} rad
1 second = $\frac{1}{60}$ min = 4·848 × 10^{-6} rad

Speed

1 revolution/minute = 0·1047 rad s^{-1}
1 mile/hour = 0·4470 m s^{-1}
1 knot = 0·5148 m s^{-1} (1 knot = 1 nautical mile/h)

Mass

1 ounce = 28·35 g
1 pound (lb) = 0·4536 kg
1 slug = 14·59 kg (1 slug = 32·17 lb)
1 hundredweight (cwt) = 50·8 kg (1 cwt = 112 lb)
1 ton = 1·016 t (1 ton = 2240 lb)

Density

1 lb/in^3 = 27·68 g cm^{-3}
1 lb/ft^3 = 16·02 kg m^{-3}
1 lb/gallon = 99·78 kg m^{-3}
1 ton/yd^3 = 1·329 t m^{-3}

Moments of mass

1 lb ft = 0·1383 kg m
1 lb ft^2 = 421·4 kg cm^2

Force

1 lb force (lbf) = 4·448 N (standard gravity)
1 poundal (pdl) = 0·1383 N (1 pdl = 0·0311 lbf)
1 ton force = 9·964 kN

Pressure (or stress)

1 lb/ft^2 = 47·88 Pa
1 lb/in^2 (psi) = 6·895 kPa
1 ton/ft^2 = 107·3 kPa
1 ton/in^2 = 15·44 MPa
1 in Hg (0°C) = 3·386 kPa (1 in Hg = 0·491 psi)
1 ft H_2O (4°C) = 2·989 kPa (1 ft H_2O = 0·434 psi)
1 atmosphere (atm) = 1·013 25 bar

Energy, etc.

1 ft lbf = 1·356 J
1 ft pdl = 42·14 mJ
1 horse-power = 745·7 W (1 hp = 550 ft lbf/s)
1 electron volt (eV) = 1·602 × 10^{-19} J
*1 degree Fahrenheit = 1 rankine (R) = $\frac{5}{9}$ K
†1 calorie = 4·187 J
†1 Btu = 1·055 kJ (1 Btu = 252 cal)
1 Chu = 1·899 kJ (1 Chu = $\frac{9}{5}$ Btu)
1 therm = 105·5 MJ (1 therm = 10^5 Btu)
1 Btu/hour = 0·293 W
1 Btu/lb = 2·325 kJ kg^{-1}

* The degree Fahrenheit is used only for the temperature scale having zero at the absolute temperature 459·67 R, on which scale the ice point of water is close to 32 °F.
† The calorie and Btu were defined in terms of heating water, and their precise values depended on the temperature limits specified. The figures given are based on agreed international equivalents.

Mathematics

Constants

Constant	Value	$\log_{10}$	ln ($\log_e$)
π	3·141 59	0·497 15	1·144 73
π^2	9·869 60	0·994 30	2·289 46
$1/\pi$	0·318 31	$\bar{1}$·502 85	−1·144 73
$1/\pi^2$	0·101 32	$\bar{1}$·005 70	−2·289 46
$\sqrt{\pi}$	1·772 45	0·248 57	0·572 36
e	2·718 28	0·434 29	1·000 00
γ (Euler constant)	0·577 22	$\bar{1}$·761 34	−0·549 54
	2	0·301 03	0·693 15
	3	0·477 12	1·098 61
	10	1·000 00	2·302 59
$\sqrt{2}$	1·414 21	0·150 51	0·346 57
$\sqrt{3}$	1·732 05	0·238 56	0·549 31
$\sqrt{10}$	3·162 28	0·500 00	1·151 29
$180/\pi$	57·295 78	1·758 12	4·048 23

Binomial coefficients

$$\binom{n}{m} = \frac{n!}{(n-m)!\,m!} = \binom{n}{n-m}$$

n \ m	0	1	2	3	4	5	6	
0	1							
1	1	1						
2	1	2	1					
3	1	3	3	1				
4	1	4	6	4	1			
5	1	5	10	10	5	1		
6	1	6	15	20	15	6	1	
7	1	7	21	35	35	21	7	
8	1	8	28	56	70	56	28	etc.
9	1	9	36	84	126	126	84	
10	1	10	45	120	210	252	210	

Large factorials can be calculated from the version of *Stirling's formula* which gives

$$\log_{10} n! \approx 0.399\,09 + (n + \tfrac{1}{2})\log_{10} n - 0.434\,294\,5n + \frac{0.0362}{n}$$

and is correct to the fifth decimal place or better for $n \geqslant 70$.

Series

$$1-\frac{1}{3}+\frac{1}{5}-\frac{1}{7}+\ldots=\frac{\pi}{4}$$

$$\frac{1}{1^2}+\frac{1}{2^2}+\frac{1}{3^2}+\ldots=\frac{\pi^2}{6}$$

$$\frac{1}{1^2}+\frac{1}{3^2}+\frac{1}{5^2}+\ldots=\frac{\pi^2}{8}$$

$$\frac{1}{1^3}-\frac{1}{3^3}+\frac{1}{5^3}-\ldots=\frac{\pi^3}{32}$$

$$1+2+3+\ldots+n=\sum_1^n r=\frac{n(n+1)}{2}$$

$$1^2+2^2+3^2+\ldots+n^2=\sum_1^n r^2=\frac{n(n+1)(2n+1)}{6}$$

$$1^3+2^3+3^3+\ldots+n^3=\sum_1^n r^3=\frac{n^2(n+1)^2}{4}$$

Arithmetic

$$a+(a+d)+(a+2d)+\ldots+\{a+(n-1)d\}$$
$$=\frac{n}{2}\{2a+(n-1)d\}$$

Geometric

$$1+x+x^2+x^3+\ldots+x^{n-1}=\frac{1-x^n}{1-x} \quad (x\neq 1)$$
$$=\frac{1}{1-x} \quad (|x|<1, n\to\infty)$$

Binomial

$$(1+x)^n=1+nx+\frac{n(n-1)}{2!}x^2+\frac{n(n-1)(n-2)}{3!}x^3$$
$$+\ldots+\binom{n}{r}x^r+\ldots$$

($|x|<1$, all real n; all x, n a positive integer)

Exponential and logarithmic

$$e^x=1+x+\frac{x^2}{2!}+\frac{x^3}{3!}+\ldots$$

$$a^x=1+x\ln a+\frac{(x\ln a)^2}{2!}+\frac{(x\ln a)^3}{3!}+\ldots$$

$$\ln(1+x)=x-\frac{x^2}{2}+\frac{x^3}{3}-\frac{x^4}{4}+\ldots \quad (|x|<1)$$

Trigonometric

$$\sin x=x-\frac{x^3}{3!}+\frac{x^5}{5!}-\ldots$$

$$\cos x=1-\frac{x^2}{2!}+\frac{x^4}{4!}-\ldots$$

$$\tan x=x+\frac{x^3}{3}+\frac{2x^5}{15}+\frac{17x^7}{315}+\ldots \quad \left(|x|<\frac{\pi}{2}\right)$$

$$\sin^{-1}x=x+\frac{x^3}{6}+\frac{1.3}{2.4}\frac{x^5}{5}+\frac{1.3.5}{2.4.6}\frac{x^7}{7}+\ldots \quad (|x|<1)$$

$$\tan^{-1}x=x-\frac{x^3}{3}+\frac{x^5}{5}-\ldots \quad (|x|<1)$$

$$\sinh x=x+\frac{x^3}{3!}+\frac{x^5}{5!}+\ldots$$

$$\cosh x=1+\frac{x^2}{2!}+\frac{x^4}{4!}+\ldots$$

$$\tanh x=x-\frac{x^3}{3}+\frac{2x^5}{15}-\frac{17x^7}{315}+\ldots \quad \left(|x|<\frac{\pi}{2}\right)$$

Maclaurin

$$f(x)=f(0)+xf'(0)+\frac{x^2}{2!}f''(0)+\frac{x^3}{3!}f'''(0)+\ldots$$

Taylor

$$f(a+h)=f(a)+hf'(a)+\frac{h^2}{2!}f''(a)+\frac{h^3}{3!}f'''(a)+\ldots$$

Fourier

A periodic function $f(x)$ with period T is represented by the exponential series

$$f(x)=c_0+c_1e^{j2\pi x/T}+c_2e^{j4\pi x/T}+\ldots$$

where

$$c_n=\frac{1}{T}\int_0^T f(x)e^{-j2\pi nx/T}\mathrm{d}x$$

Alternatively, for real $f(x)$ only,

$$f(x)=\frac{a_0}{2}+a_1\cos\frac{2\pi x}{T}+a_2\cos\frac{4\pi x}{T}+\ldots$$
$$+b_1\sin\frac{2\pi x}{T}+b_2\sin\frac{4\pi x}{T}+\ldots$$

where

$$a_n=\frac{2}{T}\int_0^T f(x)\cos\frac{2\pi nx}{T}\mathrm{d}x$$

$$b_n=\frac{2}{T}\int_0^T f(x)\sin\frac{2\pi nx}{T}\mathrm{d}x$$

In these series, terms with $n=0$ give the average of $f(x)$, those with $n=1$ its fundamental component, and those with $n\geqslant 2$ its nth harmonic.

If $f(x)=f(-x)$, f is *even*, $b_n=0$, and c_n is real.
If $f(x)=-f(-x)$, f is *odd*, $a_n=0$, and c_n is imaginary.
If $f(x)=-f\left(x+\frac{T}{2}\right)$, f has half-wave symmetry and $a_n=b_n=c_n=0$ for all even n.

Fourier series for certain waveforms

The series below are expressed in terms of the angular variable θ, the period of each waveform being 2π. Similar waveforms in any variable x with period T can be represented by the same series with the substitution $\theta = 2\pi x/T$. The origin of θ is so chosen as to make the waveforms even functions ($b_n = 0$) wherever possible. α and β are angles, k an integer.

$\theta = 0$ $\theta = 2\pi$	Series	Mean square value
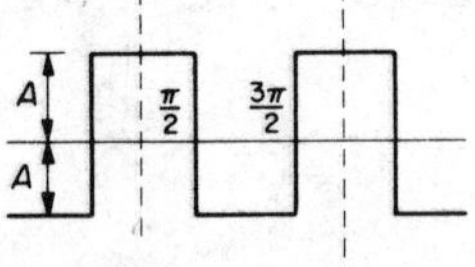	$\frac{4A}{\pi}\{\cos\theta - \frac{1}{3}\cos 3\theta + \frac{1}{5}\cos 5\theta - \ldots\}$	A^2
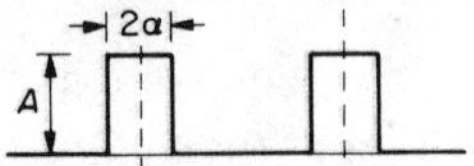	$A\left\{\frac{\alpha}{\pi} + \frac{2}{\pi}(\sin\alpha\cos\theta + \frac{1}{2}\sin 2\alpha\cos 2\theta + \frac{1}{3}\sin 3\alpha\cos 3\theta + \ldots)\right\}$	$\frac{A^2\alpha}{\pi}$
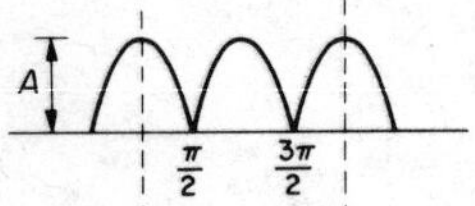	$\frac{2A}{\pi}\{1 + \frac{2}{3}\cos 2\theta - \frac{2}{15}\cos 4\theta + \frac{2}{35}\cos 6\theta - \ldots\}$	$\frac{A^2}{2}$
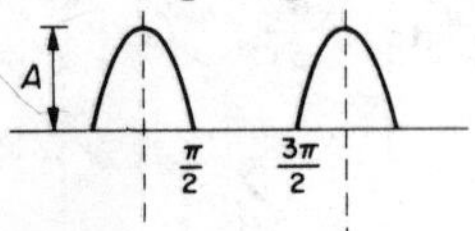	$\frac{A}{\pi}\left\{1 + \frac{\pi}{2}\cos\theta + \frac{2}{3}\cos 2\theta - \frac{2}{15}\cos 4\theta + \ldots\right\}$	$\frac{A^2}{4}$
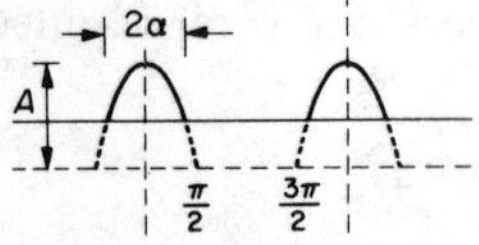	$\frac{A}{\pi}\{(\sin\alpha - \alpha\cos\alpha) + (\alpha - \frac{1}{2}\sin 2\alpha)\cos\theta + (\sin\alpha + \frac{1}{3}\sin 3\alpha - \cos\alpha\sin 2\alpha)\cos 2\theta + (\frac{1}{2}\sin 2\alpha + \frac{1}{4}\sin 4\alpha - \frac{2}{3}\cos\alpha\sin 3\alpha)\cos 3\theta + \ldots\}$	$\frac{A^2}{2\pi}\{\alpha - \frac{3}{2}\sin 2\alpha + 2\alpha\cos^2\alpha\}$
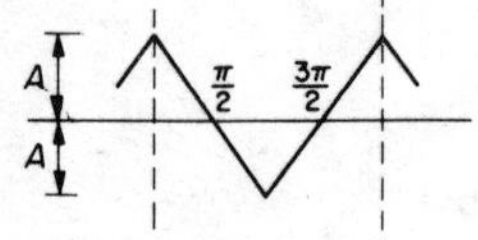	$\frac{8A}{\pi^2}\{\cos\theta + \frac{1}{9}\cos 3\theta + \frac{1}{25}\cos 5\theta + \ldots\}$	$\frac{A^2}{3}$
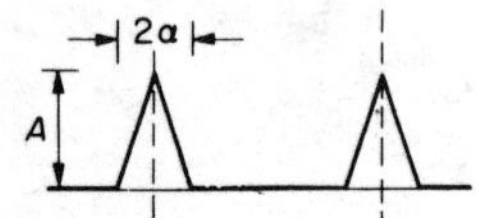	$\frac{A}{\pi\alpha}\left\{\frac{\alpha^2}{2} + 4\sin^2\frac{\alpha}{2}\cos\theta + \sin^2\alpha\cos 2\theta + \frac{4}{9}\sin^2\frac{3\alpha}{2}\cos 3\theta + \ldots\right\}$	$\frac{A^2\alpha}{3\pi}$
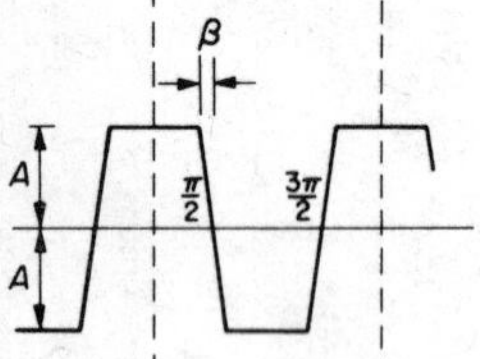	$\frac{4A}{\pi\beta}\{\sin\beta\cos\theta - \frac{1}{9}\sin 3\beta\cos 3\theta + \frac{1}{25}\sin 5\beta\cos 5\theta - \ldots\}$	$A^2\left(1 - \frac{4}{3}\frac{\beta}{\pi}\right)$
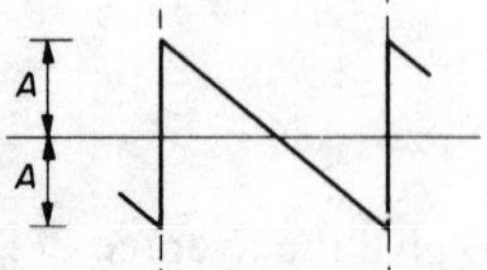	$\frac{2A}{\pi}\{\sin\theta + \frac{1}{2}\sin 2\theta + \frac{1}{3}\sin 3\theta + \ldots\}$	$\frac{A^2}{3}$
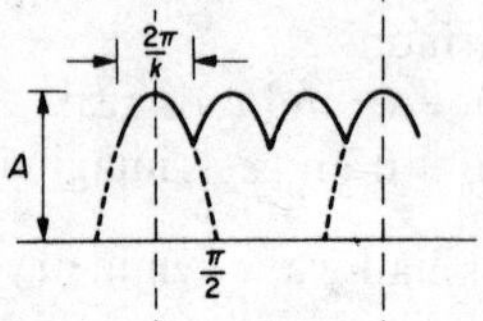	$\frac{Ak}{\pi}\sin\frac{\pi}{k}\left\{1 + \frac{2}{k^2-1}\cos k\theta - \frac{2}{4k^2-1}\cos 2k\theta + \frac{2}{9k^2-1}\cos 3k\theta - \ldots\right\}$	

$f(x) = \frac{a_0}{2} + \sum_{n=1}^{\infty}\left(a_n\cos\frac{n\pi x}{L} + b_n\sin\frac{n\pi x}{L}\right)$ period $2L$

$a_0 = \frac{1}{L}\int_{-L}^{L} f(x)\,dx$ mean of $f(x)$ over a period

$a_n = \frac{1}{L}\int_{-L}^{L} f(x)\cos\frac{n\pi x}{L}\,dx$ $n = 0, 1, 2 \ldots$

Trigonometric, hyperbolic and exponential functions

De Moivre's theorem is:

$$(\cos x + \mathrm{j} \sin x)^n = \cos nx + \mathrm{j} \sin nx$$

$$\mathrm{e}^{\mathrm{j}x} = \cos x + \mathrm{j} \sin x$$

$$\mathrm{e}^{x} = \cosh x + \sinh x$$

$$\sin x = \frac{\mathrm{e}^{\mathrm{j}x} - \mathrm{e}^{-\mathrm{j}x}}{2\mathrm{j}}; \quad \cos x = \frac{\mathrm{e}^{\mathrm{j}x} + \mathrm{e}^{-\mathrm{j}x}}{2}$$

$$\sinh x = \frac{\mathrm{e}^{x} - \mathrm{e}^{-x}}{2}; \quad \cosh x = \frac{\mathrm{e}^{x} + \mathrm{e}^{-x}}{2}$$

$$\sin \mathrm{j}x = \mathrm{j} \sinh x; \quad \cos \mathrm{j}x = \cosh x$$

$$\sinh \mathrm{j}x = \mathrm{j} \sin x; \quad \cosh \mathrm{j}x = \cos x$$

$$\sin^{-1} \mathrm{j}x = \mathrm{j} \sinh^{-1} x; \quad \cos^{-1} x = -\mathrm{j} \cosh^{-1} x$$

$$\sinh^{-1} \mathrm{j}x = \mathrm{j} \sin^{-1} x; \quad \cosh^{-1} x = \mathrm{j} \cos^{-1} x$$

$$\sinh^{-1} x = \ln \{x + \sqrt{(x^2 + 1)}\};$$

$$\cosh^{-1} x = \ln \{x + \sqrt{(x^2 - 1)}\}$$

$$\tanh^{-1} x = \tfrac{1}{2} \ln \frac{1 + x}{1 - x}$$

Trigonometric relations

$$\sin(-\mathrm{A}) = -\sin \mathrm{A}; \quad \cos(-\mathrm{A}) = \cos \mathrm{A}$$

$$\sin^2 \mathrm{A} + \cos^2 \mathrm{A} = 1$$

$$\tan^2 \mathrm{A} + 1 = \sec^2 \mathrm{A}$$

$$1 + \cot^2 \mathrm{A} = \mathrm{cosec}^2 \mathrm{A}$$

$$\sin(\mathrm{A} \pm \mathrm{B}) = \sin \mathrm{A} \cos \mathrm{B} \pm \cos \mathrm{A} \sin \mathrm{B}$$

$$\cos(\mathrm{A} \pm \mathrm{B}) = \cos \mathrm{A} \cos \mathrm{B} \mp \sin \mathrm{A} \sin \mathrm{B}$$

$$\tan(\mathrm{A} \pm \mathrm{B}) = \frac{\tan \mathrm{A} \pm \tan \mathrm{B}}{1 \mp \tan \mathrm{A} \tan \mathrm{B}}$$

$$\sin^2 \mathrm{A} = \tfrac{1}{2}(1 - \cos 2\mathrm{A})$$

$$\cos^2 \mathrm{A} = \tfrac{1}{2}(1 + \cos 2\mathrm{A})$$

$$\tan \mathrm{A} = \frac{\sin 2\mathrm{A}}{1 + \cos 2\mathrm{A}}$$

$$a \sin \mathrm{A} + b \cos \mathrm{A} = (a^2 + b^2)^{1/2} \sin\left(\mathrm{A} + \tan^{-1}\frac{b}{a}\right)$$

$$\sin \mathrm{A} \pm \sin \mathrm{B} = 2 \sin \frac{\mathrm{A} \pm \mathrm{B}}{2} \cos \frac{\mathrm{A} \mp \mathrm{B}}{2}$$

$$\cos \mathrm{A} + \cos \mathrm{B} = 2 \cos \frac{\mathrm{A} + \mathrm{B}}{2} \cos \frac{\mathrm{A} - \mathrm{B}}{2}$$

$$\cos \mathrm{A} - \cos \mathrm{B} = -2 \sin \frac{\mathrm{A} + \mathrm{B}}{2} \sin \frac{\mathrm{A} - \mathrm{B}}{2}$$

$$2 \sin \mathrm{A} \cos \mathrm{B} = \sin(\mathrm{A} + \mathrm{B}) + \sin(\mathrm{A} - \mathrm{B})$$

$$2 \cos \mathrm{A} \cos \mathrm{B} = \cos(\mathrm{A} + \mathrm{B}) + \cos(\mathrm{A} - \mathrm{B})$$

$$2 \sin \mathrm{A} \sin \mathrm{B} = \cos(\mathrm{A} - \mathrm{B}) - \cos(\mathrm{A} + \mathrm{B})$$

$$\left.\begin{array}{l} \dfrac{a}{\sin \mathrm{A}} = \dfrac{b}{\sin \mathrm{B}} = \dfrac{c}{\sin \mathrm{C}} \\ a^2 = b^2 + c^2 - 2bc \cos \mathrm{A} \end{array}\right\} \text{in a triangle ABC}$$

Hyperbolic relations

$$\sinh(-\mathrm{A}) = -\sinh \mathrm{A}; \quad \cosh(-\mathrm{A}) = \cosh \mathrm{A}$$

$$\cosh^2 \mathrm{A} - \sinh^2 \mathrm{A} = 1$$

$$1 - \tanh^2 \mathrm{A} = \mathrm{sech}^2 \mathrm{A}$$

$$\coth^2 \mathrm{A} - 1 = \mathrm{cosech}^2 \mathrm{A}$$

$$\sinh(\mathrm{A} \pm \mathrm{B}) = \sinh \mathrm{A} \cosh \mathrm{B} \pm \cosh \mathrm{A} \sinh \mathrm{B}$$

$$\cosh(\mathrm{A} \pm \mathrm{B}) = \cosh \mathrm{A} \cosh \mathrm{B} \pm \sinh \mathrm{A} \sinh \mathrm{B}$$

$$\tanh(\mathrm{A} \pm \mathrm{B}) = \frac{\tanh \mathrm{A} \pm \tanh \mathrm{B}}{1 \pm \tanh \mathrm{A} \tanh \mathrm{B}}$$

$$\sinh^2 \mathrm{A} = \tfrac{1}{2}(\cosh 2\mathrm{A} - 1)$$

$$\cosh^2 \mathrm{A} = \tfrac{1}{2}(\cosh 2\mathrm{A} + 1)$$

$$\tanh \mathrm{A} = \frac{\sinh 2\mathrm{A}}{\cosh 2\mathrm{A} + 1}$$

$$\sinh \mathrm{A} \pm \sinh \mathrm{B} = 2 \sinh \frac{\mathrm{A} \pm \mathrm{B}}{2} \cosh \frac{\mathrm{A} \mp \mathrm{B}}{2}$$

$$\cosh \mathrm{A} + \cosh \mathrm{B} = 2 \cosh \frac{\mathrm{A} + \mathrm{B}}{2} \cosh \frac{\mathrm{A} - \mathrm{B}}{2}$$

$$\cosh \mathrm{A} - \cosh \mathrm{B} = 2 \sinh \frac{\mathrm{A} + \mathrm{B}}{2} \sinh \frac{\mathrm{A} - \mathrm{B}}{2}$$

$$2 \sinh \mathrm{A} \cosh \mathrm{B} = \sinh(\mathrm{A} + \mathrm{B}) + \sinh(\mathrm{A} - \mathrm{B})$$

$$2 \cosh \mathrm{A} \cosh \mathrm{B} = \cosh(\mathrm{A} + \mathrm{B}) + \cosh(\mathrm{A} - \mathrm{B})$$

$$2 \sinh \mathrm{A} \sinh \mathrm{B} = \cosh(\mathrm{A} + \mathrm{B}) - \cosh(\mathrm{A} - \mathrm{B})$$

Differentials

In the following, u, v and w are functions of x and a is a constant.

$f(x)$	$f'(x)$
uv	$u\dfrac{dv}{dx} + v\dfrac{du}{dx}$
uvw	$uv\dfrac{dw}{dx} + vw\dfrac{du}{dx} + wu\dfrac{dv}{dx}$
$\dfrac{u}{v}$	$\dfrac{1}{v^2}\left(v\dfrac{du}{dx} - u\dfrac{dv}{dx}\right)$
$f(u, v)$	$\dfrac{\partial f}{\partial u}\dfrac{du}{dx} + \dfrac{\partial f}{\partial v}\dfrac{dv}{dx}$
ax^n	anx^{n-1}
e^{ax}	$a\,e^{ax}$
a^x	$a^x \ln a$
x^x	$x^x(1 + \ln x)$
$\ln x$	$1/x$
$\log_a x$	$\dfrac{1}{x}\log_a e$

$f(x)$	$f'(x)$
$\sin x$	$\cos x$
$\cos x$	$-\sin x$
$\tan x$	$\sec^2 x$
$\operatorname{cosec} x$	$-\cot x \operatorname{cosec} x$
$\sec x$	$\tan x \sec x$
$\cot x$	$-\operatorname{cosec}^2 x$
$\sinh x$	$\cosh x$
$\cosh x$	$\sinh x$
$\tanh x$	$\operatorname{sech}^2 x$
$\operatorname{cosech} x$	$-\coth x \operatorname{cosech} x$
$\operatorname{sech} x$	$-\tanh x \operatorname{sech} x$
$\coth x$	$-\operatorname{cosech}^2 x$
$\sin^{-1}(x/a)$	$(a^2 - x^2)^{-1/2}$
$\cos^{-1}(x/a)$	$-(a^2 - x^2)^{-1/2}$
$\tan^{-1}(x/a)$	$a/(a^2 + x^2)$
$\operatorname{cosec}^{-1}(x/a)$	$-a/x(x^2 - a^2)^{1/2}$
$\sec^{-1}(x/a)$	$a/x(x^2 - a^2)^{1/2}$
$\cot^{-1}(x/a)$	$-a/(a^2 + x^2)$
$\sinh^{-1}(x/a)$	$(a^2 + x^2)^{-1/2}$
$\cosh^{-1}(x/a)$	$(x^2 - a^2)^{-1/2}$
$\tanh^{-1}(x/a)$	$a/(a^2 - x^2)$
$\operatorname{cosech}^{-1}(x/a)$	$-a/x(x^2 + a^2)^{1/2}$
$\operatorname{sech}^{-1}(x/a)$	$-a/x(a^2 - x^2)^{1/2}$
$\coth^{-1}(x/a)$	$-a/(x^2 - a^2)$

Indefinite integrals

The constant of integration is omitted in each case.

$f(x)$	$\int f(x)\,dx$	$f(x)$	$\int f(x)\,dx$
$(a + bx)^n \ (n \neq -1)$	$\dfrac{(a + bx)^{n+1}}{b(n + 1)}$	$\ln x$	$x(\ln x - 1)$
$(a + bx)^{-1}$	$\dfrac{1}{b}\ln(a + bx)$	$(\ln x)^2$	$x\{(\ln x)^2 - 2\ln x + 2\}$
$\dfrac{x}{ax + b}$	$\dfrac{ax + b - b\ln(ax + b)}{a^2}$	$x^n \ln ax \ (n \neq -1)$	$\dfrac{x^{n+1}}{n + 1}\left(\ln ax - \dfrac{1}{n + 1}\right)$
a^x	$a^x/\ln a$	$(\ln ax)^n/x \ (n \neq -1)$	$(\ln ax)^{n+1}/(n + 1)$
xa^x	$\dfrac{a^x x}{\ln a} - \dfrac{a^x}{(\ln a)^2}$	$1/x \ln x$	$\ln(\ln x)$
$x\,e^{ax}$	$e^{ax}(ax - 1)/a^2$		
$(a + b\,e^{cx})^{-1}$	$\dfrac{x}{a} - \dfrac{1}{ac}\ln(a + b\,e^{cx})$		

$f(x)$	$\int f(x)\,\mathrm{d}x$	$f(x)$	$\int f(x)\,\mathrm{d}x$
$(x^2+a^2)^{-1}$	$\frac{1}{a}\tan^{-1}(x/a)$		
$(x^2-a^2)^{-1}$	$\frac{1}{2a}\ln\left\lvert\frac{x-a}{x+a}\right\rvert$	$(x^2+a^2)^{-1/2}$	$\sinh^{-1}(x/a)$
		$(x^2-a^2)^{-1/2}$	$\cosh^{-1}(x/a)$
$(x^2+a^2)^{1/2}$	$\frac{1}{2}\{x(x^2+a^2)^{1/2}+a^2\sinh^{-1}(x/a)\}$	$(a^2-x^2)^{-1/2}$	$\sin^{-1}(x/a)$
$(x^2-a^2)^{1/2}$	$\frac{1}{2}\{x(x^2-a^2)^{1/2}-a^2\cosh^{-1}(x/a)\}$	$a/x(x^2-a^2)^{1/2}$	$\sec^{-1}(x/a)$
$(a^2-x^2)^{1/2}$	$\frac{1}{2}\{x(a^2-x^2)^{1/2}+a^2\sin^{-1}(x/a)\}$		
$\sin x$	$-\cos x$	$\sinh x$	$\cosh x$
$\cos x$	$\sin x$	$\cosh x$	$\sinh x$
$\tan x$	$-\ln\cos x$	$\tanh x$	$\ln\cosh x$
$\operatorname{cosec} x$	$\ln\tan(x/2)$	$\operatorname{cosech} x$	$\ln\tanh(x/2)$
$\sec x$	$\ln(\sec x+\tan x)$	$\operatorname{sech} x$	$\tan^{-1}(\sinh x)$
$\cot x$	$\ln\sin x$	$\coth x$	$\ln\sinh x$
$\sin^2 x$	$\frac{1}{2}(x-\frac{1}{2}\sin 2x)$	$\sinh^2 x$	$\frac{1}{2}(-x+\frac{1}{2}\sinh 2x)$
$\cos^2 x$	$\frac{1}{2}(x+\frac{1}{2}\sin 2x)$	$\cosh^2 x$	$\frac{1}{2}(x+\frac{1}{2}\sinh 2x)$
$\tan^2 x$	$\tan x-x$	$\tanh^2 x$	$x-\tanh x$
$\operatorname{cosec}^2 x$	$-\cot x$	$\operatorname{cosech}^2 x$	$-\coth x$
$\sec^2 x$	$\tan x$	$\operatorname{sech}^2 x$	$\tanh x$
$\cot^2 x$	$-x-\cot x$	$\coth^2 x$	$x-\coth x$
$x\sin x$	$\sin x-x\cos x$	$x\sinh x$	$x\cosh x-\sinh x$
$x\cos x$	$\cos x+x\sin x$	$x\cosh x$	$x\sinh x-\cosh x$
$\sin^{-1} x$	$x\sin^{-1}x+(1-x^2)^{1/2}$	$\sinh^{-1} x$	$x\sinh^{-1}x-(x^2+1)^{1/2}$
$\cos^{-1} x$	$x\cos^{-1}x-(1-x^2)^{1/2}$	$\cosh^{-1} x$	$x\cosh^{-1}x-(x^2-1)^{1/2}$
$\tan^{-1} x$	$x\tan^{-1}x-\frac{1}{2}\ln(1+x^2)$	$\tanh^{-1} x$	$x\tanh^{-1}x+\frac{1}{2}\ln(1-x^2)$

Definite integrals

Legendre's normal *elliptic integrals* include:

$$F(\theta,k)=\int_0^\theta \frac{\mathrm{d}\theta}{(1-k^2\sin^2\theta)^{1/2}} \quad \text{(first kind)}$$

$$E(\theta,k)=\int_0^\theta (1-k^2\sin^2\theta)^{1/2}\,\mathrm{d}\theta \quad \text{(second kind)}$$

The 'complete' form of these is:

$$F(\pi/2,k)=K(k)=\int_0^{\pi/2}\frac{\mathrm{d}\theta}{(1-k^2\sin^2\theta)^{1/2}}$$

$$=\frac{\pi}{2}\left\{1+(\tfrac{1}{2})^2k^2+\left(\frac{1.3}{2.4}\right)^2k^4+\ldots\right\}(k^2<1)$$

$$E(\pi/2,k)=E(k)=\int_0^{\pi/2}(1-k^2\sin^2\theta)^{1/2}\,\mathrm{d}\theta$$

$$=\frac{\pi}{2}\left\{1-(\tfrac{1}{2})^2k^2-\left(\frac{1.3}{2.4}\right)^2\frac{k^4}{3}-\ldots\right\}(k^2<1)$$

The *error function* is:

$$\operatorname{erf} x=\frac{2}{\sqrt{\pi}}\int_0^x \mathrm{e}^{-u^2}\mathrm{d}u=\frac{2}{\sqrt{\pi}}\left(x-\frac{x^3}{3}+\frac{1}{2!}\frac{x^5}{5}-\frac{1}{3!}\frac{x^7}{7}+\ldots\right)$$

The *gamma function* is:

$$\Gamma(n)=\int_0^\infty x^{n-1}\,\mathrm{e}^{-x}\,\mathrm{d}x=\int_0^1\left(\ln\frac{1}{x}\right)^{n-1}\mathrm{d}x \quad (\mathrm{Re}\,n>0)$$

If n is a positive integer,

$$\Gamma(n) = (n-1)!$$

Also

$$\Gamma(\tfrac{1}{2}) = \sqrt{\pi}$$

The *beta function* is:

$$B(p,q) = \int_0^1 x^{p-1}(1-x)^{q-1}\,dx = \frac{\Gamma(p)\Gamma(q)}{\Gamma(p+q)} \quad (\mathrm{Re}\; p, q > 0)$$

Stirling's formula can be expressed as:

$$\ln \Gamma(n) \sim (n - \tfrac{1}{2})\ln n - n + \ln(\sqrt{2\pi})$$

Euler's constant is:

$$\gamma = -\int_0^\infty e^{-x} \ln x\,dx = 0{\cdot}577\,22$$

Other definite integrals

$$\int_0^\infty \frac{x^{m-1}}{1+x^n}\,dx = (\pi/n)\operatorname{cosec}(m\pi/n) \quad (0 < m < n)$$

$$\int_0^\infty e^{-ax}\,dx = \frac{1}{a} \quad (\mathrm{Re}\; a > 0)$$

$$\int_0^\infty x^n e^{-ax}\,dx = \frac{\Gamma(n+1)}{a^{n+1}} \quad (n > -1, \mathrm{Re}\; a > 0)$$

$$\int_0^\infty x\,e^{-x^2}\,dx = \tfrac{1}{2}$$

$$\int_0^\infty x^2 e^{-x^2}\,dx = \frac{\sqrt{\pi}}{4}$$

$$\int_0^\infty \frac{\sin ax}{x}\,dx = \begin{cases} \pi/2 & (a > 0) \\ 0 & (a = 0) \\ -\pi/2 & (a < 0) \end{cases} \quad (a \text{ real})$$

$$\int_0^\infty \frac{\tan x}{x}\,dx = \frac{\pi}{2}$$

$$\int_0^\infty e^{-ax} \sin bx\,dx = \frac{b}{a^2+b^2} \quad (a > 0)$$

$$\int_0^\infty e^{-ax} \cos bx\,dx = \frac{a}{a^2+b^2} \quad (a > 0)$$

$$\int_0^\pi \sin ax \cos ax\,dx = 0$$

$$\int_0^\pi \sin ax \sin bx\,dx = \int_0^\pi \cos ax \cos bx\,dx = 0$$

$(a, b$ integers; $a \neq b)$

$$\int_0^{\pi/2} \sin^m x \cos^n x\,dx =$$

$$\frac{(m-1)(m-3)\ldots(2 \text{ or } 1)\,(n-1)(n-3)\ldots(2 \text{ or } 1)}{(m+n)(m+n-2)\ldots(2 \text{ or } 1)} \times C$$

$(m, n$ integers; $C = \dfrac{\pi}{2}$ for m and n even,

$C = 1$ for m or n odd)

$$\int_0^\pi \frac{\cos nu\,du}{\cos u - \cos x} = \frac{\pi \sin nx}{\sin x} \quad (n = 1, 2, 3 \ldots) \text{ (principal value)}$$

Fourier transform

The Fourier transform of a function of time $x(t)$ may be written as

$$X(f) = \int_{-\infty}^{\infty} x(t) e^{-j2\pi ft}\,dt$$

in terms of cyclic frequency f, or as

$$X(\omega) = \int_{-\infty}^{\infty} x(t) e^{-j\omega t}\,dt$$

in terms of angular frequency ω or $2\pi f$. (Although in general $X(\omega) = X(2\pi f) \neq X(f)$, the same symbol is conventionally used for both forms since they differ only in scale.) The inverse Fourier transform is then

$$x(t) = \int_{-\infty}^{\infty} X(f) e^{j2\pi ft}\,df = \frac{1}{2\pi}\int_{-\infty}^{\infty} X(\omega) e^{j\omega t}\,d\omega$$

Properties of the Fourier transform

Symmetry

If $x(t)$ is real, then $X(f) = X^*(-f)$
If $x(t)$ is even, then $X(f) = X(-f)$
If $x(t)$ is odd, then $X(f) = -X(-f)$
If $x(t)$ is real and even, so is $X(f)$
If $x(t)$ is real and odd, $X(f)$ is imaginary and odd.
(In all these cases $X(\omega)$ has the same properties as $X(f)$.)

Duality

If $X(f)$ or $X(\omega)$ is the transform of $x(t)$, then $x(-f)$ or $2\pi x(-\omega)$ is the transform of $X(t)$.

Convolution

If $x(t) * y(t)$ represents the convolution

$$\int_{-\infty}^{\infty} x(\tau)y(t-\tau)\mathrm{d}\tau$$

then the transform of $x(t)*y(t)$ is $X(f)Y(f)$ or $X(\omega)Y(\omega)$ and the transform of $x(t)y(t)$ is $X(f)*Y(f)$ or $\frac{1}{2\pi}X(\omega)*Y(\omega)$.

Parseval–Rayleigh theorem

$$\int_{-\infty}^{\infty} |x(t)|^2\mathrm{d}t = \int_{-\infty}^{\infty} |X(f)|^2\mathrm{d}f = \frac{1}{2\pi}\int_{-\infty}^{\infty} |X(\omega)|^2\mathrm{d}\omega$$

Power spectral density

The power spectral density of $x(t)$ is

$$W(f) = \lim_{T\to\infty} \frac{1}{T}|X(f)|^2$$

where $X(f)$ is the Fourier transform of $x(t)$ in the interval $-T/2 < t < T/2$.

Autocorrelation

The autocorrelation function of $x(t)$ is the even function

$$R_x(\tau) = \lim_{T\to\infty} \frac{1}{T}\int_{-T/2}^{T/2} x(t)x(t+\tau)\mathrm{d}t$$

and $R_x(0)$ is the mean square value of $x(t)$. $R_x(\tau)$ is the inverse Fourier transform of the power spectral density of $x(t)$, i.e.

$$R_x(\tau) = \int_{-\infty}^{\infty} W(f)\mathrm{e}^{\mathrm{j}2\pi f\tau}\mathrm{d}f$$

and

$$W(f) = \int_{-\infty}^{\infty} R_x(\tau)\mathrm{e}^{-\mathrm{j}2\pi f\tau}\mathrm{d}\tau.$$

White noise is a random quantity having $W(f)$ constant and $R_x(\tau)$ an impulse at $\tau = 0$.

Fourier transforms of various functions

$x(t)$	$X(f)$	$X(\omega)$
$x(t-\tau)$	$\mathrm{e}^{-\mathrm{j}2\pi f\tau}X(f)$	$\mathrm{e}^{-\mathrm{j}\omega\tau}X(\omega)$
$x^{(n)}(t)$	$(\mathrm{j}2\pi f)^n X(f)$	$(\mathrm{j}\omega)^n X(\omega)$
$\int_{-\infty}^{t} x(t)\mathrm{d}t$	$\frac{1}{\mathrm{j}2\pi f}X(f) + \frac{1}{2}X(0)\delta(f)$	$\frac{1}{\mathrm{j}\omega}X(\omega) + \pi X(0)\delta(\omega)$
k, a constant	$k\delta(f)$	$2\pi k\delta(\omega)$
$\mathrm{e}^{\mathrm{j}\omega_0 t}$	$\delta(f-f_0)$	$2\pi\delta(\omega-\omega_0)$
$\mathrm{e}^{\mathrm{j}\omega_0 t}x(t)$	$X(f-f_0)$	$X(\omega-\omega_0)$
$\cos\omega_0 t$	$\frac{1}{2}\{\delta(f-f_0)+\delta(f+f_0)\}$	$\pi\{\delta(\omega-\omega_0)+\delta(\omega+\omega_0)\}$
$\sin\omega_0 t$	$\mathrm{j}\frac{1}{2}\{-\delta(f-f_0)+\delta(f+f_0)\}$	$\mathrm{j}\pi\{-\delta(\omega-\omega_0)+\delta(\omega+\omega_0)\}$
$x(t)\cos\omega_0 t$	$\frac{1}{2}\{X(f-f_0)+X(f+f_0)\}$	$\frac{1}{2}\{X(\omega-\omega_0)+X(\omega+\omega_0)\}$

$x(t)$	$X(f)$	$X(\omega)$
Unit impulse $\delta(t)$ 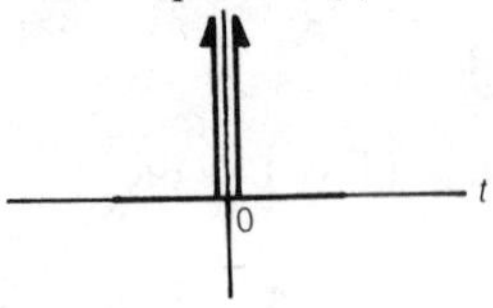	1	1
Unit step $u(t)$ 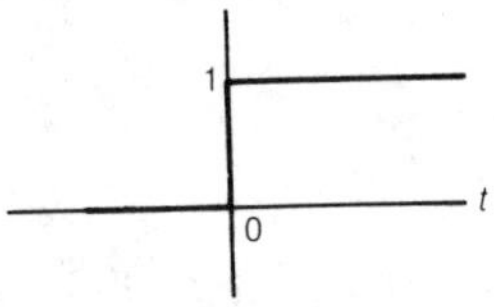	$\frac{1}{2}\delta(f) + \dfrac{1}{j2\pi f}$	$\pi\delta(\omega) + \dfrac{1}{j\omega}$
Sign function $\epsilon(t)$ 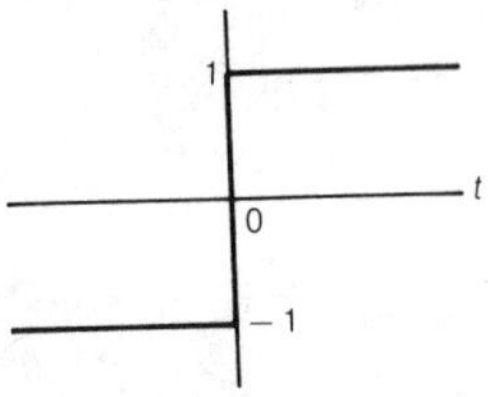	$\dfrac{1}{j\pi f}$	$\dfrac{2}{j\omega}$
Rectangular pulse $\Pi(t/\tau)$ 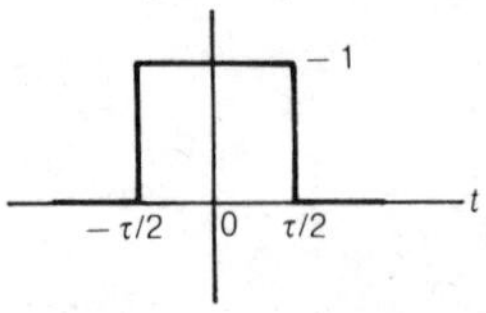	$\dfrac{\sin \pi f\tau}{\pi f} = \tau \operatorname{sinc} f\tau$ 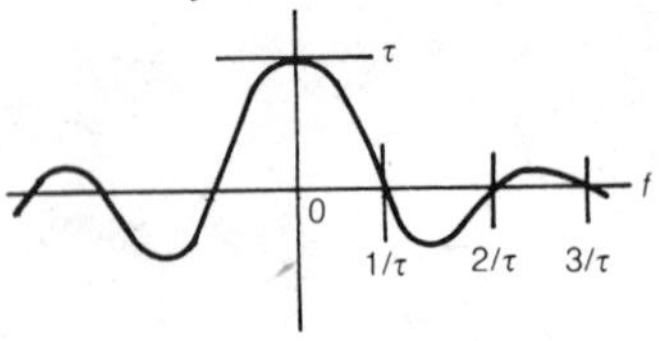	$\dfrac{2\sin(\omega\tau/2)}{\omega} = \tau \operatorname{sinc} \dfrac{\omega\tau}{2\pi}$ 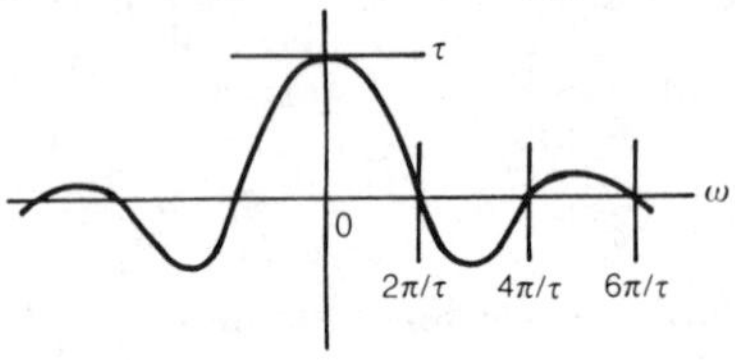
Triangular pulse $\Lambda(t/\tau)$ 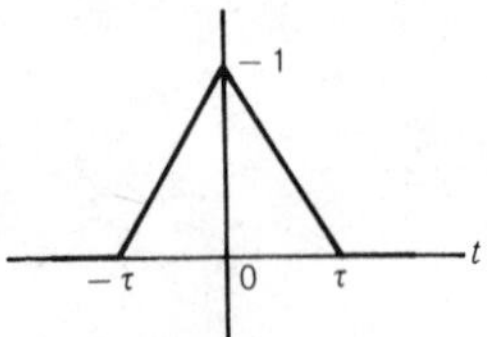	$\dfrac{\sin^2 \pi f\tau}{\pi^2 f^2 \tau} = \tau \operatorname{sinc}^2 f\tau$	$\dfrac{4\sin^2(\omega\tau/2)}{\omega^2\tau} = \tau \operatorname{sinc}^2 \dfrac{\omega\tau}{2\pi}$
Carrier pulse $\Pi(t/\tau)\cos\omega_0 t$ 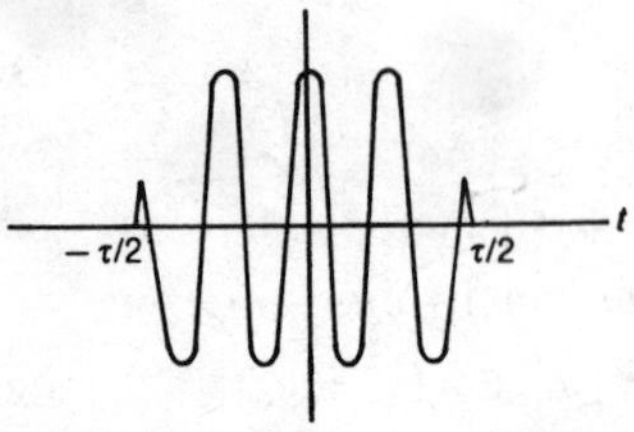	$\dfrac{\tau}{2}\{\operatorname{sinc}(f - f_0)\tau + \operatorname{sinc}(f + f_0)\tau\}$	$\dfrac{\tau}{2}\left\{\operatorname{sinc}\dfrac{(\omega-\omega_0)\tau}{2\pi} + \operatorname{sinc}\dfrac{(\omega+\omega_0)\tau}{2\pi}\right\}$

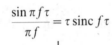

$x(t)$	$X(f)$	$X(\omega)$
Exponential pulse $u(t)\,\mathrm{e}^{-at}$ $(a>0)$ 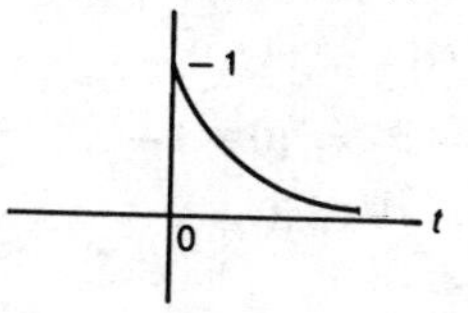	$\dfrac{1}{a+\mathrm{j}2\pi f}$	$\dfrac{1}{a+\mathrm{j}\omega}$
Two-sided exponential pulse $\mathrm{e}^{-a\lvert t\rvert}$ $(a>0)$ 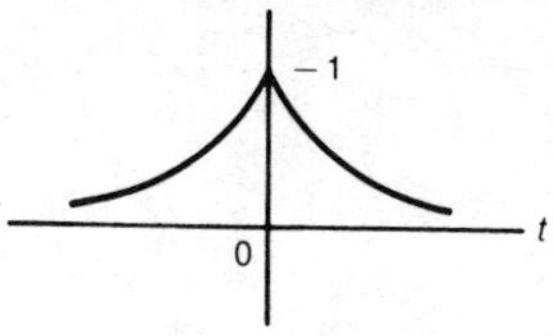	$\dfrac{2a}{a^2+4\pi^2 f^2}$	$\dfrac{2a}{a^2+\omega^2}$
Gaussian pulse $\mathrm{e}^{-\pi(t/\tau)^2}$ 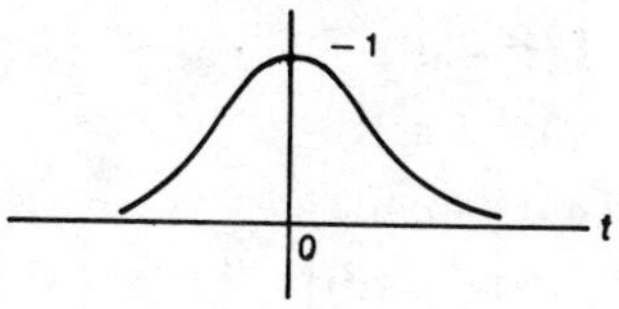	$\tau\,\mathrm{e}^{-\pi(f\tau)^2}$	$\tau\,\mathrm{e}^{-(\omega\tau)^2/4\pi}$
Cosine-squared pulse $\Pi(t/\tau)\cos^2(\pi t/2\tau)$ 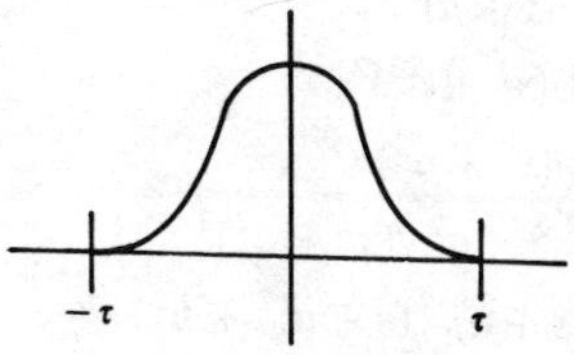	$\dfrac{\pi^2\sin 2\pi f\tau}{2\pi f(\pi^2-4\pi^2 f^2\tau^2)}$	$\dfrac{\pi^2\sin\omega\tau}{\omega(\pi^2-\omega^2\tau^2)}$
Impulse train $\sum_{-\infty}^{\infty}\delta(t-nT)$ 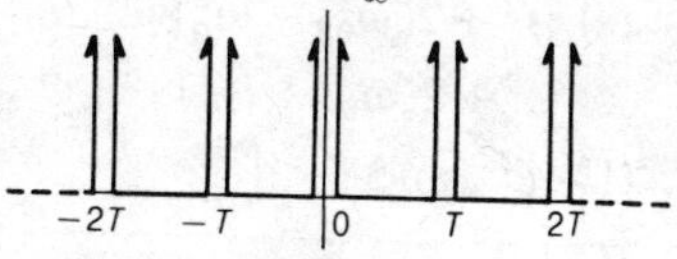	$\dfrac{1}{T}\sum_{-\infty}^{\infty}\delta\left(f-\dfrac{n}{T}\right)$	$\dfrac{2\pi}{T}\sum_{-\infty}^{\infty}\delta\left(\omega-\dfrac{2\pi n}{T}\right)$

Discrete Fourier transform

If $x(n)$ is the nth of a sequence of N samples of a function of time x, then the Fourier transform of these has N discrete values of which the mth is given by

$$X(m)=\frac{1}{N}\sum_{n=0}^{N-1}x(n)\,\mathrm{e}^{-\mathrm{j}2\pi nm/N}$$

The inverse Fourier transform of $X(m)$ is then

$$x(n)=\sum_{m=0}^{N-1}X(m)\,\mathrm{e}^{\mathrm{j}2\pi nm/N}$$

Laplace transform

The Laplace transform of a function $f(t)$ is

$$F(s)=\int_0^{\infty} f(t)\,\mathrm{e}^{-st}\,\mathrm{d}t$$

and the inverse transform of $F(s)$ for $t>0$ is then

$$f(t)=\frac{1}{2\pi\mathrm{j}}\int_{\sigma-\mathrm{j}\infty}^{\sigma+\mathrm{j}\infty} F(s)\,\mathrm{e}^{st}\,\mathrm{d}s$$

where σ is a real constant greater than the real part of each singularity of $F(s)$.

Properties of the Laplace transform

Convolution

If $F(s), G(s)$ are the Laplace transforms of $f(t), g(t)$ then $F(s)G(s)$ is the transform of the convolution of f and g from 0 to t, i.e. of

$$f*g=\int_0^t f(\tau)g(t-\tau)\,\mathrm{d}\tau.$$

Initial-value theorem

$$\lim_{s\to\infty} sF(s)=f(0)$$

Final-value theorem

$$\lim_{s\to 0} sF(s)=\lim_{t\to\infty} f(t)$$

Laplace transforms of various functions

Function	Transform
$f(t-\tau)$	$\mathrm{e}^{-s\tau}F(s)$
$\mathrm{e}^{-at}f(t)$	$F(s+a)$
$f'(t)$	$sF(s)-f(0+)$
$f''(t)$	$s^2F(s)-sf(0+)-f'(0+)$
$f^{(n)}(t)$	$s^nF(s)-s^{n-1}f(0+)$ $-s^{n-2}f'(0+)-\ldots$ $-f^{(n-1)}(0+)$
$\int_0^t f(t)\mathrm{d}t$	$F(s)/s$
Unit impulse at $t=0, \delta(t)$	1
Unit step at $t=0, u(t)$	$1/s$
$t^{n-1}/(n-1)!$	$1/s^n$
e^{-at}	$1/(s+a)$
$(1/a)(1-e^{-at})$	$1/\{s(s+a)\}$
$\cos at$	$s/(s^2+a^2)$
$\cosh at$	$s/(s^2-a^2)$
$\sin at$	$a/(s^2+a^2)$
$\sinh at$	$a/(s^2-a^2)$
$(1/a^2)(1-\cos at)$	$1/\{s(s^2+a^2)\}$
$(1/a^3)(at-\sin at)$	$1/\{s^2(s^2+a^2)\}$
$t\,\mathrm{e}^{-at}$	$1/(s+a)^2$
$\mathrm{e}^{-at}(1-at)$	$s/(s+a)^2$
$(1/2a^3)(\sin at-at\cos at)$	$1/(s^2+a^2)^2$
$(t/2a)\sin at$	$s/(s^2+a^2)^2$
$t\cos at$	$(s^2-a^2)/(s^2+a^2)^2$
$\mathrm{e}^{-at}\cos bt$	$(s+a)/\{(s+a)^2+b^2\}$
$\mathrm{e}^{-at}\sin bt$	$b/\{(s+a)^2+b^2\}$
$(\mathrm{e}^{-bt}-\mathrm{e}^{-at})/(a-b)$	$1/(s+a)(s+b)$
$(a\,\mathrm{e}^{-at}-b\,\mathrm{e}^{-bt})/(a-b)$	$s/(s+a)(s+b)$
*$\mathrm{e}^{-\alpha t}\sin\beta t/\beta$	$1/(s^2+2\zeta\omega_0 s+\omega_0^2)$
*$\mathrm{e}^{-\alpha t}\{\cos\beta t-(\alpha/\beta)\sin\beta t\}$	$s/(s^2+2\zeta\omega_0 s+\omega_0^2)$

*where $\alpha=\zeta\omega_0$, $\beta=\omega_0\sqrt{(1-\zeta^2)}$, and $\zeta<1$.

The z-transform

If $f(iT)$ represents samples of a function $f(t)$ taken at intervals T $(i=0,1,2\ldots)$ then the z-transform of f is

$$F(z)=\sum_{i=0}^{\infty} z^{-i}f(iT)$$

and the inverse z-transform of F is

$$f(iT)=\frac{1}{2\pi\mathrm{j}}\oint z^{i-1}F(z)\mathrm{d}z.$$

In these the complex variable z is related to the complex variable s of the Laplace transform by

$$z=\mathrm{e}^{sT}.$$

Properties of the z-transform

Convolution

If $F(z)$ and $G(z)$ are the z-transforms of $f(iT)$ and $g(iT)$ then $F(z)G(z)$ is the transform of

$$f*g=\sum_{k=0}^{i} f(k)g(i-k)$$

Initial-value theorem

$$\lim_{z\to\infty} F(z)=f(0)$$

Final-value theorem

$$\lim_{z\to 1}(z-1)F(z)=\lim_{i\to\infty} f(iT)$$

z-transforms of various functions

Function sampled	*Transform*
$f(t-kT)$ (k is an integer)	$z^{-k}F(z)$
Unit impulse at $t=0, \delta(t)$	1
Unit step at $t=0, u(t)$	$z/(z-1)$
Unit ramp at $t=0, f(t)=t$	$Tz/(z-1)^2$
Unit acceleration at $t=0, f(t)=t^2/2$	$T^2z(z+1)/2(z-1)^3$
e^{-at}	$z/(z-e^{-aT})$
$\cos at$	$\dfrac{z(z-\cos aT)}{(z^2-2z\cos aT+1)}$
$\sin at$	$\dfrac{z\sin aT}{(z^2-2z\cos aT+1)}$

Complex variable

Cauchy-Riemann relations

If $z=x+jy$ and the function $f(z)=u+jv$, then for $f(z)$ to be analytic it is necessary that

$$\frac{\partial u}{\partial x}=\frac{\partial v}{\partial y};\quad \frac{\partial u}{\partial y}=-\frac{\partial v}{\partial x}.$$

Cauchy's theorem

If $f(z)$ is analytic in a closed region bounded by a contour C,

$$\oint_C f(z)\,dz=0.$$

Cauchy's integral

If a is a point inside C, $f(z)$ being analytic within and on C,

$$f(a)=\frac{1}{2\pi j}\oint_C \frac{f(z)}{z-a}\,dz.$$

Also,

$$f^{(n)}(a)=\frac{n!}{2\pi j}\oint_C \frac{f(z)}{(z-a)^{n+1}}\,dz.$$

Residue theorem

If $f(z)$ is analytic within and on C except at poles $a, b, c\ldots$ enclosed by C,

$$\oint_C f(z)\,dz=2\pi j(A+B+C\ldots)$$

where $A, B, C\ldots$ are the residues of the poles.

The Nyquist criterion

A consequence of the residue theorem is the following. If $f(z)$ is analytic within and on C except for P poles and Z zeros (a pole or zero of order n being counted n times) within C, then

$$\theta_c=2\pi(Z-P),$$

where θ_c is the change in the argument of $f(z)$ for one circuit of C; or

$$N=Z-P,$$

where N is the number of times $f(z)$ encircles its origin counter-clockwise for one counter-clockwise circuit of C. This is the basis of the Nyquist stability criterion.

Algebraic equations

The *quadratic* equation $ax^2+bx+c=0$ has roots

$$x=\frac{-b\pm\sqrt{(b^2-4ac)}}{2a}$$

which are:

real and unequal if $b^2>4ac$,

real, equal and given by $-b/2a$ if $b^2=4ac$,

complex and conjugate if $b^2<4ac$.

The *cubic* equation $x^3+ax^2+bx+c=0$ is reduced to the form $y^3+py+q=0$, in which

$$p=-\frac{a^2}{3}+b;\quad q=2\left(\frac{a}{3}\right)^3-\frac{ab}{3}+c$$

by the substitution $x = y - a/3$. The roots are obtained from

$$y = A + B, \quad -\tfrac{1}{2}(A + B) \pm \mathrm{j}\frac{\sqrt{3}}{2}(A - B)$$

where

$$A = \sqrt[3]{\left\{-\frac{q}{2} + \sqrt{\left(\frac{q^2}{4} + \frac{p^3}{27}\right)}\right\}};$$

$$B = \sqrt[3]{\left\{-\frac{q}{2} - \sqrt{\left(\frac{q^2}{4} + \frac{p^3}{27}\right)}\right\}}.$$

If

$\frac{q^2}{4} + \frac{p^3}{27} > 0$, one root is real and two are complex and conjugate,
$= 0$, all roots are real and two are equal,
< 0, all roots are real and different.

The general equation

$$a_0x^n + a_1x^{n-1} + \ldots + a_{n-1}x + a_n = 0$$

of degree n has n roots (real, or complex and conjugate in pairs) of which at least one is real for odd n. Their sum is $-a_1/a_0$ and the sum of their products taken r at a time is $(-1)^r a_r/a_0$.

The Routh-Hurwitz criterion

The number of roots of the general equation which are positive or have positive real parts is the number of sign changes in the sequence

$D_0, D_1, D_1D_2, D_2D_3, \ldots$, where

$$D_0, D_1, D_2, D_3, \ldots = a_0, a_1, \begin{vmatrix} a_1 & a_0 \\ a_3 & a_2 \end{vmatrix}, \begin{vmatrix} a_1 & a_0 & 0 \\ a_3 & a_2 & a_1 \\ a_5 & a_4 & a_3 \end{vmatrix}, \ldots$$

All roots have negative real parts if there is no sign change and no coefficient is zero.

Simultaneous linear equations

The set of n equations in n unknowns

$$\sum_{k=1}^{n} a_{ik}x_k = b_i \quad (i = 1, 2, 3, \ldots n)$$

has a unique solution if the determinant of the coefficients $\det[a_{ik}]$ or Δ is non-zero; the solution is given by *Cramer's rule*:

$$x_k = \Delta_k/\Delta \quad (k = 1, 2, 3, \ldots n)$$

where Δ_k is obtained by replacing the kth column of Δ by the column of b_i.

Differential equations

Bessel's equation of order ν is

$$z^2\frac{\mathrm{d}^2w}{\mathrm{d}z^2} + z\frac{\mathrm{d}w}{\mathrm{d}z} + (z^2 - \nu^2)w = 0$$

and its solutions include the following Bessel functions.

First kind, order ν:

$$J_\nu(z) = \sum_{r=0}^{\infty} (-1)^r\left(\frac{z}{2}\right)^{\nu+2r} \Big/ r!\,\Gamma(\nu + r + 1).$$

First kind, order n (an integer):

$$J_n(z) = \sum_{r=0}^{\infty} (-1)^r\left(\frac{z}{2}\right)^{n+2r} \Big/ r!(n + r)!$$

Second kind, order ν (non-integral):

$$Y_\nu(z) = \{\cos\nu\pi\, J_\nu(z) - J_{-\nu}(z)\}/\sin\nu\pi.$$

Second kind, order n (an integer):

$$Y_n(z) = \left[\frac{\partial}{\partial\nu}\{\cos\nu\pi\, J_\nu(z) - J_{-\nu}(z)\}\Big/\frac{\partial}{\partial\nu}\sin\nu\pi\right]_{\nu=n}.$$

Third kind (Hankel functions), order ν:

$$H_\nu^{(1)}(z) = J_\nu(z) + \mathrm{j}Y_\nu(z)$$
$$H_\nu^{(2)}(z) = J_\nu(z) - \mathrm{j}Y_\nu(z)$$

Complete solutions may take the form, for any ν,

$$w = AJ_\nu(z) + BY_\nu(z)$$

or

$$w = AH_\nu^{(1)}z + BH_\nu^{(2)}(z)$$

where A and B are constants.

Legendre's equation of degree n has the form

$$(1 - z^2)\frac{\mathrm{d}^2w}{\mathrm{d}z^2} - 2z\frac{\mathrm{d}w}{\mathrm{d}z} + n(n + 1)w = 0$$

and its solutions are Legendre functions of the first and second kinds; for positive integral n the first are Legendre polynomials. Associated Legendre functions of degree n and order m are solutions to equations of the form

$$(1-z^2)\frac{d^2w}{dz^2} - 2z\frac{dw}{dz} + \left\{n(n+1) - \frac{m^2}{1-z^2}\right\}w = 0$$

Laguerre's equation of degree n has the form

$$z\frac{d^2w}{dz^2} + (1-z)\frac{dw}{dz} + nw = 0.$$

Its solutions for positive integral n are the Laguerre polynomials

$$L_n(z) = \frac{e^z}{n!}\frac{d^n(z^n e^{-z})}{dz^n}.$$

Chebyshev polynomials are solutions to equations of the form

$$(1-z^2)\frac{d^2w}{dz^2} - z\frac{dw}{dz} + n^2w = 0$$

for positive integral n, including:

First kind $T_n(z) = \cos(n\cos^{-1}z)$

Second kind $U_n(z) = \sin(n\cos^{-1}z)$.

Mathieu's equation takes the form

$$\frac{d^2w}{dz^2} + (a - 16q\cos 2z)w = 0$$

in which a, q are real numbers.

Riccati's equation has the general form

$$\frac{dy}{dx} = ay^2 + by + c$$

in which a, b, c may be functions of x.

Cauchy's equation has the form

$$x^2\frac{d^2y}{dx^2} + ax\frac{dy}{dx} + by = 0$$

where a, b are constants; its solution is

$$y = Ax^{m_1} + Bx^{m_2}$$

where m_1, m_2 are the roots of

$$m^2 + (a-1)m + b = 0.$$

The general form of the *wave equation* is

$$\nabla^2 y - a\frac{\partial^2 y}{\partial t^2} - b\frac{\partial y}{\partial t} - cy = 0$$

and has solutions in the form of attenuated travelling waves. If $b = c = 0$, the solutions are undamped travelling waves of phase velocity $1/\sqrt{a}$. If $a = c = 0$, there results the equation of *diffusion* or *heat conduction* which has solutions of exponential form. If $a = b = c = 0$, the equation becomes

$$\nabla^2 y = 0$$

which is *Laplace's equation*; its solutions give the spatial variation of a potential y whose gradient is a vector field of zero divergence.

Poisson's equation is

$$\nabla^2 y = \rho$$

in which ρ may be a function of position; its solutions give the potential y of a field whose divergence at any point is ρ.

Vector analysis

For two vectors **A**, **B** with angle θ between them:

Scalar product: $\mathbf{A}.\mathbf{B} = \mathbf{B}.\mathbf{A} = AB\cos\theta$

Vector product: $\mathbf{A}\times\mathbf{B} = -\mathbf{B}\times\mathbf{A}$

Scalar triple product: $\mathbf{A}\times\mathbf{B}.\mathbf{C} = \mathbf{B}\times\mathbf{C}.\mathbf{A} = \mathbf{C}\times\mathbf{A}.\mathbf{B}$

Vector triple product: $\mathbf{A}\times(\mathbf{B}\times\mathbf{C}) = (\mathbf{A}.\mathbf{C})\mathbf{B} - (\mathbf{A}.\mathbf{B})\mathbf{C}$

The vector product has magnitude $AB\sin\theta$ and is normal to the plane containing **A** and **B**.

For unit vectors **i**, **j**, **k** on right-handed orthogonal axes:

$\mathbf{i}.\mathbf{i} = \mathbf{j}.\mathbf{j} = \mathbf{k}.\mathbf{k} = 1$

$\mathbf{i}.\mathbf{j} = \mathbf{j}.\mathbf{k} = \mathbf{k}.\mathbf{i} = 0$

$\mathbf{i}\times\mathbf{i} = \mathbf{j}\times\mathbf{j} = \mathbf{k}\times\mathbf{k} = 0$

$\mathbf{i}\times\mathbf{j} = \mathbf{k} = -(\mathbf{j}\times\mathbf{i})$ etc.

In *Cartesian coordinates:*

$$\mathbf{A}.\mathbf{B} = A_xB_x + A_yB_y + A_zB_z$$

$$\mathbf{A}\times\mathbf{B} = \mathbf{i}(A_yB_z - B_yA_z) + \mathbf{j}(A_zB_x - B_zA_x) + \mathbf{k}(A_xB_y - B_xA_y)$$

$$= \begin{vmatrix} \mathbf{i} & \mathbf{j} & \mathbf{k} \\ A_x & A_y & A_z \\ B_x & B_y & B_z \end{vmatrix}$$

$d\vec{S}$: $h_u h_v\, du\, dv\, \hat{u}\times\hat{v}$ or $r\,dr\,d\theta\,\hat{k}$ or (ends) $r\,dr\,d\phi\,\hat{k}$ or (curved sides) $r\,dz\,d\phi\,\hat{e}_r$ or $r^2\sin\theta\,dr\,d\theta\,d\phi\,\hat{e}_r$

dV : $h_u h_v h_w\, du\, dv\, dw\, \hat{u}\cdot(\hat{v}\times\hat{w})$ or / or $r\,dr\,d\phi\,dz$ or — or $r^2\sin\theta\,dr\,d\theta\,d\phi$

$$\mathbf{A} \times \mathbf{B}.\mathbf{C} = A_x(B_yC_z - C_yB_z) + A_y(B_zC_x - C_zB_x) + A_z(B_xC_y - C_xB_y)$$

$$= \begin{vmatrix} A_x & A_y & A_z \\ B_x & B_y & B_z \\ C_x & C_y & C_z \end{vmatrix}$$

$$\nabla \equiv \mathbf{i}\frac{\partial}{\partial x} + \mathbf{j}\frac{\partial}{\partial y} + \mathbf{k}\frac{\partial}{\partial z}$$

(In the following, V is a scalar field, $\mathbf{F}$ a vector field.)

$$\text{grad } V = \nabla V = \mathbf{i}\frac{\partial V}{\partial x} + \mathbf{j}\frac{\partial V}{\partial y} + \mathbf{k}\frac{\partial V}{\partial z}$$

$$\text{div } \mathbf{F} = \nabla.\mathbf{F} = \frac{\partial F_x}{\partial x} + \frac{\partial F_y}{\partial y} + \frac{\partial F_z}{\partial z}$$

$$\text{curl } \mathbf{F} = \nabla \times \mathbf{F} = \mathbf{i}\left(\frac{\partial F_z}{\partial y} - \frac{\partial F_y}{\partial z}\right) + \mathbf{j}\left(\frac{\partial F_x}{\partial z} - \frac{\partial F_z}{\partial x}\right) + \mathbf{k}\left(\frac{\partial F_y}{\partial x} - \frac{\partial F_x}{\partial y}\right)$$

$$\nabla^2 \equiv \frac{\partial^2}{\partial x^2} + \frac{\partial^2}{\partial y^2} + \frac{\partial^2}{\partial z^2}$$

$$\text{div grad } V = \nabla^2 V = \frac{\partial^2 V}{\partial x^2} + \frac{\partial^2 V}{\partial y^2} + \frac{\partial^2 V}{\partial z^2}.$$

$$\nabla^2\mathbf{F} = \mathbf{i}\nabla^2 F_x + \mathbf{j}\nabla^2 F_y + \mathbf{k}\nabla^2 F_z$$

In *spherical coordinates* (unit vectors $\mathbf{u}_r$, $\mathbf{u}_\theta$, $\mathbf{u}_\phi$):

$$\text{grad } V = \mathbf{u}_r\frac{\partial V}{\partial r} + \mathbf{u}_\theta\frac{1}{r}\frac{\partial V}{\partial \theta} + \mathbf{u}_\phi\frac{1}{r\sin\theta}\frac{\partial V}{\partial \phi}$$

$$\text{div } \mathbf{F} = \frac{1}{r^2}\frac{\partial}{\partial r}(r^2F_r) + \frac{1}{r\sin\theta}\frac{\partial}{\partial \theta}(F_\theta\sin\theta) + \frac{1}{r\sin\theta}\frac{\partial F_\phi}{\partial \phi}$$

$$\text{curl } \mathbf{F} = \mathbf{u}_r\frac{1}{r\sin\theta}\left[\frac{\partial}{\partial\theta}(F_\phi\sin\theta) - \frac{\partial F_\theta}{\partial\phi}\right] + \mathbf{u}_\theta\frac{1}{r}\left[\frac{1}{\sin\theta}\frac{\partial F_r}{\partial\phi} - \frac{\partial}{\partial r}(rF_\phi)\right] + u_\phi\frac{1}{r}\left[\frac{\partial}{\partial r}(rF_\theta) - \frac{\partial F_r}{\partial\theta}\right]$$

$$\nabla^2 V = \frac{1}{r^2}\frac{\partial}{\partial r}\left(r^2\frac{\partial V}{\partial r}\right) + \frac{1}{r^2\sin\theta}\frac{\partial}{\partial\theta}\left(\sin\theta\frac{\partial V}{\partial\theta}\right) + \frac{1}{r^2\sin^2\theta}\frac{\partial^2 V}{\partial\phi^2}.$$

In *cylindrical coordinates* (unit vectors $\mathbf{u}_r$, $\mathbf{u}_\phi$, $\mathbf{u}_z$):

$$\text{grad } V = \mathbf{u}_r\frac{\partial V}{\partial r} + \mathbf{u}_\phi\frac{1}{r}\frac{\partial V}{\partial\phi} + \mathbf{u}_z\frac{\partial V}{\partial z}$$

$$\text{div } \mathbf{F} = \frac{1}{r}\frac{\partial}{\partial r}(rF_r) + \frac{1}{r}\frac{\partial F_\phi}{\partial\phi} + \frac{\partial F_z}{\partial z}$$

$$\text{curl } \mathbf{F} = \mathbf{u}_r\left(\frac{1}{r}\frac{\partial F_z}{\partial\phi} - \frac{\partial F_\phi}{\partial z}\right) + \mathbf{u}_\phi\left(\frac{\partial F_r}{\partial z} - \frac{\partial F_z}{\partial r}\right) + \mathbf{u}_z\frac{1}{r}\left[\frac{\partial}{\partial r}(rF_\phi) - \frac{\partial F_r}{\partial\phi}\right]$$

$$\nabla^2 V = \frac{1}{r}\frac{\partial}{\partial r}\left(r\frac{\partial V}{\partial r}\right) + \frac{1}{r^2}\frac{\partial^2 V}{\partial\phi^2} + \frac{\partial^2 V}{\partial z^2}.$$

General vector identities

$$\text{curl grad } V = \nabla \times \nabla V = 0$$

$$\text{div curl } \mathbf{F} = \nabla.\nabla \times \mathbf{F} = 0$$

$$\text{curl curl } \mathbf{F} = \nabla \times \nabla \times \mathbf{F} = \text{grad div } \mathbf{F} - \nabla^2\mathbf{F}$$

$$\text{grad}(V_1V_2) = V_1 \text{ grad } V_2 + V_2 \text{ grad } V_1$$

$$\text{grad}(\mathbf{F}_1.\mathbf{F}_2) = (\mathbf{F}_1.\nabla)\mathbf{F}_2 + (\mathbf{F}_2.\nabla)\mathbf{F}_1 + \mathbf{F}_1 \times \text{curl } \mathbf{F}_2 + \mathbf{F}_2 \times \text{curl } \mathbf{F}_1$$

$$\text{div}(V\mathbf{F}) = V \text{ div } \mathbf{F} + \mathbf{F}.\text{grad } V$$

$$\text{div}(\mathbf{F}_1 \times \mathbf{F}_2) = \mathbf{F}_2.\text{curl } \mathbf{F}_1.\text{curl } \mathbf{F}_2$$

$$\text{curl}(V\mathbf{F}) = V \text{ curl } \mathbf{F} - \mathbf{F} \times \text{grad } V$$

$$\text{curl}(\mathbf{F}_1 \times \mathbf{F}_2) = \mathbf{F}_1 \text{ div } \mathbf{F}_2 - \mathbf{F}_2 \text{ div } \mathbf{F}_1 + (\mathbf{F}_2.\nabla)\mathbf{F}_1 - (\mathbf{F}_1.\nabla)\mathbf{F}_2$$

A vector field of zero divergence is said to be *solenoidal.* If the line integral of $\mathbf{F}$ around any closed path is zero, then $\mathbf{F}$ has zero curl, can always be expressed as grad V and is said to be *lamellar, conservative* or *irrotational.*

Gauss's divergence theorem

If $d\tau$ is an element of a volume T bounded by a surface S of which $d\mathbf{S}$ is an element, then

$$\iiint_T \text{div } \mathbf{F}\, d\tau = \iint_S \mathbf{F}.d\mathbf{S}.$$

Stokes's theorem

If $d\mathbf{S}$ is an element of a surface S bounded by a closed curve C of which $d\mathbf{l}$ is an element, then

$$\iint_S \text{curl } \mathbf{F}.d\mathbf{S} = \int_C \mathbf{F}.d\mathbf{l}.$$

Matrices

If $A_{m,n}$ represents a matrix A of order $m \times n$ (i.e. having m rows and n columns) with the element a_{jk} in the jth row and kth column, then:

(i) $A_{m,n} + B_{m,n} = C_{m,n} = B_{m,n} + A_{m,n}$
and $c_{jk} = a_{jk} + b_{jk}$.

(Only matrices of the same order may be added or subtracted.)

(ii) $\lambda A_{m,n} = B_{m,n}$
and $b_{jk} = \lambda a_{jk}$, where λ is a scalar.

(iii) $A_{m,n}B_{n,p} = C_{m,p}$

and $c_{jk} = \sum_{l=1}^{n} a_{jl}b_{lk}$.

(Two matrices can be multiplied only if they are *conformable*, i.e. if the first has as many columns as the second has rows; in general $AB \neq BA$.)

(iv) The *transpose* of $A_{m,n}$ is

$B_{n,m} = A^T$

and $b_{jk} = a_{kj}$.

(It follows that $(A + B)^T = A^T + B^T$ and $(AB)^T = B^T A^T$.)

(v) A is a *square matrix* if $m = n$

A is a *row matrix* if $m = 1$

A is a *column matrix* if $n = 1$.

Square matrices

A square matrix A is:

symmetric if $A^T = A$

skew-symmetric if $A^T = -A$

diagonal if $a_{jk} = 0$ $(j \neq k)$.

A *unit matrix* U is a diagonal matrix with $u_{jk} = 1$ $(j = k)$.

The *determinant* of a square matrix A of order n has the value

$$\det A = |A| = \sum_{k=1}^{n} a_{jk}C_{jk} \quad \text{for any } j$$

$$= \sum_{j=1}^{n} a_{jk}C_{jk} \quad \text{for any } k$$

where

$$C_{jk} = (-1)^{j+k}M_{jk}$$

and M_{jk} is the determinant of order $n-1$ obtained by deleting the jth row and kth column of A. The determinants C_{jk} and M_{jk} are the *cofactor* and *minor* respectively of the element a_{jk}. If $|A| = 0$ then A is a *singular* matrix.

The *adjoint* or adjugate matrix adj A of a square matrix A is the transpose of the matrix of the cofactors of A. The *inverse* A^{-1} of a non-singular square matrix A is a square matrix of the same order such that

$$AA^{-1} = U$$

and is given by

$$A^{-1} = \frac{\operatorname{adj} A}{|A|}$$

If A and B are square non-singular matrices of the same order, then

$$(AB)^{-1} = B^{-1}A^{-1}.$$

A is *orthogonal* if $A^{-1} = A^T$.

The Jacobian

If u, v are functions of the variables x, y then the Jacobian of u and v with respect to x and y is the determinant

$$J = \begin{vmatrix} \frac{\partial u}{\partial x} & \frac{\partial u}{\partial y} \\ \frac{\partial v}{\partial x} & \frac{\partial v}{\partial y} \end{vmatrix} = \frac{\partial(u, v)}{\partial(x, y)}.$$

For n functions u_i of n variables x_i the Jacobian is similarly

$$J = \frac{\partial(u_1, u_2, \ldots u_n)}{\partial(x_1, x_2, \ldots x_n)}$$

The functions are independent if $J \neq 0$.

Matrix representation of vectors

An n-dimensional vector may be represented as a row matrix of order $1 \times n$ or as a column matrix of order $n \times 1$. The column matrix for a vector **A** may be written $\{A\}$. The scalar product of **A** and **B** is then

$$\mathbf{A} \cdot \mathbf{B} = \{A\}^T\{B\} = \{B\}^T\{A\}.$$

The vector product of two three-dimensional vectors **A** and **B** in Cartesian coordinates is

$$\mathbf{A} \times \mathbf{B} = [A]\{B\}$$

where $[A]$ is the skew-symmetric matrix

$$\begin{bmatrix} 0 & -A_z & A_y \\ A_z & 0 & -A_x \\ -A_y & A_x & 0 \end{bmatrix}.$$

Rotation of axes

If a vector is represented by $\{R\}$ in a system of Cartesian coordinates $OXYZ$, and by $\{r\}$ in a second system $Oxyz$ having the same origin O, then

$$\{r\} = [C]\{R\}$$

where $[C]$ is the *rotation matrix*

$$\begin{bmatrix} l_{xX} & l_{xY} & l_{xZ} \\ l_{yX} & l_{yY} & l_{yZ} \\ l_{zX} & l_{zY} & l_{zZ} \end{bmatrix}$$

in which $l_{xX} = \cos(xOX)$, xOX being the angle between Ox and OX, etc.

Properties of plane curves and figures

Pappus's theorems

(i) The surface area generated by a curve of length l revolving about an axis is

$$A = 2\pi l\bar{y}$$

where $\bar{y}$ is the perpendicular distance of the centroid of the curve from the axis.

(ii) The volume generated by a plane surface of area A rotati about an axis is

$$V = 2\pi A\bar{y}$$

where $\bar{y}$ is the perpendicular distance of the centroid of A from the axis.

Conic sections

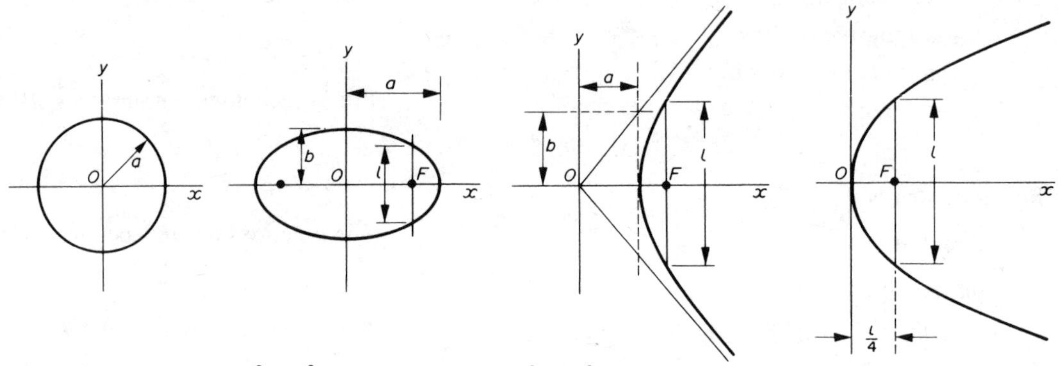

	Circle	Ellipse	Hyperbola	Parabola
Cartesian equation	$x^2 + y^2 = a^2$	$\frac{x^2}{a^2} + \frac{y^2}{b^2} = 1$	$\frac{x^2}{a^2} - \frac{y^2}{b^2} = 1$	$y^2 = lx$
Eccentricity ϵ	0	$\left(1 - \frac{b^2}{a^2}\right)^{1/2} < 1$	$\left(1 + \frac{b^2}{a^2}\right)^{1/2} > 1$	1
Focal distance OF	0	$a\epsilon$	$a\epsilon$	$l/4$
Latus rectum l	$2a$	$2b^2/a$	$2b^2/a$	
Circumference	$2\pi a$	$4aE \approx 2\pi\sqrt{\left(\frac{a^2 + b^2}{2}\right)}$		
Enclosed area	πa^2	πab		$h^3/6l$*
Polar equation, origin O	$r = a$	$r^2 = b^2/(1 - \epsilon^2 \cos^2 \theta)$	$r^2 = -b^2/(1 - \epsilon^2 \cos^2 \theta)$	$r = l \cos\theta/(1 - \cos^2\theta)$
Polar equation, origin F	$r = a$	$r = l/2(1 - \epsilon \cos\theta)$	$r = l/2(1 - \epsilon\cos\theta)$	$r = l/2(1 - \cos\theta)$

* Area enclosed by curve and vertical chord of length h.

Other curves

Catenary

$$y = a \cosh(x/a)$$

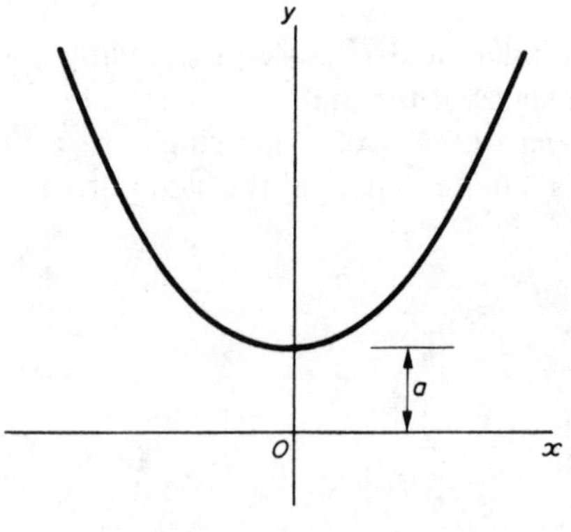

Cycloids

$x = a\theta - b \sin \theta$

$y = a - b \cos \theta$

(i) $a = b$ (arc $8a$, area $3\pi a^2$)

(ii) $a < b$ (prolate)

(iii) $a > b$ (curtate)

b is the generating radius on the circle of radius a.

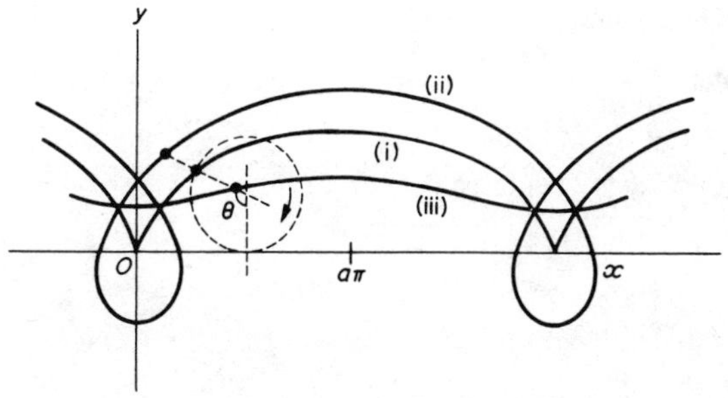

Epicycloids

$$x = (a + b) \cos \phi - b \cos \left(\frac{a + b}{b}\right) \phi$$

$$y = (a + b) \sin \phi - b \sin \left(\frac{a + b}{b}\right) \phi$$

(i) $0 < b < a$

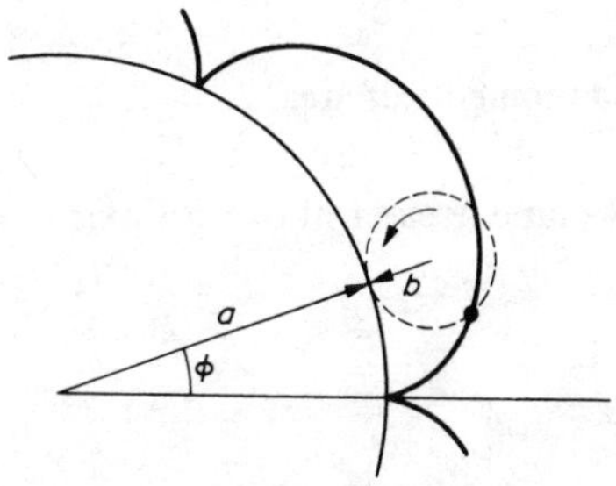

(ii) $a = b$: *cardioid*

polar equation: $r = 2a(1 - \cos \theta)$

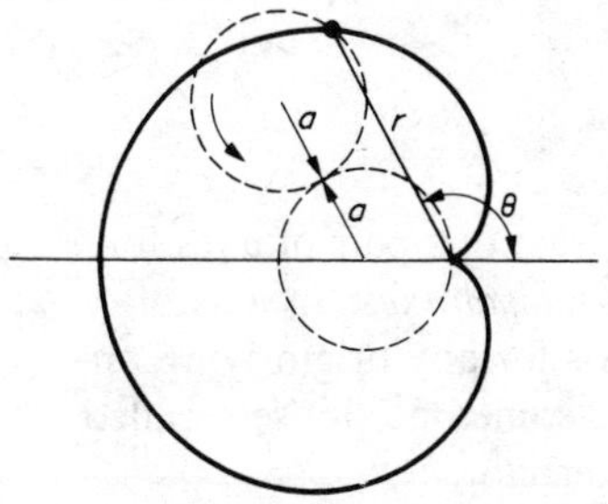

(iii) $b \to -b$: *hypocycloid*

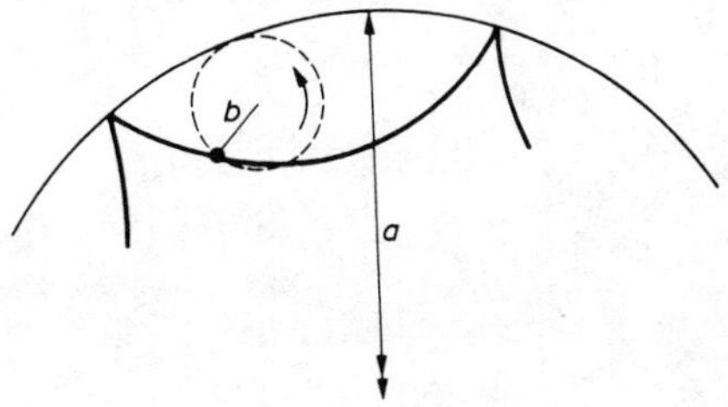

Logarithmic spiral

$r = a\, e^{b\theta}$

Area between radii $r_1, r_2 = \dfrac{r_2^2 - r_1^2}{4b}$

Length between radii $r_1, r_2 = \dfrac{(r_2 - r_1)\sqrt{(b^2 + 1)}}{b}$

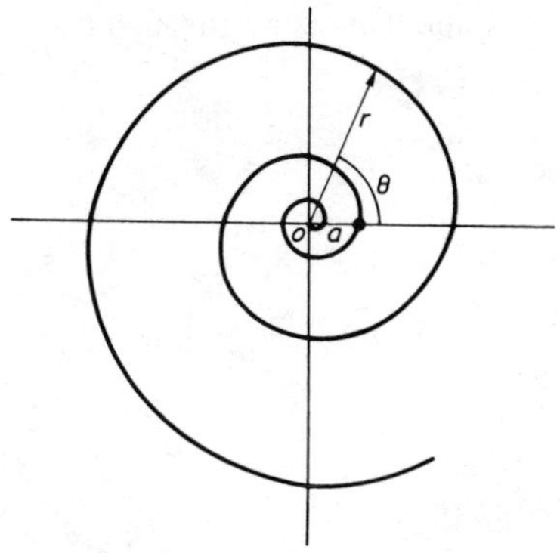

Archimedean spiral

$r = a\theta$
Area $= a^2\theta^3/6$
Length $\rightarrow a\theta^2/2$ for large θ

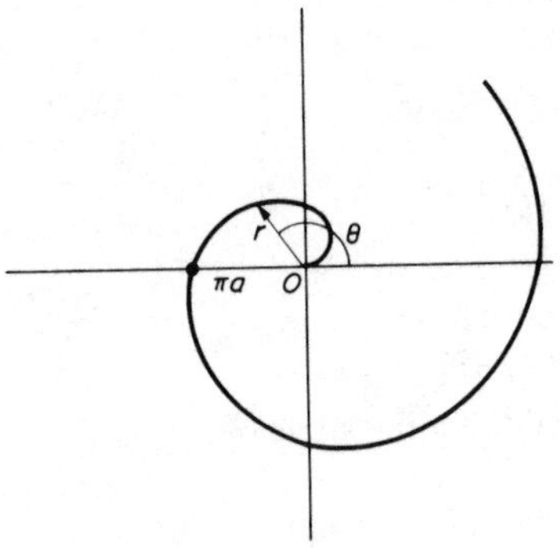

Areas, centroids and second moments of area

The second moment of a plane area A about an axis x in its plane is

$$I_{xx} = \int y^2 \mathrm{d}A = Ak_x^2$$

where k_x is the radius of gyration about x. The product moment of area is

$$I_{xy} = \int xy\mathrm{d}A.$$

For any origin there is at least one pair of axes for which $I_{xy} = 0$. These are *principal axes*; an axis of symmetry is a principal axis for any origin lying on it. The above moments are sometimes loosely called moment and product of inertia.

The centroid of A is defined by the coordinates

$$\bar{x} = \frac{1}{A}\int_A x\mathrm{d}A, \quad \bar{y} = \frac{1}{A}\int_A y\mathrm{d}A$$

Polar moment

The polar second moment for any point O in the plane of A is the second moment of A about an axis z through O normal to A. It is given by

$$I_{zz} = I_{xx} + I_{yy}$$

for any pair of orthogonal axes x, y in A with origin O.

Parallel-axis theorem

If I is the second moment of area about any axis through the centroid of A, then that for a parallel axis at a perpendicular distance d is

$$I' = I + Ad^2$$

and if I is the product moment for any pair of Cartesian axes through the centroid, that for a parallel pair at distances a, b is

$$I' = I + Aab$$

Rotation of axes

If I_{xx}, I_{yy} and I_{xy} are values for axes x,y, then the corresponding values for axes x',y' at an angle α to x,y and having the same origin are

$$I_{x'x'} = \tfrac{1}{2}(I_{xx} + I_{yy}) + \tfrac{1}{2}(I_{xx} - I_{yy}) \cos 2\alpha - I_{xy} \sin 2\alpha$$

$$I_{y'y'} = \tfrac{1}{2}(I_{xx} + I_{yy}) + \tfrac{1}{2}(I_{yy} - I_{xx}) \cos 2\alpha + I_{xy} \sin 2\alpha$$

$$I_{x'y'} = \tfrac{1}{2}(I_{xx} - I_{yy}) \sin 2\alpha + I_{xy} \cos 2\alpha.$$

In the following table, $\bar{x}, \bar{y}$ are the coordinates of the centroid C with respect to the origin O; I_{xx}, etc., are moments for axes *through* C in the directions x, y.

Figure and area, A	$\bar{x}, \bar{y}$	I_{xx}	I_{yy}	I_{xy}
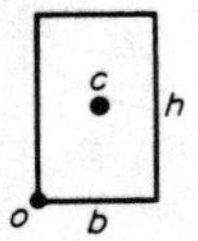 bh rectangle	$\frac{b}{2}, \frac{h}{2}$	$\frac{Ah^2}{12}$	$\frac{Ab^2}{12}$	0
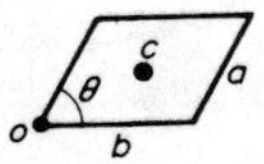 $ab \sin \theta$ parallelogram	$\frac{b + a \cos \theta}{2}, \frac{a \sin \theta}{2}$	$\frac{A}{12} (a \sin \theta)^2$	$\frac{A}{12} (b^2 + a^2 \cos^2 \theta)$	$\frac{A}{12} a^2 \sin \theta \cos \theta$
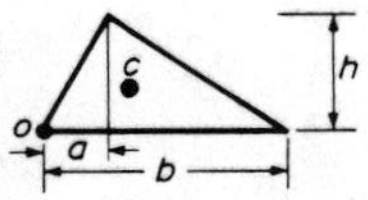 $\frac{bh}{2}$ triangle	$\frac{a+b}{3}, \frac{h}{3}$	$\frac{Ah^2}{18}$	$\frac{A}{18} (b^2 - ab + a^2)$	$\frac{Ah}{36} (2a - b)$
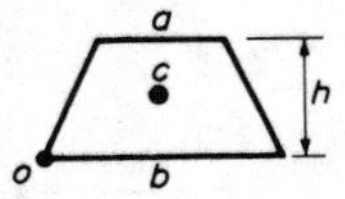 $\frac{h}{2}(a + b)$ trapezium	$\bar{y} = \frac{h(2a + b)}{3(a + b)}$	$\frac{Ah^2(a^2 + 4ab + b^2)}{18(a + b)^2}$		
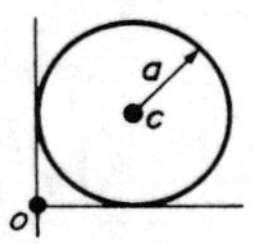 πa^2 circle	a, a	$\frac{Aa^2}{4}$	$\frac{Aa^2}{4}$	0
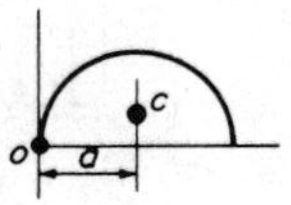 $\frac{\pi a^2}{2}$ semicircle	$a, \frac{4a}{3\pi}$	$\frac{Aa^2(9\pi^2 - 64)}{36\pi^2}$	$\frac{Aa^2}{4}$	0
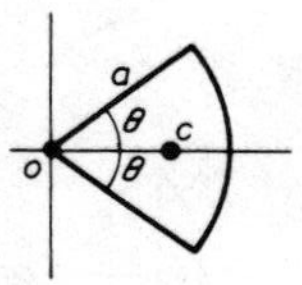 $a^2\theta$ sector of circle	$\frac{2a \sin \theta}{3\theta}, 0$	$\frac{Aa^2(\theta - \sin \theta \cos \theta)}{4\theta}$	$A \left\{ \frac{a^2(\theta + \sin \theta \cos \theta)}{4\theta} - \bar{x}^2 \right\}$	0

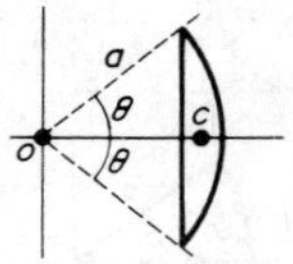	$a^2(\theta - \frac{1}{2}\sin 2\theta)$ segment of circle	$\dfrac{2a\sin^3\theta}{3(\theta - \frac{1}{2}\sin 2\theta)}, 0$	$\dfrac{Aa^2}{4}\left\{1 - \dfrac{\sin^2\theta\sin 2\theta}{3(\theta - \frac{1}{2}\sin 2\theta)}\right\}$	$\dfrac{Aa^2}{4}\left\{1 + \dfrac{\sin^2\theta\sin 2\theta}{\theta - \frac{1}{2}\sin 2\theta}\right\} - A\bar{x}^2$	0
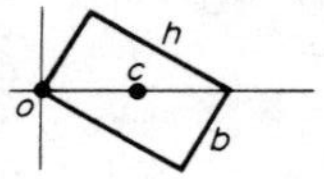	bh rectangle	$\frac{1}{2}\sqrt{(b^2 + h^2)}, 0$	$\dfrac{Ab^2h^2}{6(h^2 + b^2)}$	$\dfrac{A(h^4 + b^4)}{12(h^2 + b^2)}$	$\dfrac{Abh(h^2 - b^2)}{12(h^2 + b^2)}$
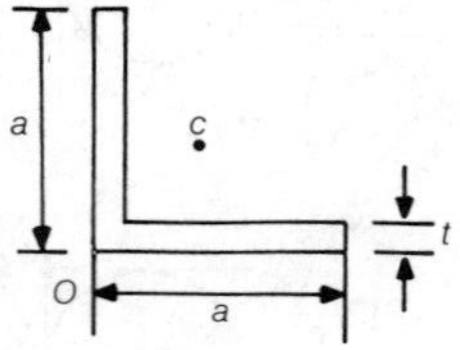	$2at$ angle section ($t \ll a$)	$\dfrac{a}{4}, \dfrac{a}{4}$	$\dfrac{5Aa^2}{48}$	$\dfrac{5Aa^2}{48}$	$\dfrac{Aa^2}{16}$
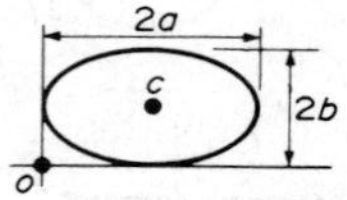	πab ellipse	a, b	$\dfrac{Ab^2}{4}$	$\dfrac{Aa^2}{4}$	0
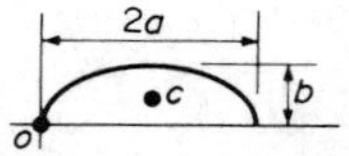	$\dfrac{\pi ab}{2}$ semi-ellipse	$a, \dfrac{4b}{3\pi}$	$\dfrac{Ab^2(9\pi^2 - 64)}{36\pi^2}$	$\dfrac{Aa^2}{4}$	0
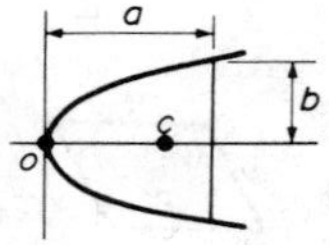	$\dfrac{4ab}{3}$ parabola	$\dfrac{3a}{5}, 0$	$\dfrac{Ab^2}{5}$	$\dfrac{12Aa^2}{175}$	0
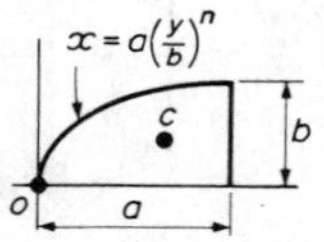	$\dfrac{nab}{n+1}$	$\dfrac{(n+1)a}{2n+1}, \dfrac{(n+1)b}{2(n+2)}$			
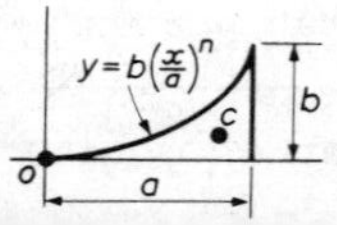	$\dfrac{ab}{n+1}$	$\dfrac{(n+1)a}{n+2}, \dfrac{(n+1)b}{2(2n+1)}$			

Moments of inertia, etc., of rigid bodies

In the following, ρ is mass density, m total mass and V volume.

The moment of inertia of a rigid body about the x-axis of a Cartesian set is

$$I_{xx} = \int_V \rho(y^2 + z^2)\mathrm{d}V$$

and the product of inertia for axes x and y is

$$I_{xy} = \int_V \rho xy\mathrm{d}V.$$

The centre of mass is defined by the coordinates

$$\bar{x} = \frac{1}{m}\int_V x\rho\mathrm{d}V$$

The inertia tensor or matrix is

$$[I] = \begin{bmatrix} I_{xx} & -I_{xy} & -I_{xz} \\ -I_{yz} & I_{yy} & -I_{yz} \\ -I_{zx} & -I_{zy} & I_{zz} \end{bmatrix}$$

For any origin there exists at least one set of axes for which the products of inertia are all zero. These are *principal axes*, and the corresponding moments I_{xx} etc. are the principal moments of inertia for that origin.

If two axes, say x and y, lie in a plane of mass symmetry, then only the product I_{xy} can be non-zero. If there are two orthogonal planes of symmetry, their intersection is a principal axis for any origin lying on it.

Parallel-axis theorem

If I is the moment of inertia about any axis through the centre of mass C, then that for a parallel axis at a perpendicular distance d is

$$I' = I + md^2$$

and if I is the product of inertia for any Cartesian pair through C, that for a parallel pair at distances a, b is

$$I' = I + mab$$

Rotation of axes

If I is the inertia matrix for certain axes, the matrix for a new set having the same origin and a rotation matrix $[C]$ is

$$[I'] = [C]\,[I]\,[C]^T$$

For any origin, the sum of the moments $I_{xx} + I_{yy} + I_{zz}$ is invariant.

In the following table for homogeneous bodies $\bar{x}, \bar{y}, \bar{z}$ are the coordinates of the centre of mass C with respect to the origin O; I_{xx}, etc., are the (principal) moments for axes *through* C in the directions x, y, z; A is the area of external curved surfaces only and V is the volume. Shells are assumed to have uniform thickness t.

Body 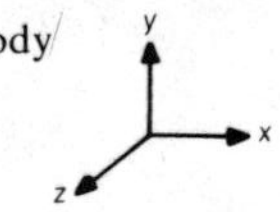	A	V	$\bar{x}, \bar{y}, \bar{z}$	I_{xx}	I_{yy}	I_{zz}
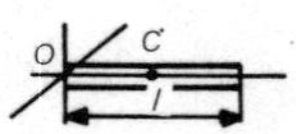 uniform rod	small	small	$\frac{l}{2}, 0, 0$	0	$\frac{ml^2}{12}$	$\frac{ml^2}{12}$
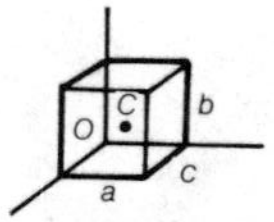 rectangular prism		abc	$\frac{a}{2}, \frac{b}{2}, \frac{c}{2}$	$\frac{m(b^2+c^2)}{12}$	$\frac{m(c^2+a^2)}{12}$	$\frac{m(a^2+b^2)}{12}$
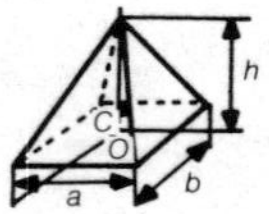 right rectangular pyramid		$\frac{abh}{3}$	$0, \frac{h}{4}, 0$	$\frac{m(4b^2+3h^2)}{80}$	$\frac{m(a^2+b^2)}{20}$	$\frac{m(4a^2+3h^2)}{80}$
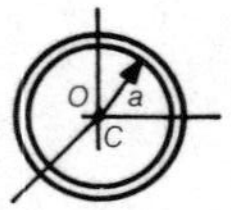 uniform hoop	small	small	$0, 0, 0$	$\frac{ma^2}{2}$	$\frac{ma^2}{2}$	ma^2
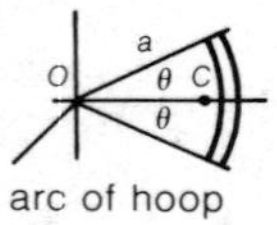 arc of hoop	small	small	$\frac{a \sin\theta}{\theta}, 0, 0$	$\frac{ma^2(\theta - \sin\theta\cos\theta)}{2\theta}$	$ma^2\left(\frac{1}{2} + \frac{\sin 2\theta}{4\theta} - \frac{\sin^2\theta}{\theta^2}\right)$	$ma^2\left(1 - \frac{\sin^2\theta}{\theta^2}\right)$
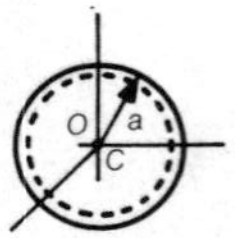 spherical shell	$4\pi a^2$	At	$0, 0, 0$	$\frac{2ma^2}{3}$	$\frac{2ma^2}{3}$	$\frac{2ma^2}{3}$
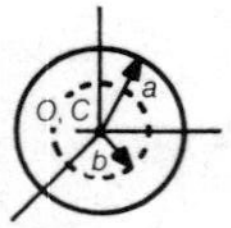 hollow sphere	$4\pi a^2$	$\frac{4\pi}{3}(a^3 - b^3)$	$0, 0, 0$	$\frac{2m(a^5-b^5)}{5(a^3-b^3)}$	$\frac{2m(a^5-b^5)}{5(a^3-b^3)}$	$\frac{2m(a^5-b^5)}{5(a^3-b^3)}$

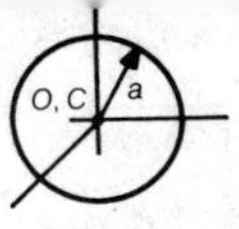 sphere	$4\pi a^2$	$\frac{4\pi a^3}{3}$	$0,0,0$	$\frac{2ma^2}{5}$	$\frac{2ma^2}{5}$	$\frac{2ma^2}{5}$
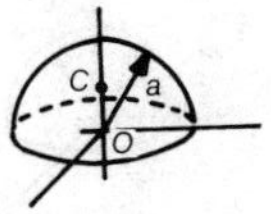 hemisphere	$2\pi a^2$	$\frac{2\pi a^3}{3}$	$0,\frac{3a}{8},0$	$\frac{83ma^2}{320}$	$\frac{2ma^2}{5}$	$\frac{83ma^2}{320}$
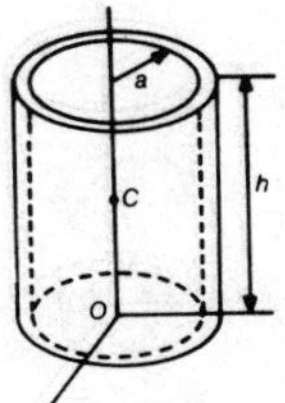 cylindrical shell	$2\pi ah$	At	$0,\frac{h}{2},0$	$\frac{m(6a^2+h^2)}{12}$	ma^2	$\frac{m(6a^2+h^2)}{12}$
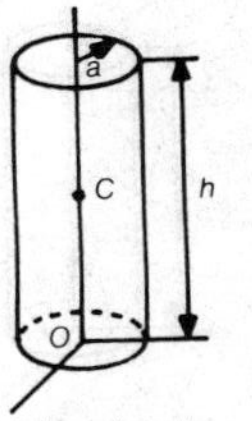 right circular cylinder	$2\pi ah$	$\pi a^2 h$	$0,\frac{h}{2},0$	$\frac{m(3a^2+h^2)}{12}$	$\frac{ma^2}{2}$	$\frac{m(3a^2+h^2)}{12}$
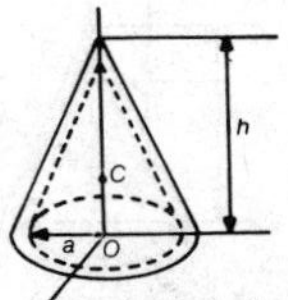 conical shell	$\pi a\sqrt{a^2+h^2}$	At	$0,\frac{h}{3},0$	$\frac{m(9a^2+2h^2)}{36}$	$\frac{ma^2}{2}$	$\frac{m(9a^2+2h^2)}{36}$

Body 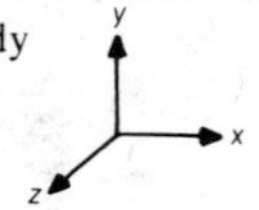	A	V	$\bar{x}, \bar{y}, \bar{z}$	I_{xx}	I_{yy}	I_{zz}
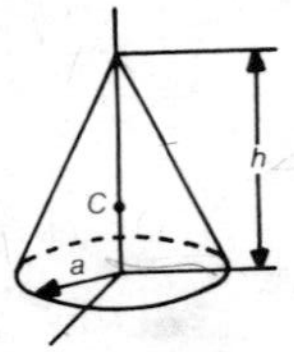 right circular cone	$\pi a\sqrt{a^2+h^2}$	$\frac{\pi a^2 h}{3}$	$0, \frac{h}{4}, 0$	$\frac{3m(4a^2+h^2)}{80}$	$\frac{3ma^2}{10}$	$\frac{3m(4a^2+h^2)}{80}$
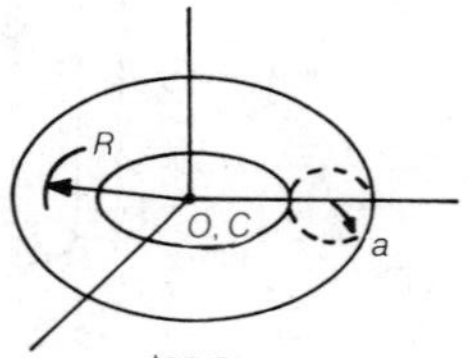 torus	$4\pi^2 Ra$	$2\pi^2 Ra^2$	$0, 0, 0$	$\frac{m(5a^2+4R^2)}{8}$	$\frac{m(3a^2+4R^2)}{4}$	$\frac{m(5a^2+4R^2)}{8}$
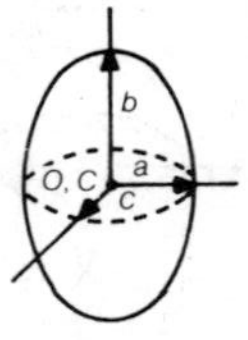 ellipsoid		$\frac{4\pi abc}{3}$	$0, 0, 0$	$\frac{m(b^2+c^2)}{5}$	$\frac{m(c^2+a^2)}{5}$	$\frac{m(a^2+b^2)}{5}$
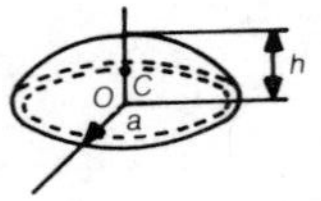 segment of spherical shell*	$2\pi rh$	At	$0, \frac{h}{2}, 0$	$\frac{mh(6a-h)}{12}$	$\frac{mh(3a-h)}{3}$	$\frac{mh(6a-h)}{12}$
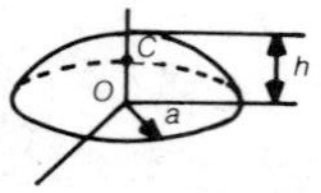 segment of sphere*	$2\pi rh$	$\pi h^2\left(r-\frac{h}{3}\right)$	$0, \frac{h(4a-h)}{4(3a-h)}, 0$			

*radius r

Numerical analysis

Solution of algebraic equation $f(x) = 0$.

(i) Newton's method:

$$x_{n+1} = x_n - f(x_n)/f'(x_n)$$

(ii) Secant method:

$$x_{n+1} = \frac{-x_n f(x_{n-1}) + x_{n-1} f(x_n)}{f(x_n) - f(x_{n-1})}$$

where x_n is the nth estimate.

Approximations to derivatives

$$f'(x) = \frac{f(x+h) - f(x-h)}{2h}$$

$$f''(x) = \frac{f(x+h) - 2f(x) + f(x-h)}{h^2}$$

$$f'''(x) = \frac{f(x+2h) - 3f(x+h) + 3f(x-h) - f(x-2h)}{2h^3}$$

$$f^{(4)}(x) = \frac{f(x+2h) - 4f(x+h) + 6f(x) - 4f(x-h) + f(x-2h)}{h^4}$$

where h is an increment in x.

Numerical integration by equal intervals h

(i) Trapezoidal rule:

$$\int_{x_0}^{x_1} y(x)\,dx = \frac{h}{2}(y_0 + y_1) - 0(h^3 y_0''/12)$$

(ii) Simpson's rule:

$$\int_{x_0}^{x_2} y(x)\,dx = \frac{h}{3}(y_0 + 4y_1 + y_2) - 0(h^5 y_1^{(4)}/90)$$

where $x_n = x_0 + nh, y_n = y(x_n)$.

Everett's interpolation formula for a table of $y(x)$

If $x = x_0 + s(x_1 - x_0)$ and $p = 1 - s$, then

$$y(x) \approx \binom{p}{1} y_0 + \binom{p+1}{3} \delta^2 y_0 + \ldots$$

$$+ \binom{s}{1} y_1 + \binom{s+1}{3} \delta^2 y_1 + \ldots$$

where $\delta^2 y_0 = y_1 - 2y_0 + y_{-1}$, etc.

Smoothing

Third-order, five-point: a least-squares cubic for five successive points $y_{-2} \ldots y_2$ is fixed by the points

$$y_0^1 = y_0 - \tfrac{3}{35}\delta^4 y_0 \text{(similarly } y_1^1, y_2^1, \ldots)$$

$$\left.\begin{aligned} y_{-1}^1 &= y_{-1} + \tfrac{2}{35}\delta^4 y_0 \\ y_{-2}^1 &= y_{-2} - \tfrac{1}{70}\delta^4 y_0 \end{aligned}\right\} \text{end points}$$

Gaussian integration (second order)

$$\int_{-1}^{1} f(x)\,dx = f\left(-\frac{1}{\sqrt{3}}\right) + f\left(\frac{1}{\sqrt{3}}\right)$$

Integration of ordinary differential equations $\dfrac{dy}{dx} = f(x, y)$

(i) *Runge-Kutta*

2nd order: $y_{n+1} = y_n + \dfrac{h}{2}[f(x_n, y_n)$

$+ f\{x_n + h, y_n + hf(x_n, y_n)\}]$

4th order: $y_{n+1} = y_n + \tfrac{1}{6}(k_1 + 2k_2 + 2k_3 + k_4)$

where

$$k_1 = hf(x_n, y_n)$$

$$k_2 = hf\left(x_n + \frac{h}{2}, y_n + \frac{k_1}{2}\right)$$

$$k_3 = hf\left(x_n + \frac{h}{2}, y_n + \frac{k_2}{2}\right)$$

$$k_4 = hf(x_n + h, y_n + k_3)$$

(ii) *Adams-Bashforth*

Predictor:

$$y_{n+1} = y_n + \frac{h}{24}\{55f(x_n, y_n) - 59f(x_{n-1}, y_{n-1})$$

$$+ 37f(x_{n-2}, y_{n-2}) - 9f(x_{n-3}, y_{n-3})\}$$

Corrector:

$$y_{n+1}^k = y_n + \frac{h}{24}\{9f(x_{n+1}, y_{n+1}^{k-1}) + 19f(x_n, y_n)$$

$$- 5f(x_{n-1}, y_{n-1}) + f(x_{n-2}, y_{n-2})\}$$

Here $x_n = x_0 + nh$, y_n^k is the kth estimate of $y(x_n)$.

Statistics

The variance of n values of a variable x is

$$s^2 = \frac{1}{n-1} \sum_{j=1}^{n} (x_j - \bar{x})^2$$

and the standard deviation is s.

If the value of a variable x has a probability function of density $f(x)$ then the mean of x is

$$\mu = \int_{-\infty}^{\infty} x f(x)\, dx$$

the variance of the distribution of x is

$$\sigma^2 = \int_{-\infty}^{\infty} (x - \mu)^2 f(x)\, dx$$

and σ is the standard deviation.

Distributions

If a certain event has a probability p of occurring in each of n independent trials, then the occurrence of x events has the probability

$$f(x) = \binom{n}{x} p^x (1-p)^{n-x}$$

x has the mean $\mu = np$ and $f(x)$ has the variance $np(1-p)$. This is the *binomial distribution.*

The limit of the binomial distribution for $p \to 0$, $n \to \infty$ is the *Poisson distribution* with probability function

$$f(x) = \frac{\mu^x}{x!} e^{-\mu} \quad (x = 0, 1, 2 \ldots)$$

when np is defined as the mean, μ. The variance is then

$$\sigma^2 = \mu$$

The Gauss or *normal distribution* is defined by the probability function

$$f(x) = \frac{1}{\sigma\sqrt{(2\pi)}} e^{-(x-\mu)^2/2\sigma^2}$$

in which μ is the mean of x and σ the standard deviation.

The probability that in any one trial the variable will assume a value $\leqslant x$ is the *distribution function*

$$F(x) = \int_{-\infty}^{x} f(t)\, dt$$

For the normal distribution with $\mu = 0$, $\sigma^2 = 1$ this becomes

$$\Phi(x) = \frac{1}{\sqrt{(2\pi)}} \int_{-\infty}^{x} e^{-t^2/2}\, dt$$

corresponding to a probability function (or frequency curve)

$$\phi(x) = \frac{1}{\sqrt{(2\pi)}} e^{-x^2/2}$$

The probability that $a < x < b$ for a variable x with mean μ and standard deviation σ is then

$$p(a < x < b) = \Phi\left(\frac{b-\mu}{\sigma}\right) - \Phi\left(\frac{a-\mu}{\sigma}\right)$$

the probability that $(\mu - n\sigma) < x < (\mu + n\sigma)$ is

$$p_{n\sigma} = \Phi(n) - \Phi(-n)$$

For $n = 2$ this gives $p_{2\sigma} = 0{\cdot}955$, for $n = 3$ $p_{3\sigma} = 0{\cdot}997$.

Percentage points of the *t*-distribution

This table gives percentage points $t_\nu(P)$ defined by the equation

$$\frac{P}{100} = \frac{1}{\sqrt{\nu\pi}} \frac{\Gamma(\frac{1}{2}\nu+\frac{1}{2})}{\Gamma(\frac{1}{2}\nu)} \int_{t_\nu(P)}^{\infty} \frac{dt}{(1+t^2/\nu)^{\frac{1}{2}(\nu+1)}}.$$

Let X_1 and X_2 be independent random variables having a normal distribution with zero mean and unit variance and a χ^2-distribution with ν degrees of freedom respectively; then $t = X_1/\sqrt{X_2/\nu}$ has Student's *t*-distribution with ν degrees of freedom, and the probability that $t \geqslant t_\nu(P)$ is $P/100$. The lower percentage points are given by symmetry as $-t_\nu(P)$, and the probability that $|t| \geqslant t_\nu(P)$ is $2P/100$.

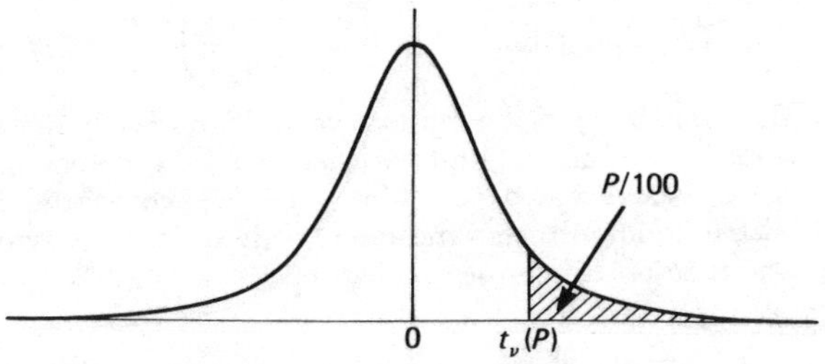

The limiting distribution of t as ν tends to infinity is the normal distribution with zero mean and unit variance. When ν is large interpolation in ν should be harmonic.

P	40	30	25	20	15	10	5	2·5	1	0·5	0·1	0·05
$\nu = 1$	0·3249	0·7265	1·0000	1·3764	1·963	3·078	6·314	12·71	31·82	63·66	318·3	636·6
2	·2887	·6172	0·8165	1·0607	·386	1·886	2·920	4·303	6·965	9·925	22·33	31·60
3	·2767	·5844	0·7649	0·9785	·250	1·638	2·353	3·182	4·541	5·841	10·21	12·92
4	·2707	·5686	0·7407	0·9410	·190	1·533	2·132	2·776	3·747	4·604	7·173	8·610
5	0·2672	0·5594	0·7267	0·9195	1·156	1·476	2·015	2·571	3·365	4·032	5·893	6·869
6	·2648	·5534	·7176	·9057	·134	·440	1·943	·447	3·143	3·707	5·208	5·959
7	·2632	·5491	·7111	·8960	·119	·415	1·895	·365	2·998	3·499	4·785	5·408
8	·2619	·5459	·7064	·8889	·108	·397	1·860	·306	2·896	3·355	4·501	5·041
9	·2610	·5435	·7027	·8834	·100	·383	1·833	·262	2·821	3·250	4·297	4·781
10	0·2602	0·5415	0·6998	0·8791	1·093	1·372	1·812	2·228	2·764	3·169	4·144	4·587
11	·2596	·5399	·6974	·8755	·088	·363	·796	·201	·718	3·106	4·025	·437
12	·2590	·5386	·6955	·8726	·083	·356	·782	·179	·681	3·055	3·930	·318
13	·2586	·5375	·6938	·8702	·079	·350	·771	·160	·650	3·012	3·852	·221
14	·2582	·5366	·6924	·8681	·076	·345	·761	·145	·624	2·977	3·787	·140
15	0·2579	0·5357	0·6912	0·8662	1·074	1·341	1·753	2·131	2·602	2·947	3·733	4·073
16	·2576	·5350	·6901	·8647	·071	·337	·746	·120	·583	·921	·686	4·015
17	·2573	·5344	·6892	·8633	·069	·333	·740	·110	·567	·898	·646	3·965
18	·2571	·5338	·6884	·8620	·067	·330	·734	·101	·552	·878	·610	3·922
19	·2569	·5333	·6876	·8610	·066	·328	·729	·093	·539	·861	·579	3·883
20	0·2567	0·5329	0·6870	0·8600	1·064	1·325	1·725	2·086	2·528	2·845	3·552	3·850
21	·2566	·5325	·6864	·8591	·063	·323	·721	·080	·518	·831	·527	·819
22	·2564	·5321	·6858	·8583	·061	·321	·717	·074	·508	·819	·505	·792
23	·2563	·5317	·6853	·8575	·060	·319	·714	·069	·500	·807	·485	·768
24	·2562	·5314	·6848	·8569	·059	·318	·711	·064	·492	·797	·467	·745
25	0·2561	0·5312	0·6844	0·8562	1·058	1·316	1·708	2·060	2·485	2·787	3·450	3·725
26	·2560	·5309	·6840	·8557	·058	·315	·706	·056	·479	·779	·435	·707
27	·2559	·5306	·6837	·8551	·057	·314	·703	·052	·473	·771	·421	·690
28	·2558	·5304	·6834	·8546	·056	·313	·701	·048	·467	·763	·408	·674
29	·2557	·5302	·6830	·8542	·055	·311	·699	·045	·462	·756	·396	·659
30	0·2556	0·5300	0·6828	0·8538	1·055	1·310	1·697	2·042	2·457	2·750	3·385	3·646
32	·2555	·5297	·6822	·8530	·054	·309	·694	·037	·449	·738	·365	·622
34	·2553	·5294	·6818	·8523	·052	·307	·691	·032	·441	·728	·348	·601
36	·2552	·5291	·6814	·8517	·052	·306	·688	·028	·434	·719	·333	·582
38	·2551	·5288	·6810	·8512	·051	·304	·686	·024	·429	·712	·319	·566
40	0·2550	0·5286	0·6807	0·8507	1·050	1·303	1·684	2·021	2·423	2·704	3·307	3·551
50	·2547	·5278	·6794	·8489	·047	·299	·676	2·009	·403	·678	·261	·496
60	·2545	·5272	·6786	·8477	·045	·296	·671	2·000	·390	·660	·232	·460
120	·2539	·5258	·6765	·8446	·041	·289	·658	1·980	·358	·617	·160	·373
∞	0·2533	0·5244	0·6745	0·8416	1·036	1·282	1·645	1·960	2·326	2·576	3·090	3·291

The normal distribution function

The function tabulated is $\Phi(x) = \dfrac{1}{\sqrt{2\pi}} \displaystyle\int_{-\infty}^{x} e^{-\frac{1}{2}t^2}\, dt$. $\Phi(x)$ is the probability that a random variable, normally distributed with zero mean and unit variance, will be less than or equal to x. When $x < 0$ use $\Phi(x) = 1 - \Phi(-x)$, as the normal distribution with zero mean and unit variance is symmetric about zero.

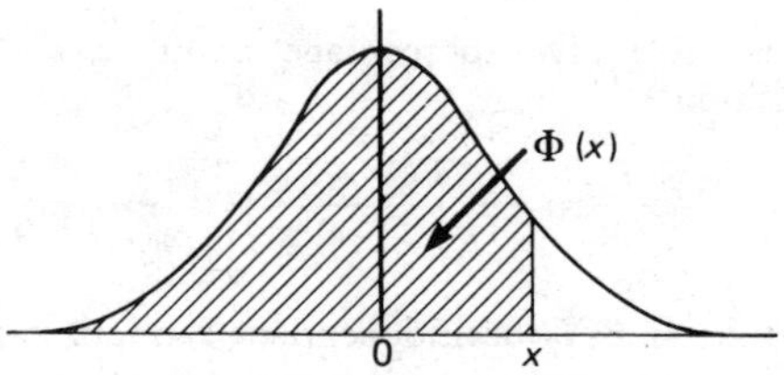

x	$\Phi(x)$	x	$\Phi(x)$	x	$\Phi(x)$	x	$\Phi(x)$	x	$\Phi(x)$	x	$\Phi(x)$
0·00	0·5000	0·40	0·6554	0·80	0·7881	1·20	0·8849	1·60	0·9452	2·00	0·97725
·01	·5040	·41	·6591	·81	·7910	·21	·8869	·61	·9463	·01	·97778
·02	·5080	·42	·6628	·82	·7939	·22	·8888	·62	·9474	·02	·97831
·03	·5120	·43	·6664	·83	·7967	·23	·8907	·63	·9484	·03	·97882
·04	·5160	·44	·6700	·84	·7995	·24	·8925	·64	·9495	·04	·97932
0·05	0·5199	0·45	0·6736	0·85	0·8023	1·25	0·8944	1·65	0·9505	2·05	0·97982
·06	·5239	·46	·6772	·86	·8051	·26	·8962	·66	·9515	·06	·98030
·07	·5279	·47	·6808	·87	·8078	·27	·8980	·67	·9525	·07	·98077
·08	·5319	·48	·6844	·88	·8106	·28	·8997	·68	·9535	·08	·98124
·09	·5359	·49	·6879	·89	·8133	·29	·9015	·69	·9545	·09	·98169
0·10	0·5398	0·50	0·6915	0·90	0·8159	1·30	0·9032	1·70	0·9554	2·10	0·98214
·11	·5438	·51	·6950	·91	·8186	·31	·9049	·71	·9564	·11	·98257
·12	·5478	·52	·6985	·92	·8212	·32	·9066	·72	·9573	·12	·98300
·13	·5517	·53	·7019	·93	·8238	·33	·9082	·73	·9582	·13	·98341
·14	·5557	·54	·7054	·94	·8264	·34	·9099	·74	·9591	·14	·98382
0·15	0·5596	0·55	0·7088	0·95	0·8289	1·35	0·9115	1·75	0·9599	2·15	0·98422
·16	·5636	·56	·7123	·96	·8315	·36	·9131	·76	·9608	·16	·98461
·17	·5675	·57	·7157	·97	·8340	·37	·9147	·77	·9616	·17	·98500
·18	·5714	·58	·7190	·98	·8365	·38	·9162	·78	·9625	·18	·98537
·19	·5753	·59	·7224	·99	·8389	·39	·9177	·79	·9633	·19	·98574
0·20	0·5793	0·60	0·7257	1·00	0·8413	1·40	0·9192	1·80	0·9641	2·20	0·98610
·21	·5832	·61	·7291	·01	·8438	·41	·9207	·81	·9649	·21	·98645
·22	·5871	·62	·7324	·02	·8461	·42	·9222	·82	·9656	·22	·98679
·23	·5910	·63	·7357	·03	·8485	·43	·9236	·83	·9664	·23	·98713
·24	·5948	·64	·7389	·04	·8508	·44	·9251	·84	·9671	·24	·98745
0·25	0·5987	0·65	0·7422	1·05	0·8531	1·45	0·9265	1·85	0·9678	2·25	0·98778
·26	·6026	·66	·7454	·06	·8554	·46	·9279	·86	·9686	·26	·98809
·27	·6064	·67	·7486	·07	·8577	·47	·9292	·87	·9693	·27	·98840
·28	·6103	·68	·7517	·08	·8599	·48	·9306	·88	·9699	·28	·98870
·29	·6141	·69	·7549	·09	·8621	·49	·9319	·89	·9706	·29	·98899
0·30	0·6179	0·70	0·7580	1·10	0·8643	1·50	0·9332	1·90	0·9713	2·30	0·98928
·31	·6217	·71	·7611	·11	·8665	·51	·9345	·91	·9719	·31	·98956
·32	·6255	·72	·7642	·12	·8686	·52	·9357	·92	·9726	·32	·98983
·33	·6293	·73	·7673	·13	·8708	·53	·9370	·93	·9732	·33	·99010
·34	·6331	·74	·7704	·14	·8729	·54	·9382	·94	·9738	·34	·99036
0·35	0·6368	0·75	0·7734	1·15	0·8749	1·55	0·9394	1·95	0·9744	2·35	0·99061
·36	·6406	·76	·7764	·16	·8770	·56	·9406	·96	·9750	·36	·99086
·37	·6443	·77	·7794	·17	·8790	·57	·9418	·97	·9756	·37	·99111
·38	·6480	·78	·7823	·18	·8810	·58	·9429	·98	·9761	·38	·99134
·39	·6517	·79	·7852	·19	·8830	·59	·9441	·99	·9767	·39	·99158
0·40	0·6554	0·80	0·7881	1·20	0·8849	1·60	0·9452	2·00	0·9772	2·40	0·99180

The normal distribution function (continued)

x	$\Phi(x)$	x	$\Phi(x)$	x	$\Phi(x)$	x	$\Phi(x)$	x	$\Phi(x)$	x	$\Phi(x)$
2·40	0·99180	2·55	0·99461	2·70	0·99653	2·85	0·99781	3·00	0·99865	3·15	0·99918
·41	·99202	·56	·99477	·71	·99664	·86	·99788	·01	·99869	·16	·99921
·42	·99224	·57	·99492	·72	·99674	·87	·99795	·02	·99874	·17	·99924
·43	·99245	·58	·99506	·73	·99683	·88	·99801	·03	·99878	·18	·99926
·44	·99266	·59	·99520	·74	·99693	·89	·99807	·04	·99882	·19	·99929
2·45	0·99286	2·60	0·99534	2·75	0·99702	2·90	0·99813	3·05	0·99886	3·20	0·99931
·46	·99305	·61	·99547	·76	·99711	·91	·99819	·06	·99889	·21	·99934
·47	·99324	·62	·99560	·77	·99720	·92	·99825	·07	·99893	·22	·99936
·48	·99343	·63	·99573	·78	·99728	·93	·99831	·08	·99896	·23	·99938
·49	·99361	·64	·99585	·79	·99736	·94	·99836	·09	·99900	·24	·99940
2·50	0·99379	2·65	0·99598	2·80	0·99744	2·95	0·99841	3·10	0·99903	3·25	0·99942
·51	·99396	·66	·99609	·81	·99752	·96	·99846	·11	·99906	·26	·99944
·52	·99413	·67	·99621	·82	·99760	·97	·99851	·12	·99910	·27	·99946
·53	·99430	·68	·99632	·83	·99767	·98	·99856	·13	·99913	·28	·99948
·54	·99446	·69	·99643	·84	·99774	·99	·99861	·14	·99916	·29	·99950
2·55	0·99461	2·70	0·99653	2·85	0·99781	3·00	0·99865	3·15	0·99918	3·30	0·99952

The critical table below gives on the left the range of values of x for which $\Phi(x)$ takes the value on the right, correct to the last figure given; in critical cases, take the upper of the two values of $\Phi(x)$ indicated.

			0·9994		0·99990		0·99995
3·075		3·263		3·731		3·916	
	0·9990		0·9995		0·99991		0·99996
3·105		3·320		3·759		3·976	
	0·9991		0·9996		0·99992		0·99997
3·138		3·389		3·791		4·055	
	0·9992		0·9997		0·99993		0·99998
3·174		3·480		3·826		4·173	
	0·9993		0·9998		0·99994		0·99999
3·215		3·615		3·867		4·417	
	0·9994		0·9999		0·99995		1·00000

When $x > 3{\cdot}3$ the formula $1-\Phi(x) \doteqdot \frac{e^{-\frac{1}{2}x^2}}{x\sqrt{2\pi}}\left[1-\frac{1}{x^2}+\frac{3}{x^4}-\frac{15}{x^6}+\frac{105}{x^8}\right]$ is very accurate, with relative error less than $945/x^{10}$.

Percentage points of the normal distribution

This table gives percentage points $x(P)$ defined by the equation

$$\frac{P}{100} = \frac{1}{\sqrt{2\pi}}\int_{x(P)}^{\infty} e^{-\frac{1}{2}t^2}\,dt.$$

If X is a variable, normally distributed with zero mean and unit variance, $P/100$ is the probability that $X \geqslant x(P)$. The lower P per cent points are given by symmetry as $-x(P)$, and the probability that $|X| \geqslant x(P)$ is $2P/100$.

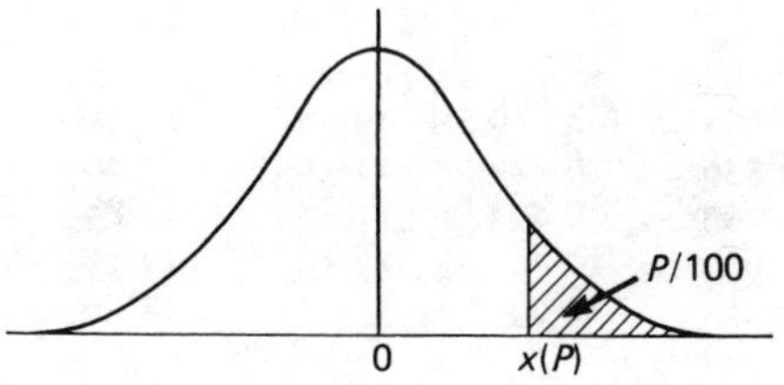

P	$x(P)$	P	$x(P)$	P	$x(P)$	P	$x(P)$	P	$x(P)$	P	$x(P)$
50	0·0000	5·0	1·6449	3·0	1·8808	2·0	2·0537	1·0	2·3263	0·10	3·0902
45	0·1257	4·8	1·6646	2·9	1·8957	1·9	2·0749	0·9	2·3656	0·09	3·1214
40	0·2533	4·6	1·6849	2·8	1·9110	1·8	2·0969	0·8	2·4089	0·08	3·1559
35	0·3853	4·4	1·7060	2·7	1·9268	1·7	2·1201	0·7	2·4573	0·07	3·1947
30	0·5244	4·2	1·7279	2·6	1·9431	1·6	2·1444	0·6	2·5121	0·06	3·2389
25	0·6745	4·0	1·7507	2·5	1·9600	1·5	2·1701	0·5	2·5758	0·05	3·2905
20	0·8416	3·8	1·7744	2·4	1·9774	1·4	2·1973	0·4	2·6521	0·01	3·7190
15	1·0364	3·6	1·7991	2·3	1·9954	1·3	2·2262	0·3	2·7478	0·005	3·8906
10	1·2816	3·4	1·8250	2·2	2·0141	1·2	2·2571	0·2	2·8782	0·001	4·2649
5	1·6449	3·2	1·8522	2·1	2·0335	1·1	2·2904	0·1	3·0902	0·0005	4·4172

Percentage points of the χ^2-distribution

This table gives percentage points $\chi^2_\nu(P)$ defined by the equation

$$\frac{P}{100} = \frac{1}{2^{\nu/2}\,\Gamma(\frac{\nu}{2})}\int_{\chi^2_\nu(P)}^{\infty} x^{\frac{1}{2}\nu-1}\,e^{-\frac{1}{2}x}\,dx.$$

If X is a variable distributed as χ^2 with ν degrees of freedom, $P/100$ is the probability that $X \geqslant \chi^2_\nu(P)$.

For $\nu > 100$, $\sqrt{2X}$ is approximately normally distributed with mean $\sqrt{2\nu-1}$ and unit variance.

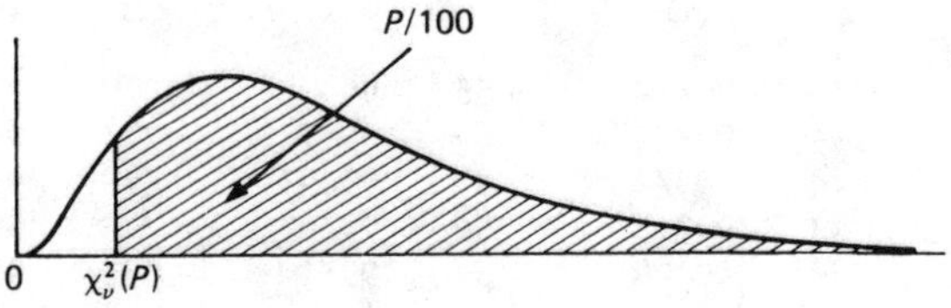

(The above shape applies for $\nu \geqslant 3$ only. When $\nu < 3$ the mode is at the origin.)

P	99·95	99·9	99·5	99	97·5	95	90	80	70	60
$\nu = 1$	0·0^{6}3927	0·0^{5}1571	0·0^{4}3927	0·0^{3}1571	0·0^{3}9821	0·003932	0·01579	0·06418	0·1485	0·2750
2	0·001000	0·002001	0·01003	0·02010	0·05064	0·1026	0·2107	0·4463	0·7133	1·022
3	0·01528	0·02430	0·07172	0·1148	0·2158	0·3518	0·5844	1·005	1·424	1·869
4	0·06392	0·09080	0·2070	0·2971	0·4844	0·7107	1·064	1·649	2·195	2·753
5	0·1581	0·2102	0·4117	0·5543	0·8312	1·145	1·610	2·343	3·000	3·655
6	0·2994	0·3811	0·6757	0·8721	1·237	1·635	2·204	3·070	3·828	4·570
7	0·4849	0·5985	0·9893	1·239	1·690	2·167	2·833	3·822	4·671	5·493
8	0·7104	0·8571	1·344	1·646	2·180	2·733	3·490	4·594	5·527	6·423
9	0·9717	1·152	1·735	2·088	2·700	3·325	4·168	5·380	6·393	7·357
10	1·265	1·479	2·156	2·558	3·247	3·940	4·865	6·179	7·267	8·295
11	1·587	1·834	2·603	3·053	3·816	4·575	5·578	6·989	8·148	9·237
12	1·934	2·214	3·074	3·571	4·404	5·226	6·304	7·807	9·034	10·18
13	2·305	2·617	3·565	4·107	5·009	5·892	7·042	8·634	9·926	11·13
14	2·697	3·041	4·075	4·660	5·629	6·571	7·790	9·467	10·82	12·08
15	3·108	3·483	4·601	5·229	6·262	7·261	8·547	10·31	11·72	13·03
16	3·536	3·942	5·142	5·812	6·908	7·962	9·312	11·15	12·62	13·98
17	3·980	4·416	5·697	6·408	7·564	8·672	10·09	12·00	13·53	14·94
18	4·439	4·905	6·265	7·015	8·231	9·390	10·86	12·86	14·44	15·89
19	4·912	5·407	6·844	7·633	8·907	10·12	11·65	13·72	15·35	16·85
20	5·398	5·921	7·434	8·260	9·591	10·85	12·44	14·58	16·27	17·81
21	5·896	6·447	8·034	8·897	10·28	11·59	13·24	15·44	17·18	18·77
22	6·404	6·983	8·643	9·542	10·98	12·34	14·04	16·31	18·10	19·73
23	6·924	7·529	9·260	10·20	11·69	13·09	14·85	17·19	19·02	20·69
24	7·453	8·085	9·886	10·86	12·40	13·85	15·66	18·06	19·94	21·65
25	7·991	8·649	10·52	11·52	13·12	14·61	16·47	18·94	20·87	22·62
26	8·538	9·222	11·16	12·20	13·84	15·38	17·29	19·82	21·79	23·58
27	9·093	9·803	11·81	12·88	14·57	16·15	18·11	20·70	22·72	24·54
28	9·656	10·39	12·46	13·56	15·31	16·93	18·94	21·59	23·65	25·51
29	10·23	10·99	13·12	14·26	16·05	17·71	19·77	22·48	24·58	26·48
30	10·80	11·59	13·79	14·95	16·79	18·49	20·60	23·36	25·51	27·44
32	11·98	12·81	15·13	16·36	18·29	20·07	22·27	25·15	27·37	29·38
34	13·18	14·06	16·50	17·79	19·81	21·66	23·95	26·94	29·24	31·31
36	14·40	15·32	17·89	19·23	21·34	23·27	25·64	28·73	31·12	33·25
38	15·64	16·61	19·29	20·69	22·88	24·88	27·34	30·54	32·99	35·19
40	16·91	17·92	20·71	22·16	24·43	26·51	29·05	32·34	34·87	37·13
50	23·46	24·67	27·99	29·71	32·36	34·76	37·69	41·45	44·31	46·86
60	30·34	31·74	35·53	37·48	40·48	43·19	46·46	50·64	53·81	56·62
70	37·47	39·04	43·28	45·44	48·76	51·74	55·33	59·90	63·35	66·40
80	44·79	46·52	51·17	53·54	57·15	60·39	64·28	69·21	72·92	76·19
90	52·28	54·16	59·20	61·75	65·65	69·13	73·29	78·56	82·51	85·99
100	59·90	61·92	67·33	70·06	74·22	77·93	82·36	87·95	92·13	95·81

Percentage points of the χ^2-distribution (continued)

P	50	40	30	20	10	5	2·5	1	0·5	0·1	0·05
$\nu = 1$	0·4549	0·7083	1·074	1·642	2·706	3·841	5·024	6·635	7·879	10·83	12·12
2	1·386	1·833	2·408	3·219	4·605	5·991	7·378	9·210	10·60	13·82	15·20
3	2·366	2·946	3·665	4·642	6·251	7·815	9·348	11·34	12·84	16·27	17·73
4	3·357	4·045	4·878	5·989	7·779	9·488	11·14	13·28	14·86	18·47	20·00
5	4·351	5·132	6·064	7·289	9·236	11·07	12·83	15·09	16·75	20·52	22·11
6	5·348	6·211	7·231	8·558	10·64	12·59	14·45	16·81	18·55	22·46	24·10
7	6·346	7·283	8·383	9·803	12·02	14·07	16·01	18·48	20·28	24·32	26·02
8	7·344	8·351	9·524	11·03	13·36	15·51	17·53	20·09	21·95	26·12	27·87
9	8·343	9·414	10·66	12·24	14·68	16·92	19·02	21·67	23·59	27·88	29·67
10	9·342	10·47	11·78	13·44	15·99	18·31	20·48	23·21	25·19	29·59	31·42
11	10·34	11·53	12·90	14·63	17·28	19·68	21·92	24·72	26·76	31·26	33·14
12	11·34	12·58	14·01	15·81	18·55	21·03	23·34	26·22	28·30	32·91	34·82
13	12·34	13·64	15·12	16·98	19·81	22·36	24·74	27·69	29·82	34·53	36·48
14	13·34	14·69	16·22	18·15	21·06	23·68	26·12	29·14	31·32	36·12	38·11
15	14·34	15·73	17·32	19·31	22·31	25·00	27·49	30·58	32·80	37·70	39·72
16	15·34	16·78	18·42	20·47	23·54	26·30	28·85	32·00	34·27	39·25	41·31
17	16·34	17·82	19·51	21·61	24·77	27·59	30·19	33·41	35·72	40·79	42·88
18	17·34	18·87	20·60	22·76	25·99	28·87	31·53	34·81	37·16	42·31	44·43
19	18·34	19·91	21·69	23·90	27·20	30·14	32·85	36·19	38·58	43·82	45·97
20	19·34	20·95	22·77	25·04	28·41	31·41	34·17	37·57	40·00	45·31	47·50
21	20·34	21·99	23·86	26·17	29·62	32·67	35·48	38·93	41·40	46·80	49·01
22	21·34	23·03	24·94	27·30	30·81	33·92	36·78	40·29	42·80	48·27	50·51
23	22·34	24·07	26·02	28·43	32·01	35·17	38·08	41·64	44·18	49·73	52·00
24	23·34	25·11	27·10	29·55	33·20	36·42	39·36	42·98	45·56	51·18	53·48
25	24·34	26·14	28·17	30·68	34·38	37·65	40·65	44·31	46·93	52·62	54·95
26	25·34	27·18	29·25	31·79	35·56	38·89	41·92	45·64	48·29	54·05	56·41
27	26·34	28·21	30·32	32·91	36·74	40·11	43·19	46·96	49·64	55·48	57·86
28	27·34	29·25	31·39	34·03	37·92	41·34	44·46	48·28	50·99	56·89	59·30
29	28·34	30·28	32·46	35·14	39·09	42·56	45·72	49·59	52·34	58·30	60·73
30	29·34	31·32	33·53	36·25	40·26	43·77	46·98	50·89	53·67	59·70	62·16
32	31·34	33·38	35·66	38·47	42·58	46·19	49·48	53·49	56·33	62·49	65·00
34	33·34	35·44	37·80	40·68	44·90	48·60	51·97	56·06	58·96	65·25	67·80
36	35·34	37·50	39·92	42·88	47·21	51·00	54·44	58·62	61·58	67·99	70·59
38	37·34	39·56	42·05	45·08	49·51	53·38	56·90	61·16	64·18	70·70	73·35
40	39·34	41·62	44·16	47·27	51·81	55·76	59·34	63·69	66·77	73·40	76·09
50	49·33	51·89	54·72	58·16	63·17	67·50	71·42	76·15	79·49	86·66	89·56
60	59·33	62·13	65·23	68·97	74·40	79·08	83·30	88·38	91·95	99·61	102·7
70	69·33	72·36	75·69	79·71	85·53	90·53	95·02	100·4	104·2	112·3	115·6
80	79·33	82·57	86·12	90·41	96·58	101·9	106·6	112·3	116·3	124·8	128·3
90	89·33	92·76	96·52	101·1	107·6	113·1	118·1	124·1	128·3	137·2	140·8
100	99·33	102·9	106·9	111·7	118·5	124·3	129·6	135·8	140·2	149·4	153·2

Properties of matter

Physical constants

Universal gas constant	R_0	$8{\cdot}315 \times 10^3\,\mathrm{J\,kmol^{-1}\,K^{-1}}$*
Boltzmann constant	k	$1{\cdot}38 \times 10^{-23}\,\mathrm{J\,K^{-1}}$
		$8{\cdot}617 \times 10^{-5}\,\mathrm{eV\,K^{-1}}$
Universal gravitational constant	G	$6{\cdot}67 \times 10^{-11}\,\mathrm{N\,m^2\,kg^{-2}}$
Mean radius of earth	R	6371 km
Mass of earth	M	$5{\cdot}976 \times 10^{24}\,\mathrm{kg}$
Gravitational acceleration (standard gravity)	g	$9{\cdot}807\,\mathrm{m\,s^{-2}}$
Stefan-Boltzmann constant	σ	$5{\cdot}670 \times 10^{-8}\,\mathrm{W\,m^{-2}\,K^{-4}}$
Avogadro number	N	$6{\cdot}022 \times 10^{26}\,\mathrm{kmol^{-1}}$
Loschmidt number		$2{\cdot}687 \times 10^{25}\,\mathrm{m^{-3}}$
Molar volume†		$22{\cdot}41\,\mathrm{m^3\,kmol^{-1}}$
Planck constant	h	$6{\cdot}626 \times 10^{-34}\,\mathrm{J\,s}$
		$4{\cdot}136 \times 10^{-15}\,\mathrm{eV\,s}$
Atomic mass unit (a.m.u.)		$1{\cdot}661 \times 10^{-27}\,\mathrm{kg}$
Velocity of light in vacuum	c	$2{\cdot}998 \times 10^8\,\mathrm{m\,s^{-1}}$
Absolute permittivity of free space	ϵ_0	$8{\cdot}854 \times 10^{-12}\,\mathrm{F\,m^{-1}}$
Absolute permeability of free space	μ_0	$4\pi \times 10^{-7}\,\mathrm{H\,m^{-1}}$
Charge of an electron	e	$1{\cdot}602 \times 10^{-19}\,\mathrm{C}$
Rest mass of an electron	m_e	$9{\cdot}110 \times 10^{-31}\,\mathrm{kg}$
Charge/mass ratio of an electron	e/m_e	$1{\cdot}759 \times 10^{11}\,\mathrm{C\,kg^{-1}}$
Mass of a proton	m_p	$1{\cdot}673 \times 10^{-27}\,\mathrm{kg}$
Impedance of free space		$376{\cdot}7\,\Omega$
Bohr magneton	μ_B	$9{\cdot}274 \times 10^{-24}\,\mathrm{A\,m^2}$
Wavelength of 1 eV photon		$1{\cdot}240\,\mu\mathrm{m}$
Faraday constant	F	$9{\cdot}649 \times 10^7\,\mathrm{C\,kmol^{-1}}$

* One kilomole of a substance is that quantity which contains as many of its (specified) particles as there are atoms in 12 kg of carbon 12.
† of a perfect gas at 0° C, 1 atm (1·013 25 bar).

The periodic table

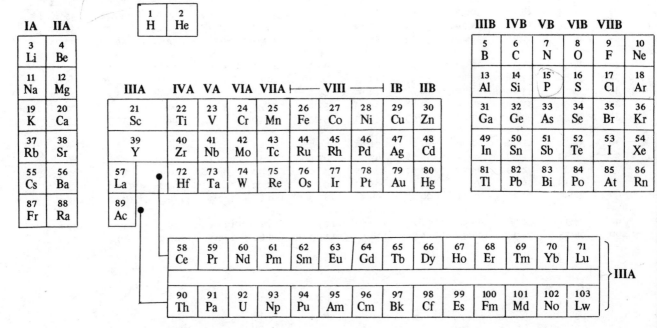

1 H	2 He

IA	IIA
3 Li	4 Be
11 Na	12 Mg
19 K	20 Ca
37 Rb	38 Sr
55 Cs	56 Ba
87 Fr	88 Ra

IIIA	IVA	VA	VIA	VIIA	VIII	VIII	VIII	IB	IIB
21 Sc	22 Ti	23 V	24 Cr	25 Mn	26 Fe	27 Co	28 Ni	29 Cu	30 Zn
39 Y	40 Zr	41 Nb	42 Mo	43 Tc	44 Ru	45 Rh	46 Pd	47 Ag	48 Cd
57 La	72 Hf	73 Ta	74 W	75 Re	76 Os	77 Ir	78 Pt	79 Au	80 Hg
89 Ac									

IIIB	IVB	VB	VIB	VIIB	
5 B	6 C	7 N	8 O	9 F	10 Ne
13 Al	14 Si	15 P	16 S	17 Cl	18 Ar
31 Ga	32 Ge	33 As	34 Se	35 Br	36 Kr
49 In	50 Sn	51 Sb	52 Te	53 I	54 Xe
81 Tl	82 Pb	83 Bi	84 Po	85 At	86 Rn

IIIA:

58 Ce	59 Pr	60 Nd	61 Pm	62 Sm	63 Eu	64 Gd	65 Tb	66 Dy	67 Ho	68 Er	69 Tm	70 Yb	71 Lu
90 Th	91 Pa	92 U	93 Np	94 Pu	95 Am	96 Cm	97 Bk	98 Cf	99 Es	100 Fm	101 Md	102 No	103 Lw

Atomic properties of the elements

(for free neutral atoms in the ground state)

Z	Atomic number
AW	Atomic weight in a.m.u. ($_6C^{12}$ = 12·000)
V_i	First ionization potential in eV
K, L, M, N, O, P, Q	Principal quantum number = 1, 2, 3, 4, 5, 6, 7
s, p, d, f, g, h	Azimuthal quantum number = 0, 1, 2, 3, 4, 5

				K	L		M			N				O				
Z	Element	AW	V_i	1s	2s	2p	3s	3p	3d	4s	4p	4d	4f	5s	5p	5d	5f	5g
1	H Hydrogen	1·008	13·5	1														
2	He Helium	4·003	24·5	2														
3	Li Lithium	6·941	5·4	2	1													
4	Be Beryllium	9·012	9·3	2	2													
5	B Boron	10·811	8·3	2	2	1												
6	C Carbon	12·011	11·2	2	2	2												
7	N Nitrogen	14·007	14·5	2	2	3												
8	O Oxygen	15·999	13·6	2	2	4												
9	F Fluorine	18·998	17·3	2	2	5												
10	Ne Neon	20·179	21·5	2	2	6												
11	Na Sodium	22·990	5·1	2	2	6	1											
12	Mg Magnesium	24·305	7·6	2	2	6	2											
13	Al Aluminium	26·982	6·0	2	2	6	2	1										
14	Si Silicon	28·086	8·1	2	2	6	2	2										
15	P Phosphorus	30·974	10·9	2	2	6	2	3										
16	S Sulphur	32·066	10·3	2	2	6	2	4										
17	Cl Chlorine	35·453	13·0	2	2	6	2	5										
18	Ar Argon	39·948	15·7	2	2	6	2	6										
19	K Potassium	39·098	4·3	2	2	6	2	6		1								
20	Ca Calcium	40·078	6·1	2	2	6	2	6		2								
21	Sc Scandium	44·956	6·7	2	2	6	2	6	1	2								
22	Ti Titanium	47·88	6·8	2	2	6	2	6	2	2								
23	V Vanadium	50·942	6·7	2	2	6	2	6	3	2								
24	Cr Chromium	51·996	6·7	2	2	6	2	6	5	1								
25	Mn Manganese	54·938	7·4	2	2	6	2	6	5	2								
26	Fe Iron	55·847	7·8	2	2	6	2	6	6	2								
27	Co Cobalt	58·933	7·8	2	2	6	2	6	7	2								
28	Ni Nickel	58·69	7·6	2	2	6	2	6	8	2								
29	Cu Copper	63·546	7·7	2	2	6	2	6	10	1								
30	Zn Zinc	65·39	9·4	2	2	6	2	6	10	2								
31	Ga Gallium	69·723	6·0	2	2	6	2	6	10	2	1							
32	Ge Germanium	72·59	8·1	2	2	6	2	6	10	2	2							
33	As Arsenic	74·922	10·5	2	2	6	2	6	10	2	3							
34	Se Selenium	78·96	9·7	2	2	6	2	6	10	2	4							
35	Br Bromine	79·904	11·8	2	2	6	2	6	10	2	5							
36	Kr Krypton	83·80	13·9	2	2	6	2	6	10	2	6							
37	Rb Rubidium	85·468	4·2	2	2	6	2	6	10	2	6			1				
38	Sr Strontium	87·62	5·7	2	2	6	2	6	10	2	6			2				
39	Y Yttrium	88·906	6·5	2	2	6	2	6	10	2	6	1		2				
40	Zr Zirconium	91·224	6·9	2	2	6	2	6	10	2	6	2		2				
41	Nb Niobium	92·906		2	2	6	2	6	10	2	6	4		1				
42	Mo Molybdenum	95·94	7·4	2	2	6	2	6	10	2	6	5		1				
43	Tc Technetium			2	2	6	2	6	10	2	6	6		1				
44	Ru Ruthenium	101·07	7·7	2	2	6	2	6	10	2	6	7		1				
45	Rh Rhodium	102·906	7·7	2	2	6	2	6	10	2	6	8		1				
46	Pd Palladium	106·42	8·3	2	2	6	2	6	10	2	6	10						
47	Ag Silver	107·868	7·5	2	2	6	2	6	10	2	6	10		1				
48	Cd Cadmium	112·41	9·0	2	2	6	2	6	10	2	6	10		2				
49	In Indium	114·82	5·8	2	2	6	2	6	10	2	6	10		2	1			
50	Sn Tin	118·710	7·3	2	2	6	2	6	10	2	6	10		2	2			
51	Sb Antimony	121·75	8·5	2	2	6	2	6	10	2	6	10		2	3			
52	Te Tellurium	127·60	9·0	2	2	6	2	6	10	2	6	10		2	4			
53	I Iodine	126·905	10·6	2	2	6	2	6	10	2	6	10		2	5			
54	Xe Xenon	131·29	12·1	2	2	6	2	6	10	2	6	10		2	6			

					K	L	M	N				O				P						Q
Z	Element		AW	V_i				4s	4p	4d	4f	5s	5p	5d	5g	6s	6p	6d	6f	6g	6h	7s
55	Cs	Caesium	132·905	3·9	2	8	18	2	6	10		2	6			1						
56	Ba	Barium	137·33	5·2	2	8	18	2	6	10		2	6			2						
57	La	Lanthanum	138·906	5·6	2	8	18	2	6	10		2	6	1		2						
58	Ce	Cerium	140·12	6·5	2	8	18	2	6	10	2	2	6			2						
59	Pr	Praseodymium	140·908	5·8	2	8	18	2	6	10	3	2	6			2						
60	Nd	Neodymium	144·240	6·3	2	8	18	2	6	10	4	2	6			2						
61	Pm	Promethium			2	8	18	2	6	10	5	2	6			2						
62	Sm	Samarium	150·36	6·6	2	8	18	2	6	10	6	2	6			2						
63	Eu	Europium	151·96	5·6	2	8	18	2	6	10	7	2	6			2						
64	Gd	Gadolineum	157·25	6·7	2	8	18	2	6	10	7	2	6	1		2						
65	Tb	Terbium	158·925	6·7	2	8	18	2	6	10	9	2	6			2						
66	Dy	Dysprosium	162·50	6·8	2	8	18	2	6	10	10	2	6			2						
67	Ho	Holmium	164·930		2	8	18	2	6	10	11	2	6			2						
68	Er	Erbium	167·26		2	8	18	2	6	10	12	2	6			2						
69	Tm	Thulium	168·934		2	8	18	2	6	10	13	2	6			2						
70	Yb	Ytterbium	173·04	7·1	2	8	18	2	6	10	14	2	6			2						
71	Lu	Lutetium	174·967		2	8	18	2	6	10	14	2	6	1		2						
72	Hf	Hafnium	178·49		2	8	18	2	6	10	14	2	6	2		2						
73	Ta	Tantalum	180·948		2	8	18	2	6	10	14	2	6	3		2						
74	W	Tungsten	183·85	8·1	2	8	18	2	6	10	14	2	6	4		2						
75	Re	Rhenium	186·2		2	8	18	2	6	10	14	2	6	5		2						
76	Os	Osmium	190·2		2	8	18	2	6	10	14	2	6	6		2						
77	Ir	Iridium	192·22		2	8	18	2	6	10	14	2	6	9								
78	Pt	Platinum	195·08	8·9	2	8	18	2	6	10	14	2	6	9		1						
79	Au	Gold	196·967	9·2	2	8	18	2	6	10	14	2	6	10		1						
80	Hg	Mercury	200·59	10·4	2	8	18	2	6	10	14	2	6	10		2						
81	Tl	Thallium	204·383	6·1	2	8	18	2	6	10	14	2	6	10		2	1					
82	Pb	Lead	207·2	7·4	2	8	18	2	6	10	14	2	6	10		2	2					
83	Bi	Bismuth	208·98	8·0	2	8	18	2	6	10	14	2	6	10		2	3					
84	Po	Polonium			2	8	18	2	6	10	14	2	6	10		2	4					
85	At	Astatine			2	8	18	2	6	10	14	2	6	10		2	5					
86	Rn	Radon			2	8	18	2	6	10	14	2	6	10		2	6					
87	Fr	Francium			2	8	18	2	6	10	14	2	6	10		2	6					1
88	Ra	Radium			2	8	18	2	6	10	14	2	6	10		2	6					2
89	Ac	Actinium			2	8	18	2	6	10	14	2	6	10		2	6	1				2
90	Th	Thorium	232·038		2	8	18	2	6	10	14	2	6	10		2	6	2				2
91	Pa	Protactinium			2	8	18	2	6	10	14	2	6	10		2	6	3				2
92	U	Uranium	238·029		2	8	18	2	6	10	14	2	6	10		2	6	4				2
93	Np	Neptunium			2	8	18	2	6	10	14	2	6	10	5	2	6					2
94	Pu	Plutonium			2	8	18	2	6	10	14	2	6	10	5	2	6	1				2
95	Am	Americium			2	8	18	2	6	10	14	2	6	10	6	2	6	1				2
96	Cm	Curium			2	8	18	2	6	10	14	2	6	10	7	2	6	1				2
97	Bk	Berkelium			2	8	18	2	6	10	14	2	6	10	8	2	6	1				2
98	Cf	Californium			2	8	18	2	6	10	14	2	6	10	9	2	6	1				2
99	Es	Einsteinium			2	8	18	2	6	10	14	2	6	10	10	2	6	1				2
100	Fm	Fermium			2	8	18	2	6	10	14	2	6	10	11	2	6	1				2
101	Md	Mendelevium			2	8	18	2	6	10	14	2	6	10	12	2	6	1				2
102	No	Nobelium																				
103	Lw	Lawrencium																				

Physical properties of solids

CS Crystal structure: BCC body-centred cubic
FCC face-centred cubic
CPH close-packed hexagonal

ρ Mass density ($kg\,dm^{-3}$) $\approx$ specific gravity

t_m Melting point (°C)

t_b Boiling point (°C)

h_{if} Latent heat of fusion ($kJ\,kg^{-1}$)

h_{fg} Latent heat of vaporization ($kJ\,kg^{-1}$)

k Thermal conductivity at or near 0° C ($W\,m^{-1}\,K^{-1}$)

c Specific heat capacity at or near 0° C ($J\,kg^{-1}\,K^{-1}$)

α Coefficient of linear thermal expansion (K^{-1}) $\times 10^6$

ρ_e Electrical resistivity at 20° C (units as shown)

α_e Temperature coefficient of resistance, 0–100° C ($K^{-1} \times 10^3$)

ϵ_r Dielectric constant (relative permittivity) at $\lesssim$ 1 MHz, 20° C

$\tan\delta$ Loss factor at ~1 MHz, 20 °C (units of 10^{-4})

	CS	ρ	t_m	t_b	h_{if}	h_{fg}	k	c	α	ρ_e(nΩ m)	α_e
Metallic elements											
Aluminium	FCC	2·7	660	2400	387	9460	205	880	23	27	4·2
Carbon: Diamond	Diamond	3·51	transforms to graphite				70	430	1·2		
Graphite	Hexagonal	2·25	–	3652*				610			
Copper	FCC	8·96	1083	2580	205	5230	390	380	17	16·8	4·3
Gold	FCC	19·3	1063	2660	66	1750	310	145	14	23	3·9
Iron	BCC/FCC	7·9	1535	2900	270	6600	76	437	12	97	6·5
Lead	FCC	11·3	327	1750	24	850	35	126	29	206	4·3
Nickel	FCC	8·9	1453	2820	305	5850	91	444	13	68	6·8
Platinum	FCC	21·5	1769	3800	113	2400	69	125	9	106	3·9
Silicon	Diamond	2·3	1412	2355			84	700	7·6	10^4–10^7	
Silver	FCC	10·5	961	2180	105	2330	418	232	19	16	4·1
Tantalum	BCC	16·6	3000	5300	160		54	140	6	135	~3·5
Tin	Diamond/ tetragonal	7·3	232	~2500	59	2400	64	224	23	120	~4·5
Titanium	CPH/BCC	4·5	1680	3300	435		17	500	9	550	~3·5
Tungsten	BCC	19·3	3380	~6000	185		190	130	4·5	55	4·6
Zinc	CPH	7·1	420	907	110	1750	113	384	31	59	4·2

* Sublimation temperature

	ρ	t_m^*	k	c	α	ρ_e(nΩ m)	α_e
Alloys							
Aluminium 2024 (4·5% Cu)	2·8	640	147	900	22·5	~52	~2·3
Brass (70/30)	8·55	965	121	370	20	62	1·6
Cast Iron: grey	7·0	1250	180		13	500	
nodular	7·1	1150	65	461	11	150	
Constantan (60% Cu)	8·9	1280	22	410	16	490	~0·02
Manganin (84% Cu)	8·5	950	22	405	19	440	~0
Nimonic 80A (superalloy)	8·19	1400	11·2	460	12·7	1170	
Nichrome (80/20)	8·36	1420	13	430	12·5	1030	0·1
Phosphor-bronze (5% Sn)	8·85	1060	~75	380	18	105	3·5
Solder (soft) (50% Sn)	8·9	215	50	210	23	150	
Steel (mild)	~7·85	~1500	~50	~450	~11	~120	~3·0
Steel (austenitic stainless)	7·9	1500	16	~500	16	850	
Titanium-6Al-4V	4·42	1700	5·8	610	8	1680	0·4

* Alloys generally do not have a unique melting point; solid and liquid co-exist over a freezing range. The temperature t_m given here is the maximum temperature of this range.

Non-metals	ρ	t_m	k	c	α	ρ_e(MΩ m)	ϵ_r	tan δ
Ceramics and brittle materials								
Alumina	3·9	2050	39	1050	8	10^3–10^6	4·5–8·4	2–100
Brick	1·4–2·2		0·4–0·8	800	3–9	1–2		
Concrete	2·4		1·0–1·5	1100	10–14			
Dry ground	~1·6					0·01–0·1		
Glass (soda)	2·48	750*	1·0	990	8·5	5×10^4	5·8	13–100
Glass (borosilicate)	2·23	950	1·0	800	4·0			
Granite	2·7		2–4	800	6–9		7–9	
Ice†	0·92	0	2·3	2100				
Mica	2·8		~0·5	840		10^5–10^9	5–7	1–2
Porcelain	2·4	1550	0·8–1·85	1100	2·2	10^4–10^7	5·5–7	60–100
Quartz (crystal)	2·65		5–9	730	7·5–13·7	10^6–2×10^8	4·5–5	
Quartz (fused)	2·2		1·3	840	0·5	10^{10}	3·8	2
Sandstone	2·4		1·1–2·3	900	5–12		10	
Silicon carbide	3·2	2840	84	1422	4·3			
Silicon nitride	3·2	1900	17	627	3·2			
Tungsten carbide	15·7	2777	84	–	4·9			
Zirconia	5·6	2570	1·5	670	8			
Polymers, composites and natural materials								
Carbon-fibre reinforced plastic	1·5–2·0	–	0·3–1·0	–	3–5			
Epoxy resin	1·2–1.4	130–170*	0·2–0·5	1700–2000	55–90			
Glass-fibre reinforced plastic	1·5–2·5	–	0·3–1·0	–	15–20			
Nylon 6	1·14	200–220*	0·25–0·33	1600	80–130	10^4–10^7	3–7	200–1300
Paper (dry)	~1·0		0·06			10^4	1·9–2·9	20–45
Perspex	1·2	85–115*	0·19–0·23	1450	50–80		2·5–3·5	160–300
Polybutadiene (synthetic rubber)	1·5	~90*	~0·15	~2500	~600			
Polypropylene	0·91	40*	0·2	1900	100–300			
Polystyrene	1·06	80–105*	0·08–0·2	1300	60–80	10^{10}	2·4–3·5	<20
Polythene (low density)	0·91–0·94	80*	0·35	2250	160–190	10^5	2·3	2–5
Polythene (high density)	0·95–0·98	110*	0·52	2100	150–300			
PTFE	2·2		0·23–0·27	1050	90–130	10^9	2·1	<3
PVC (plasticized)	1·7	70–80*	0·16–0·19		50–250	10^4–10^7	4–6	600
PVC (unplasticized)	1·4	70–80*	0·15		50–70	10^4–10^7	4–6	600
Rubber (natural, vulcanized)	1·1–1·2	125	~0·15		~200	~10^7	2–3·5	280
Timber (along grain)	0·4–0·8		~0·15	~1600	3–5		2–9	350–600

*Softening temperature
†$h_{if} = 333\,\mathrm{kJ\,kg^{-1}}$

Mechanical properties of solids

E Young modulus (GPa)
G Shear modulus (GPa)
K Bulk modulus (GPa)
ν Poisson ratio

σ_y Proof or yield stress (MPa)
σ_f Ultimate (failure) stress or tensile strength (MPa)
ϵ_f Tensile strain to failure (%)
K_{Ic} Fracture toughness (MPa m$^{1/2}$)

Values of σ_y and σ_f, particularly, usually depend strongly on the preparation and condition of a material. The ranges given are typical but not necessarily exhaustive, and unless otherwise stated those for metals refer to drawn or wrought, rather than cast, material of commercial purity.

	E	G	K	ν	σ_y	σ_f	ϵ_f
Metallic elements in the annealed state							
Aluminium	70	25	75	0·34	15–20	40–50	50–70
Copper	130	48	138	0·34	33	210	60
Gold	79	26	171	0·42		100	30
Iron	211	82	170	0·29	80–100	350	
Lead	16	6	46	0·44		12	
Nickel	200	76	176	0·31	60	310	40
Platinum	170	61	280	0·39	14–35	140–195	30–40
Silicon	107	62		0·42	5000–9000		0
Silver	83	30	104	0·37		170	50
Tantalum	186	69	196	0·34	180	200	35–45
Tin	47	17	52	0·36	9–14	15–200	
Titanium	120	46	108	0·36	100–225	240–370	70–80
Tungsten	411	161	311	0·28	550	550–620	0
Zinc (wrought)	105	42	69	0·25		110–200	40–80
Alloys							
Aluminium 2024 (age-hardened)	72	28	75	0·33	395	475	10
Brass (70/30) (annealed)	101	37	112	0·35	115	320	67
(rolled)	101	37	112	0·35	390	460	20
Cast iron (grey)	100–145	40–58		0·26	100–260	150–400	
(nodular)	169–172	66		0·28	230–460	370–800	2–17
Constantan (60% Cu)	163	61	157	0·33	200–440	400–570	
Manganin (84% Cu)	124	47				465	
Mumetal (77% Ni)	220					500–900	
Nimonic 80A (superalloy)	214			0·35	800	1300	20
Nichrome (80/20)	186				100–400	170–900	
Phosphor-bronze (5% Sn)	100			0·38	110–670	330–750	2–50
Solder (soft) (50% Sn)	40				33	42	60
Steel: mild	210	81	160–170	0·27–0·30	240	400–500	10–20
Steel: high-yield structural	210	81	170	0·30	400	600	20
Steel: ultra high strength					1600	2000	10
Steel: austenitic stainless	190–200	74–86	–	0·25–0·29	255	660	45
Titanium-6Al-4V	115				800–900	900–1000	10–20

Hello!

Non-metals	E	ν	σ_f(tension)	σ_f(compression)	K_{Ic}
Ceramics and brittle materials					
Alumina	380	0·24	300–400	3000	3–5
Brick (grade A)	10–50			69–140	~1–2
Concrete (28 day)	10–17	0·1–0·21		27–55	0·2–0·4
Diamond	1050	0·13	–	5000–10000	
Glass (soda)	74	0·23	50	1000	0·7
Glass (borosilicate)	65	0·25	55	1200	0·8
Gallium arsenide	80	0·24	–	1000–3000	~1
Granite	40–70			90–235	~1
Ice	9·1	–	1·7	6	0·12
Porcelain	70	0·25	45	350	
Sandstone	14–55			30–135	
Silicon	107	0·42	–	5000–9000	1·1
Silicon carbide	410	0·13	200–500	2000	3–5
Silicon nitride	310	0·15	300–850	1200	4
Zirconia	200	0·30	500–830	2000	12
Polymers, composites and natural materials					
Carbon-fibre reinforced plastic	150–250	–	1000–1500	–	~60
Epoxy resin	2·1–5·5	0·38–0·4	40–85	100–200	0·6–1·0
Glass-fibre reinforced plastic	80–100	–	~1000	–	~90
Nylon 6	2–3·5		50–100	60–110	3–5
Perspex	3·3		80–90	80–140	1·6
Polybutadiene	0·004–0·1		~10		
Polypropylene	1·2–1·7		50–70		3·5
Polystyrene	3·0–3·3		35–68	10–110	2
Polythene (low density)	0·15–0·24		7–17	15–20	1–2
Polythene (high density)	0·1–1·0		20–37		2–5
PTFE	0·35		17–28		
PVC (plasticized)	~0·3		75–100	75–100	
PVC (unplasticized)	2·4–3·0		40–60		2·4
Timber (along grain)					
Pine, oak, ash	7–20		80–150	20–80	7–10
Balsa	2–5		20–40	10–30	2–3
Timber (across grain)					
Pine, oak, ash	0·7–1·5		10–40		0·7–1
Balsa	0·1–0·3		3–10		0·1–0·2

Properties of reinforcing fibres

r fibre radius (μm)
ρ mass density (kg dm^{-3})
E Young modulus (GPa)
σ_f failure stress (MPa)
α_l axial coefficient of thermal expansion (K^{-1}) × 10^6
α_r radial coefficient of thermal expansion (K^{-1}) × 10^6

	r	ρ	E	σ_f	α_l	α_r
α-alumina	10–20	3–4	200–380	1300–1700	4–9	4–9
Alumino-silicate (Saffil)	~3	2·8	103	1030		
Carbon (high modulus)	7–10	1·95	390	2200	~ −1·0	7–12
Carbon (high strength)	8–9	1·75	250	2700	~ −0·3	7–12
E-glass	10–12		76	1400–2500	4·9	4·9
Kevlar	12		130	2800–3600	−2	59
Silicon carbide (Nicalon)	10–15	2·5	170–200	5300–6500		
Silicon carbide (monofilament)	100–150	3	400–410	8000–9000		

Work functions

ϕ_t and ϕ_p are the least energies, in electron volts, to extract an electron from the element by thermionic and photoelectric emission respectively.

Element	ϕ_t	ϕ_p	Element	ϕ_t	ϕ_p	Element	ϕ_t	ϕ_p
Aluminium	4·3	4·1	Gold	5·2	4·8	Sodium	2·8	2·3
Barium	2·7	2·5	Iron	4·5	4·6	Tantalum	4·3	4·1
Bismuth	4·3	4·3	Mercury	4·5	4·5	Thorium	3·4	3·5
Caesium	2·2	1·9	Molybdenum	4·7	4·2	Titanium	4·3	4·2
Calcium	3·0	2·7	Nickel	5·2	5·0	Tungsten	4·6	4·5
Carbon	5·0	4·8	Platinum	5·7	6·3	Zinc	4·4	4·2
Chromium	4·6	4·4	Potassium	2·3	2·2			
Copper	4·7	4·3	Silver	4·3	4·7			

Properties of semiconductors

(at room temperature)

W Atomic or molecular weight
N Number density of atoms (m^{-3}) × 10^{-28}
a Lattice constant (Å)
ρ Mass density (kg dm^{-3})
m_e^*/m_0 Effective mass of electron relative to free rest mass m_0
m_h^*/m_0 Effective mass of hole (light, heavy and split-off) relative to m_0
E_g Energy gap (eV)
N_c, N_v Effective density of states in conduction and valence bands (m^{-3}) × 10^{-25}
N_i Intrinsic carrier concentration (m^{-3})
μ_e Mobility of electrons (cm^2 V^{-1} s^{-1})
μ_h Mobility of holes (cm^2 V^{-1} s^{-1})
ϵ_r Dielectric constant at low frequencies
E_B Breakdown field (V μm^{-1})

	W	N	a	ρ	m_e^*/m_0	m_h^*/m_0	E_g	N_c	N_v	N_i	μ_e	μ_h	ϵ_r	E_B
Carbon C	12·01	17·60	3·57	3·51	0·2	0·25	5·47				1 800	1 200	5·7	
Germanium Ge	72·60	4·42	5·65	5·33	0·12	0·04, 0·28, 0·08	0·67	1·04	0·60	$2{\cdot}4 \times 10^{19}$	3 900	1 900	16·3	~8
Silicon Si	28·09	5·00	5·43	2·33	0·26	0·16, 0·50, 0·24	1·11	2·8	1·04	$1{\cdot}45 \times 10^{16}$	1 500	450	11·7	~30
Aluminium arsenide AlAs	101·90	4·50	5·66	3·81	0·35		2·16				1 200	420	10·1	
Aluminium antimonide AlSb	148·73	3·42	6·14	4·22	0·1	0·5	1·6				200	500	14·4	
Gallium phosphide GaP	100·69	4·94	5·45	4·13	0·35	0·6	2·25				190	120	11·1	
Gallium arsenide GaAs	144·63	4·42	5·65	5·32	0·067	0·082, 0·5, 0·16	1·42	0·047	0·70	$1{\cdot}8 \times 10^{12}$	8 500	400	13	~40
Gallium antimonide GaSb	191·47	3·53	6·10	5·62	0·04	0·4	0·7				5 000	1 400	15·7	
Indium phosphide InP	145·79	3·96	5·87	4·79	0·07	0·8	1·3				4 600	150	12·4	
Indium arsenide InAs	189·74	3·59	6·06	5·66	0·03	0·4	0·36				33 000	460	14·6	
Indium antimonide InSb	236·58	2·94	6·48	5·80	0·013	0·4	0·17				80 000	1 250	17·7	
Cadmium sulphide CdS	144·48	4·04	5·83	4·82	0·21	0·8	2·42				340	50	5·4	
Cadmium selenide CdSe	191·36	3·65	6·05	5·8	0·13	0·5	1·8				650		10·0	
Cadmium telluride CdTe	240·01	2·96	6·48	5·9	0·13	0·35	1·6				1 100	80	10·2	
Zinc oxide ZnO	81·37	8·43	3·25, 5·21	5·7	0·38		3·3				200	180	9·0	
Zinc sulphide ZnS	97·43	5·07	5·41	4·1	0·4		3·8				150		5·2	
Lead sulphide PbS	239·3	3·82	5·94	7·6			0·37				800	1 000	17·0	
Lead selenide PbSe	286·2	3·49	6·12	8·3			0·26				1 500	1 500		
Lead telluride PbTe	334·8	2·95	6·45	8·2			0·25				1 600	750	30·0	

Properties of ferromagnetic materials

Soft materials

B_{sat} Saturated flux density (T)

$(\mu_r)_{max}$ Maximum relative permeability

ρ_e Resistivity (nΩ m)

w Hysteresis loss per cycle ($mJ\,kg^{-1}$)

	B_{sat}	$(\mu_r)_{max}$	ρ_e	w
Iron	2·15	5 000	100	30*
Iron (4% Si)	1·97	10 000	600	30*
Grain-oriented silicon-iron	2·00	45 000	550	25*
Permalloy B, Radiometal (50% Ni)	1·5	30 000	500	
Supermalloy (79% Ni, 5% Mo)	0·79	10^6	600	0·1
Permalloy C, Mumetal (77% Ni, 5% Cu)	0·65	10^5	620	
Ferrite MnZn $(Fe_2O_3)_2$	~0·4	2 500	2×10^8	1·0

* Approximate values at $B_{max} = 1·3$ T.

Hard materials

B_r Remanent flux density (T)
H_c Coercive force (kA m^{-1})
$(BH)_{max}$ is in units of kJ m^{-3}

	B_r	H_c	$(BH)_{max}$
Steel (1% C)	0·80	4·8	1·44
Steel (6% W)	1·05	5·2	2·40
Steel (35% Co)	0·90	18·5	7·20
Alnico 5 (14% Ni, 24% Co, 8% Al)	1·25	46	20
Ferroxdur, BaO $(Fe_2O_3)_6$	0·35	160	12
Fe-Co powder	0·90	82	40
Alnico 9	1·05	130	100

Superconducting materials

The critical field H_c for a superconducting material at an absolute temperature T depends on the critical field at absolute zero, H_0, according to the relation

$$H_c = H_0\{1-(T/T_c)^2\}$$

where T_c is the critical temperature. Values of H_0 and T_c for some superconductors are given below.

	T_c(K)	H_0(kA m^{-1})		T_c(K)	H_0(kA m^{-1})
Aluminium	1·19	8	Zinc	0·92	4
Gallium	1·09	4	Nb_3Sn	18·5	
Indium	3·41	23	NbN	16	
Lead	7·18	65	$Nb_3(Al_{0·8}Ge_{0·2})$	20	
Mercury, α	4·15	33	V_3Si	17	
Mercury, β	3·95	27	V_3Ga	16·8	
Niobium	9·46	156	$La_{2-x}Sr_xCuO_{4-x}$	30	
Tantalum	4·48	67	$Y_1Ba_2Cu_3O_{7-x}$	92	
Thorium	1·37	13	$Bi_2Sr_2Ca_2Cu_3O_x$	108	
Tin	3·72	25	$Tl_2Ba_2Ca_2Cu_3O_x$	125	
Vanadium	5·30	105			

Optical properties

Refractive index

The following curves show the variation with wavelength of the real part (n) and imaginary part (k, the extinction coefficient) of the refractive index.

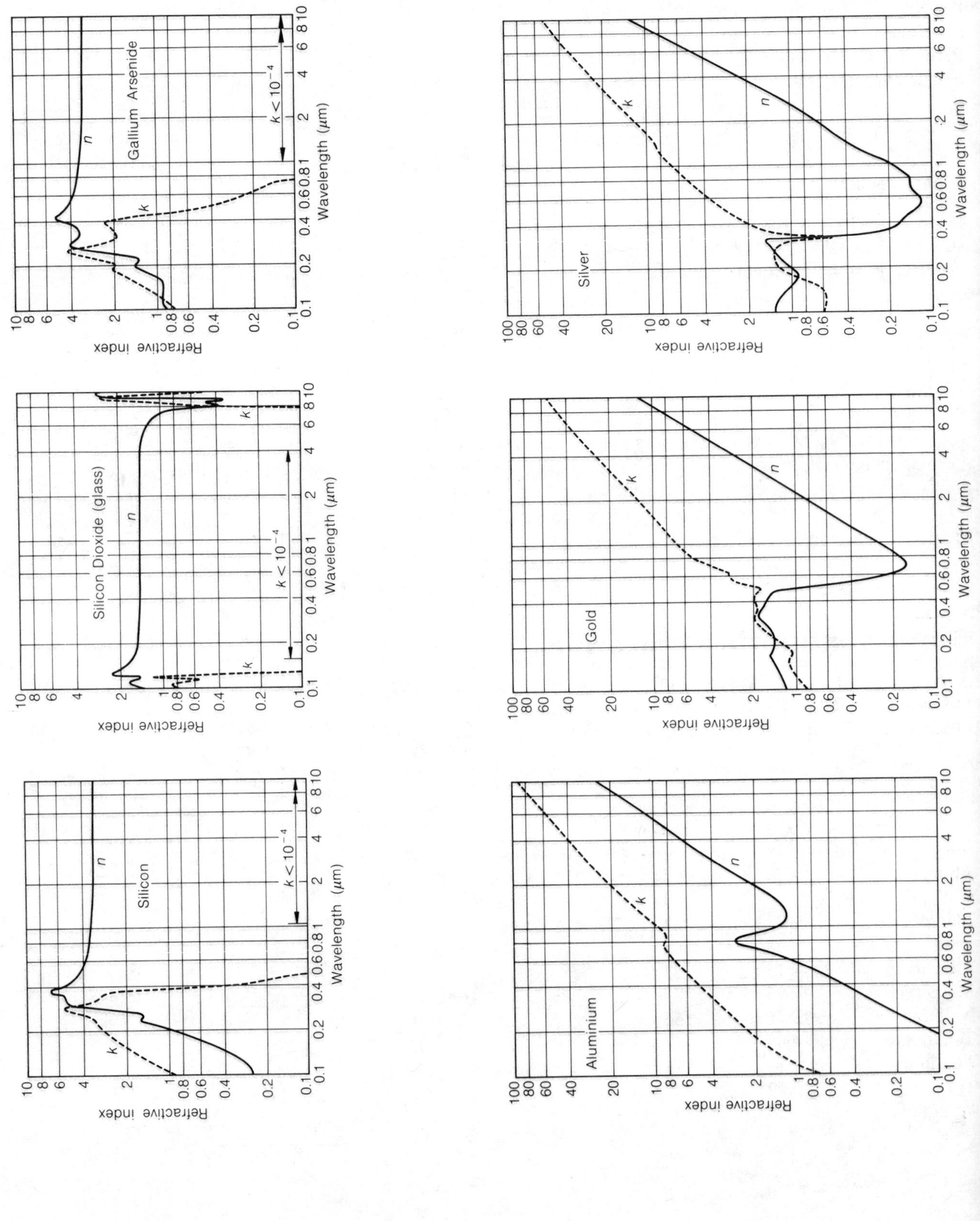

Attenuation of an optical fibre

The curve shows the attenuation of a commercial single-mode silica fibre, with germanium-doped core and matched cladding doped with phosphorus and fluorine.

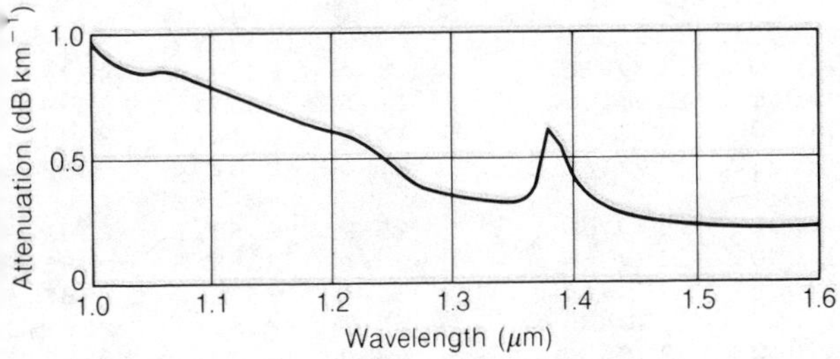

Properties of liquids

ρ Mass density (kg dm^{-3}) ≈ specific gravity
K Bulk modulus (GPa)
t_f, t_b Freezing, boiling points (° C)
h_{fg} Latent heat of boiling (kJ kg^{-1}) at ~ 1 atm
$k(t)$ Thermal conductivity (at t° C) (W m^{-1} K^{-1})
c Specific heat capacity at 20° C (J kg^{-1} K^{-1})
β Coefficient of volume thermal expansion (K^{-1} × 10^{-3})
μ Dynamic viscosity at 20° C (mPa s)
ρ_e Electrical resistivity at or near 20° C (Ωm)
ϵ_r Dielectric constant (relative permittivity) at 1 MHz, 20° C

	ρ	K	t_f	t_b	h_{fg}	$k(t)$	c	β	μ	ρ_e	ϵ_r
Mercury	13·5	25	−39	357	292	8 (20)	139	0·18	1·55	9·6 × 10^{-7}	
Sodium			98	883		85 (130)				10^{-7}*	
Water	1·00	2·3	0	100	2260	0·61 (20)	4180	0·21	1·00	5000†	81
Sea-water	1·03	2·3	−2·5	103			3930			0·2–0·3	
Mineral oil	~0·9	1-2				0·13 (100)	1700	0·7-0·9		3 × 10^{10}–10^{15}	~2·2
Carbon tetrachloride, CCl_4	1·6	1·1	−23	77	215	0·11 (20)	840	1·22	0·97		2·2
Acetone, C_3H_5OH	0·79	1·2	−95	57	560	0·16 (20)	2210	1·43	0·32		21
Ethyl alcohol, C_2H_5OH	0·79	1·3	−120	78	850	0·17 (20)	2500	1·08	1·20	3000	26
Nitrogen	0·81		−210	−196		0·15 (−203)					

* At 100°C

† Distilled water

Thermodynamic properties of fluids

t Temperature (°C)
p Absolute pressure (bar)
v Specific volume ($m^3\,kg^{-1}$) $\times 10^3$
u Specific internal energy ($kJ\,kg^{-1}$)
h Specific enthalpy ($kJ\,kg^{-1}$)
s Specific entropy ($kJ\,kg^{-1}\,K^{-1}$)
The subscripts s, f, g, and fg indicate respectively saturation, liquid, vapour, and liquid–vapour phase change.

Saturated water and steam, to 100°C

t_s	p	v_f	v_g	u_f	u_{fg}	u_g
0.01	0.006112	1.0002	206163	0.0	2375.6	2375.6
1	0.006566	1.0001	192607	4.2	2372.7	2376.9
2	0.007055	1.0001	179923	8.4	2369.9	2378.3
3	0.007575	1.0001	168169	12.6	2367.1	2379.7
4	0.008129	1.0000	157272	16.8	2364.3	2381.1
5	0.008718	1.0000	147163	21.0	2361.4	2382.4
6	0.009345	1.0000	137780	25.2	2358.6	2383.8
7	0.010012	1.0001	129064	29.4	2355.8	2385.2
8	0.010720	1.0001	120966	33.6	2352.9	2386.5
9	0.011472	1.0002	113435	37.8	2350.2	2388.0
10	0.012270	1.0003	106430	42.0	2347.3	2389.3
12	0.014014	1.0004	93835	50.4	2341.7	2392.1
14	0.015973	1.0007	82900	58.8	2336.0	2394.8
16	0.018168	1.0010	73384	67.1	2330.5	2397.6
18	0.020624	1.0013	65087	75.5	2324.8	2400.3
20	0.023366	1.0017	57838	83.9	2319.2	2403.1
22	0.026422	1.0022	51492	92.2	2313.6	2405.7
24	0.029821	1.0026	45926	100.6	2307.9	2408.5
26	0.033597	1.0032	41034	108.9	2302.3	2411.2
28	0.037782	1.0037	36728	117.3	2296.6	2413.9
30	0.042415	1.0043	32929	125.7	2291.0	2416.7
32	0.047534	1.0049	29572	134.0	2285.4	2419.4
34	0.053180	1.0056	26601	142.4	2279.7	2422.1
36	0.059400	1.0063	23967	150.7	2274.1	2424.8
38	0.066240	1.0070	21627	159.1	2268.4	2427.5
40	0.073750	1.0078	19546	167.5	2262.8	2430.2
42	0.081985	1.0086	17692	175.8	2257.1	2432.9
44	0.091001	1.0094	16036	184.2	2251.4	2435.6
46	0.100860	1.0103	14557	192.5	2245.8	2438.3
48	0.111620	1.0112	13233	200.9	2240.0	2440.9
50	0.123350	1.0121	12046	209.3	2234.3	2443.6
52	0.136130	1.0131	10980	217.6	2228.6	2446.2
54	0.150020	1.0140	10022	226.0	2222.9	2448.8
56	0.165110	1.0150	9159	234.4	2217.1	2451.5
58	0.181470	1.0161	8381	242.7	2211.4	2454.1
60	0.199200	1.0171	7679	251.1	2205.7	2456.7
62	0.218380	1.0182	7044	259.5	2199.9	2459.4
64	0.239120	1.0193	6469	267.8	2194.1	2461.9
66	0.261500	1.0205	5948	276.2	2188.4	2464.6
68	0.285630	1.0217	5476	284.6	2182.5	2467.1
70	0.311620	1.0228	5046	293.0	2176.7	2469.6
72	0.339580	1.0241	4656	301.4	2170.8	2472.2
74	0.369640	1.0253	4300	309.7	2165.1	2474.8
76	0.401910	1.0266	3976	318.1	2159.3	2477.3
78	0.436520	1.0279	3680	326.5	2153.3	2479.8
80	0.473600	1.0292	3409	334.9	2147.5	2482.3
82	0.513290	1.0305	3162	343.2	2141.6	2484.8
84	0.555730	1.0319	2935	351.6	2135.7	2487.3
86	0.601080	1.0333	2727	360.0	2129.6	2489.7
88	0.649480	1.0347	2537	368.4	2123.7	2492.2
90	0.701090	1.0361	2361	376.8	2117.7	2494.6
92	0.756080	1.0376	2200	385.3	2111.7	2497.0
94	0.814610	1.0391	2052	393.7	2105.7	2499.5
96	0.876860	1.0406	1915	402.1	2099.6	2501.8
98	0.943010	1.0421	1789	410.5	2093.7	2504.2
100	1.013250	1.0437	1673	419.0	2087.5	2506.5

h_f	h_{fg}	h_g	s_f	s_{fg}	s_g	t_s
0.0	2501.6	2501.6	0.0000	9.1575	9.1575	0.01
4.2	2499.2	2503.4	0.0153	9.1158	9.1311	1
8.4	2496.8	2505.2	0.0306	9.0741	9.1047	2
12.6	2494.5	2507.1	0.0459	9.0326	9.0785	3
16.8	2492.1	2508.9	0.0611	8.9915	9.0526	4
21.0	2489.7	2510.7	0.0762	8.9507	9.0269	5
25.2	2487.4	2512.6	0.0913	8.9102	9.0015	6
29.4	2485.0	2514.4	0.1063	8.8699	8.9762	7
33.6	2482.6	2516.2	0.1213	8.8300	8.9513	8
37.8	2480.3	2518.1	0.1362	8.7903	8.9265	9
42.0	2477.9	2519.9	0.1510	8.7510	8.9020	10
50.4	2473.2	2523.6	0.1805	8.6731	8.8536	12
58.8	2468.4	2527.2	0.2098	8.5962	8.8060	14
67.1	2463.8	2530.9	0.2388	8.5205	8.7593	16
75.5	2459.0	2534.5	0.2677	8.4458	8.7135	18
83.9	2454.3	2538.2	0.2963	8.3721	8.6684	20
92.2	2449.6	2541.8	0.3247	8.2994	8.6241	22
100.6	2444.9	2545.5	0.3530	8.2276	8.5806	24
108.9	2440.2	2549.1	0.3810	8.1569	8.5379	26
117.3	2435.4	2552.7	0.4088	8.0871	8.4959	28
125.7	2430.7	2556.4	0.4365	8.0181	8.4546	30
134.0	2426.0	2560.0	0.4640	7.9500	8.4140	32
142.4	2421.2	2563.6	0.4913	7.8827	8.3740	34
150.7	2416.5	2567.2	0.5184	7.8164	8.3348	36
159.1	2411.7	2570.8	0.5453	7.7509	8.2962	38
167.5	2406.9	2574.4	0.5721	7.6862	8.2583	40
175.8	2402.1	2577.9	0.5987	7.6222	8.2209	42
184.2	2397.3	2581.5	0.6252	7.5590	8.1842	44
192.5	2392.6	2585.1	0.6514	7.4967	8.1481	46
200.9	2387.7	2588.6	0.6776	7.4349	8.1125	48
209.3	2382.9	2592.2	0.7035	7.3741	8.0776	50
217.6	2378.1	2595.7	0.7293	7.3139	8.0432	52
226.0	2373.2	2599.2	0.7550	7.2543	8.0093	54
234.4	2368.3	2602.7	0.7804	7.1955	7.9759	56
242.7	2363.5	2606.2	0.8058	7.1373	7.9431	58
251.1	2358.6	2609.7	0.8310	7.0798	7.9108	60
259.5	2353.7	2613.2	0.8560	7.0230	7.8790	62
267.8	2348.8	2616.6	0.8809	6.9668	7.8477	64
276.2	2343.9	2620.1	0.9057	6.9111	7.8168	66
284.6	2338.9	2623.5	0.9303	6.8561	7.7864	68
293.0	2333.9	2626.9	0.9548	6.8017	7.7565	70
301.4	2328.9	2630.3	0.9792	6.7478	7.7270	72
309.7	2324.0	2633.7	1.0034	6.6945	7.6979	74
318.1	2319.0	2637.1	1.0275	6.6418	7.6693	76
326.5	2313.9	2640.4	1.0514	6.5896	7.6410	78
334.9	2308.9	2643.8	1.0753	6.5379	7.6132	80
343.3	2303.8	2647.1	1.0990	6.4868	7.5858	82
351.7	2298.7	2650.4	1.1225	6.4363	7.5588	84
360.1	2293.5	2653.6	1.1460	6.3861	7.5321	86
368.5	2288.4	2656.9	1.1693	6.3365	7.5058	88
376.9	2283.2	2660.1	1.1925	6.2874	7.4799	90
385.4	2278.0	2663.4	1.2156	6.2387	7.4543	92
393.8	2272.8	2666.6	1.2386	6.1905	7.4291	94
402.2	2267.5	2669.7	1.2615	6.1427	7.4042	96
410.6	2262.3	2672.9	1.2842	6.0954	7.3796	98
419.1	2256.9	2676.0	1.3069	6.0485	7.3554	100

Saturated water and steam, to 221 bar

For definitions see p. 48

p	t_s	v_f	v_g	u_f	u_{fg}	u_g
0.00611	0.01	1.0002	206162.9	0.0	2375.6	2375.6
0.008	3.77	1.0000	159668.5	15.9	2364.9	2380.8
0.01	6.98	1.0001	129210.7	29.3	2355.9	2385.2
0.02	17.51	1.0012	67011.6	73.5	2326.1	2399.6
0.03	24.10	1.0027	45670.0	101.0	2307.6	2408.6
0.04	28.98	1.0040	34803.3	121.4	2293.9	2415.3
0.05	32.90	1.0052	28194.5	137.8	2282.8	2420.6
0.06	36.18	1.0064	23740.6	151.5	2273.6	2425.1
0.07	39.03	1.0074	20530.4	163.4	2265.5	2428.9
0.08	41.54	1.0084	18103.8	173.9	2258.4	2432.3
0.09	43.79	1.0094	16203.4	183.3	2252.0	2435.3
0.10	45.83	1.0102	14673.7	191.8	2246.3	2438.1
0.11	47.71	1.0111	13415.2	199.7	2240.8	2440.5
0.12	49.45	1.0119	12361.0	206.9	2236.0	2442.9
0.13	51.06	1.0126	11464.9	213.7	2231.3	2445.0
0.14	52.58	1.0133	10693.4	220.0	2227.0	2447.0
0.15	54.00	1.0140	10022.1	226.0	2222.9	2448.9
0.16	55.34	1.0147	9432.4	231.6	2219.1	2450.7
0.17	56.62	1.0154	8910.3	236.9	2215.4	2452.3
0.18	57.83	1.0160	8444.6	242.0	2211.9	2453.9
0.19	58.98	1.0166	8026.6	246.8	2208.6	2455.4
0.20	60.09	1.0172	7649.2	251.5	2205.4	2456.9
0.22	62.16	1.0183	6994.6	260.1	2199.5	2459.6
0.24	64.08	1.0194	6446.2	268.2	2193.9	2462.1
0.26	65.87	1.0204	5979.9	275.7	2188.7	2464.4
0.28	67.55	1.0214	5578.4	282.7	2183.8	2466.5
0.30	69.13	1.0223	5229.0	289.3	2179.3	2468.5
0.32	70.62	1.0232	4922.0	295.6	2174.9	2470.5
0.34	72.03	1.0241	4650.1	301.5	2170.8	2472.3
0.36	73.38	1.0249	4407.6	307.1	2167.0	2474.0
0.38	74.66	1.0257	4189.8	312.5	2163.1	2475.6
0.40	75.89	1.0265	3993.2	317.7	2159.5	2477.2
0.42	77.06	1.0273	3814.8	322.6	2156.1	2478.7
0.44	78.19	1.0280	3652.2	327.3	2152.7	2480.0
0.46	79.28	1.0287	3503.2	331.9	2149.6	2481.5
0.48	80.33	1.0294	3366.3	336.3	2146.5	2482.7
0.50	81.35	1.0301	3240.1	340.5	2143.4	2484.0
0.52	82.33	1.0308	3123.3	344.6	2140.5	2485.2
0.54	83.28	1.0314	3014.8	348.6	2137.8	2486.4
0.56	84.19	1.0320	2913.9	352.4	2135.1	2487.5
0.58	85.09	1.0327	2819.7	356.2	2132.3	2488.6
0.60	85.95	1.0333	2731.7	359.8	2129.9	2489.7
0.62	86.80	1.0339	2649.1	363.4	2127.2	2490.7
0.64	87.62	1.0344	2571.5	366.8	2124.9	2491.7
0.66	88.42	1.0350	2498.5	370.2	2122.5	2492.7
0.68	89.20	1.0356	2429.7	373.5	2120.1	2493.6
0.70	89.96	1.0361	2364.7	376.7	2117.8	2494.6
0.72	90.70	1.0367	2303.1	379.8	2115.7	2495.5
0.74	91.43	1.0372	2244.8	382.9	2113.4	2496.3
0.76	92.14	1.0377	2189.5	385.8	2111.4	2497.2
0.78	92.83	1.0382	2136.9	388.8	2109.2	2498.0
0.80	93.51	1.0387	2086.9	391.6	2107.2	2498.8

h_f	h_{fg}	h_g	s_f	s_{fg}	s_g	p
0·0	2501·6	2501·6	0·0000	9·1575	9·1575	0·00611
15·9	2492·6	2508·5	0·0576	9·0008	9·0584	0·008
29·3	2485·1	2514·4	0·1060	8·8707	8·9767	0·01
73·5	2460·1	2533·6	0·2606	8·4640	8·7246	0·02
101·0	2444·6	2545·6	0·3543	8·2242	8·5785	0·03
121·4	2433·1	2554·5	0·4225	8·0530	8·4755	0·04
137·8	2423·8	2561·6	0·4763	7·9197	8·3960	0·05
151·5	2416·0	2567·5	0·5209	7·8103	8·3312	0·06
163·4	2409·2	2572·6	0·5591	7·7176	8·2767	0·07
173·9	2403·2	2577·1	0·5926	7·6369	8·2295	0·08
183·3	2397·8	2581·1	0·6224	7·5657	8·1881	0·09
191·8	2393·0	2584·8	0·6493	7·5018	8·1511	0·10
199·7	2388·4	2588·1	0·6738	7·4438	8·1176	0·11
206·9	2384·3	2591·2	0·6964	7·3908	8·0872	0·12
213·7	2380·3	2594·0	0·7172	7·3420	8·0592	0·13
220·0	2376·7	2596·7	0·7367	7·2966	8·0333	0·14
226·0	2373·2	2599·2	0·7549	7·2544	8·0093	0·15
231·6	2370·0	2601·6	0·7721	7·2147	7·9868	0·16
236·9	2366·9	2603·8	0·7883	7·1775	7·9658	0·17
242·0	2363·9	2605·9	0·8036	7·1423	7·9459	0·18
246·8	2361·1	2607·9	0·8182	7·1090	7·9272	0·19
251·5	2358·4	2609·9	0·8321	7·0773	7·9094	0·20
260·1	2353·4	2613·5	0·8581	7·0183	7·8764	0·22
268·2	2348·6	2616·8	0·8820	6·9644	7·8464	0·24
275·7	2344·2	2619·9	0·9041	6·9147	7·8188	0·26
282·7	2340·0	2622·7	0·9248	6·8684	7·7932	0·28
289·3	2336·1	2625·4	0·9441	6·8254	7·7695	0·30
295·6	2332·4	2628·0	0·9623	6·7850	7·7473	0·32
301·5	2328·9	2630·4	0·9795	6·7470	7·7265	0·34
307·1	2325·6	2632·7	0·9958	6·7111	7·7069	0·36
312·5	2322·3	2634·8	1·0113	6·6771	7·6884	0·38
317·7	2319·2	2636·9	1·0261	6·6448	7·6709	0·40
322·6	2316·3	2638·9	1·0402	6·6140	7·6542	0·42
327·3	2313·4	2640·7	1·0538	6·5845	7·6383	0·44
331·9	2310·7	2642·6	1·0667	6·5564	7·6231	0·46
336·3	2308·0	2644·3	1·0792	6·5294	7·6086	0·48
340·6	2305·4	2646·0	1·0912	6·5035	7·5947	0·50
344·7	2302·9	2647·6	1·1028	6·4786	7·5814	0·52
348·7	2300·5	2649·2	1·1140	6·4545	7·5685	0·54
352·5	2298·2	2650·7	1·1248	6·4313	7·5561	0·56
356·3	2295·8	2652·1	1·1353	6·4089	7·5442	0·58
359·9	2293·7	2653·6	1·1455	6·3872	7·5327	0·60
363·5	2291·4	2654·9	1·1553	6·3663	7·5216	0·62
366·9	2289·4	2656·3	1·1649	6·3459	7·5108	0·64
370·3	2287·3	2657·6	1·1742	6·3261	7·5003	0·66
373·6	2285·2	2658·8	1·1832	6·3070	7·4902	0·68
376·8	2283·3	2660·1	1·1921	6·2883	7·4804	0·70
379·9	2281·4	2661·3	1·2007	6·2701	7·4708	0·72
383·0	2279·4	2662·4	1·2090	6·2526	7·4616	0·74
385·9	2277·7	2663·6	1·2172	6·2353	7·4525	0·76
388·9	2275·8	2664·7	1·2252	6·2185	7·4437	0·78
391·7	2274·1	2665·8	1·2330	6·2022	7·4352	0·80

Saturated water and steam, to 221 bar

For definitions see p. 48

p	t_s	v_f	v_g	u_f	u_{fg}	u_g
0.82	94.18	1.0392	2039.2	394.4	2105.2	2499.6
0.84	94.83	1.0397	1993.8	397.2	2103.2	2500.4
0.86	95.47	1.0402	1950.4	399.9	2101.3	2501.2
0.88	96.10	1.0407	1908.9	402.5	2099.4	2501.9
0.90	96.71	1.0412	1869.1	405.1	2097.6	2502.7
0.92	97.32	1.0416	1831.1	407.7	2095.6	2503.3
0.94	97.91	1.0421	1794.6	410.2	2093.8	2504.0
0.96	98.50	1.0425	1759.6	412.6	2092.2	2504.8
0.98	99.07	1.0430	1726.0	415.0	2090.5	2505.5
1.00	99.63	1.0434	1693.7	417.4	2088.6	2506.0
1.10	102.32	1.0455	1549.2	428.7	2080.5	2509.2
1.20	104.81	1.0476	1428.1	439.3	2072.8	2512.0
1.30	107.13	1.0495	1325.0	449.1	2065.7	2514.8
1.40	109.32	1.0513	1236.3	458.3	2059.0	2517.2
1.50	111.37	1.0530	1159.0	466.9	2052.6	2519.6
1.60	113.32	1.0547	1091.1	475.2	2046.4	2521.6
1.70	115.17	1.0563	1030.9	483.0	2040.7	2523.7
1.80	116.93	1.0579	977.18	490.5	2035.1	2525.6
1.90	118.62	1.0594	928.95	497.7	2029.8	2527.5
2.00	120.23	1.0608	885.40	504.5	2024.7	2529.2
2.20	123.27	1.0636	809.80	517.4	2015.1	2532.4
2.40	126.09	1.0663	746.41	529.3	2006.0	2535.4
2.60	128.73	1.0688	692.47	540.6	1997.5	2538.2
2.80	131.21	1.0712	646.00	551.2	1989.4	2540.6
3.00	133.54	1.0735	605.53	561.1	1982.0	2543.0
3.20	135.76	1.0757	569.95	570.6	1974.7	2545.2
3.40	137.86	1.0779	538.43	579.5	1967.7	2547.2
3.60	139.87	1.0799	510.29	588.1	1961.1	2549.2
3.80	141.79	1.0819	485.02	596.4	1954.6	2551.0
4.00	143.63	1.0839	462.20	604.3	1948.5	2552.7
4.20	145.39	1.0858	441.47	611.8	1942.5	2554.4
4.40	147.09	1.0876	422.57	619.1	1936.8	2556.0
4.60	148.73	1.0894	405.26	626.2	1931.3	2557.5
4.80	150.31	1.0911	389.34	633.0	1925.8	2558.8
5.00	151.85	1.0928	374.66	639.6	1920.6	2560.2
5.20	153.33	1.0945	361.06	645.9	1915.6	2561.5
5.40	154.77	1.0961	348.44	652.2	1910.5	2562.7
5.60	156.16	1.0977	336.69	658.2	1905.8	2564.0
5.80	157.52	1.0993	325.72	664.1	1901.0	2565.1
6.00	158.84	1.1009	315.46	669.7	1896.5	2566.2
6.20	160.12	1.1024	305.84	675.3	1892.0	2567.3
6.40	161.38	1.1039	296.80	680.8	1887.5	2568.2
6.60	162.60	1.1053	288.29	686.1	1883.2	2569.2
6.80	163.79	1.1068	280.26	691.2	1879.0	2570.2
7.00	164.96	1.1082	272.68	696.3	1874.8	2571.1
7.20	166.10	1.1096	265.50	701.2	1870.8	2572.0
7.40	167.21	1.1110	258.70	706.1	1866.8	2572.9
7.60	168.30	1.1123	252.24	710.9	1862.8	2573.7
7.80	169.37	1.1137	246.10	715.4	1859.0	2574.4
8.00	170.41	1.1150	240.26	720.0	1855.3	2575.3

h_f	h_{fg}	h_g	s_f	s_{fg}	s_g	p
394.5	2272.3	2666.8	1.2407	6.1861	7.4268	0.82
397.3	2270.6	2667.9	1.2481	6.1706	7.4187	0.84
400.0	2268.9	2668.9	1.2554	6.1553	7.4107	0.86
402.6	2267.3	2669.9	1.2626	6.1404	7.4030	0.88
405.2	2265.7	2670.9	1.2696	6.1258	7.3954	0.90
407.8	2264.0	2671.8	1.2765	6.1114	7.3879	0.92
410.3	2262.4	2672.7	1.2832	6.0975	7.3807	0.94
412.7	2261.0	2673.7	1.2898	6.0838	7.3736	0.96
415.1	2259.5	2674.6	1.2963	6.0703	7.3666	0.98
417.5	2257.9	2675.4	1.3027	6.0571	7.3598	1.00
428.8	2250.8	2679.6	1.3330	5.9947	7.3277	1.10
439.4	2244.0	2683.4	1.3609	5.9375	7.2984	1.20
449.2	2237.8	2687.0	1.3868	5.8847	7.2715	1.30
458.4	2231.9	2690.3	1.4109	5.8356	7.2465	1.40
467.1	2226.3	2693.4	1.4336	5.7898	7.2234	1.50
475.4	2220.8	2696.2	1.4550	5.7467	7.2017	1.60
483.2	2215.8	2699.0	1.4752	5.7061	7.1813	1.70
490.7	2210.8	2701.5	1.4944	5.6678	7.1622	1.80
497.9	2206.1	2704.0	1.5127	5.6313	7.1440	1.90
504.7	2201.6	2706.3	1.5301	5.5967	7.1268	2.00
517.6	2193.0	2710.6	1.5628	5.5321	7.0949	2.20
529.6	2184.9	2714.5	1.5929	5.4728	7.0657	2.40
540.9	2177.3	2718.2	1.6209	5.4180	7.0389	2.60
551.5	2170.0	2721.5	1.6471	5.3669	7.0140	2.80
561.4	2163.3	2724.7	1.6717	5.3192	6.9909	3.00
570.9	2156.7	2727.6	1.6948	5.2744	6.9692	3.20
579.9	2150.4	2730.3	1.7168	5.2321	6.9489	3.40
588.5	2144.4	2732.9	1.7376	5.1921	6.9297	3.60
596.8	2138.5	2735.3	1.7575	5.1540	6.9115	3.80
604.7	2132.9	2737.6	1.7764	5.1179	6.8943	4.00
612.3	2127.5	2739.8	1.7946	5.0833	6.8779	4.20
619.6	2122.3	2741.9	1.8120	5.0502	6.8622	4.40
626.7	2117.2	2743.9	1.8287	5.0186	6.8473	4.60
633.5	2112.2	2745.7	1.8448	4.9881	6.8329	4.80
640.1	2107.4	2747.5	1.8604	4.9588	6.8192	5.00
646.5	2102.8	2749.3	1.8754	4.9305	6.8059	5.20
652.8	2098.1	2750.9	1.8899	4.9033	6.7932	5.40
658.8	2093.7	2752.5	1.9040	4.8769	6.7809	5.60
664.7	2089.3	2754.0	1.9176	4.8514	6.7690	5.80
670.4	2085.1	2755.5	1.9308	4.8267	6.7575	6.00
676.0	2080.9	2756.9	1.9437	4.8027	6.7464	6.20
681.5	2076.7	2758.2	1.9562	4.7794	6.7356	6.40
686.8	2072.7	2759.5	1.9684	4.7568	6.7252	6.60
692.0	2068.8	2760.8	1.9803	4.7347	6.7150	6.80
697.1	2064.9	2762.0	1.9918	4.7134	6.7052	7.00
702.0	2061.2	2763.2	2.0031	4.6925	6.6956	7.20
706.9	2057.4	2764.3	2.0141	4.6721	6.6862	7.40
711.7	2053.7	2765.4	2.0249	4.6522	6.6771	7.60
716.3	2050.1	2766.4	2.0354	4.6329	6.6683	7.80
720.9	2046.6	2767.5	2.0457	4.6139	6.6596	8.00

Saturated water and steam, to 221 bar

For definitions see p. 48

p	t_s	v_f	v_g	u_f	u_{fg}	u_g
8.20	171.44	1.1163	234.69	724.5	1851.6	2576.1
8.40	172.45	1.1176	229.38	729.0	1847.8	2576.7
8.60	173.43	1.1188	224.31	733.2	1844.3	2577.5
8.80	174.40	1.1201	219.46	737.5	1840.7	2578.2
9.00	175.36	1.1213	214.82	741.6	1837.2	2578.8
9.20	176.29	1.1226	210.37	745.8	1833.7	2579.5
9.40	177.21	1.1238	206.10	749.7	1830.3	2580.1
9.60	178.12	1.1250	202.01	753.7	1827.0	2580.7
9.80	179.01	1.1262	198.08	757.6	1823.7	2581.3
10.00	179.88	1.1274	194.30	761.5	1820.4	2581.9
11.00	184.06	1.1331	177.39	779.9	1804.7	2584.6
12.00	187.96	1.1386	163.21	797.0	1789.8	2586.8
13.00	191.60	1.1438	151.14	813.2	1775.7	2588.9
14.00	195.04	1.1489	140.73	828.5	1762.3	2590.8
15.00	198.28	1.1538	131.67	842.9	1749.5	2592.4
16.00	201.37	1.1586	123.70	856.6	1737.1	2593.8
17.00	204.30	1.1633	116.64	869.8	1725.3	2595.1
18.00	207.11	1.1678	110.33	882.4	1713.8	2596.2
19.00	209.79	1.1723	104.67	894.6	1702.7	2597.2
20.00	212.37	1.1766	99.55	906.2	1691.9	2598.1
21.00	214.85	1.1809	94.90	917.4	1681.5	2598.9
22.00	217.24	1.1850	90.66	928.3	1671.3	2599.6
23.00	219.55	1.1891	86.78	938.9	1661.3	2600.2
24.00	221.78	1.1932	83.21	949.0	1651.7	2600.7
25.00	223.94	1.1972	79.92	958.9	1642.2	2601.1
26.00	226.03	1.2011	76.87	968.6	1633.0	2601.6
27.00	228.06	1.2050	74.03	977.9	1623.9	2601.8
28.00	230.04	1.2088	71.40	987.1	1615.0	2602.1
29.00	231.96	1.2126	68.94	996.0	1606.3	2602.3
30.00	233.84	1.2163	66.63	1004.7	1597.8	2602.4
31.00	235.66	1.2200	64.47	1013.2	1589.2	2602.4
32.00	237.44	1.2237	62.44	1021.5	1581.0	2602.5
33.00	239.18	1.2273	60.53	1029.6	1572.9	2602.5
34.00	240.88	1.2310	58.73	1037.6	1564.8	2602.4
35.00	242.54	1.2345	57.03	1045.4	1557.0	2602.4
36.00	244.16	1.2381	55.42	1053.0	1549.2	2602.2
37.00	245.75	1.2416	53.89	1060.6	1541.4	2602.0
38.00	247.31	1.2451	52.44	1068.0	1533.9	2601.8
39.00	248.84	1.2486	51.06	1075.2	1526.4	2601.7
40.00	250.33	1.2521	49.75	1082.4	1518.9	2601.3
42.00	253.24	1.2589	47.31	1096.3	1504.4	2600.7
44.00	256.05	1.2657	45.08	1109.8	1490.1	2600.0
46.00	258.76	1.2725	43.04	1122.9	1476.1	2599.0
48.00	261.38	1.2792	41.16	1135.7	1462.5	2598.1
50.00	263.92	1.2858	39.43	1148.1	1449.0	2597.1
52.00	266.38	1.2925	37.82	1160.2	1435.8	2595.9
54.00	268.77	1.2990	36.33	1172.0	1422.6	2594.6
56.00	271.09	1.3056	34.94	1183.5	1409.8	2593.3
58.00	273.36	1.3122	33.65	1194.8	1397.1	2591.9
60.00	275.56	1.3187	32.43	1205.8	1384.6	2590.4

h_f	h_{fg}	h_g	s_f	s_{fg}	s_g	p
725.4	2043.1	2768.5	2.0558	4.5953	6.6511	8.20
729.9	2039.5	2769.4	2.0657	4.5772	6.6429	8.40
734.2	2036.2	2770.4	2.0753	4.5595	6.6348	8.60
738.5	2032.8	2771.3	2.0848	4.5421	6.6269	8.80
742.6	2029.5	2772.1	2.0941	4.5251	6.6192	9.00
746.8	2026.2	2773.0	2.1033	4.5083	6.6116	9.20
750.8	2023.0	2773.8	2.1122	4.4920	6.6042	9.40
754.8	2019.8	2774.6	2.1210	4.4759	6.5969	9.60
758.7	2016.7	2775.4	2.1297	4.4601	6.5898	9.80
762.6	2013.6	2776.2	2.1382	4.4446	6.5828	10.00
781.1	1998.6	2779.7	2.1786	4.3712	6.5498	11.00
798.4	1984.3	2782.7	2.2160	4.3034	6.5194	12.00
814.7	1970.7	2785.4	2.2509	4.2404	6.4913	13.00
830.1	1957.7	2787.8	2.2836	4.1815	6.4651	14.00
844.6	1945.3	2789.9	2.3144	4.1262	6.4406	15.00
858.5	1933.2	2791.7	2.3436	4.0740	6.4176	16.00
871.8	1921.6	2793.4	2.3712	4.0246	6.3958	17.00
884.5	1910.3	2794.8	2.3976	3.9775	6.3751	18.00
896.8	1899.3	2796.1	2.4227	3.9328	6.3555	19.00
908.6	1888.6	2797.2	2.4468	3.8899	6.3367	20.00
919.9	1878.3	2798.2	2.4699	3.8488	6.3187	21.00
930.9	1868.2	2799.1	2.4921	3.8094	6.3015	22.00
941.6	1858.2	2799.8	2.5136	3.7714	6.2850	23.00
951.9	1848.5	2800.4	2.5342	3.7348	6.2690	24.00
961.9	1839.0	2800.9	2.5542	3.6995	6.2537	25.00
971.7	1829.7	2801.4	2.5736	3.6652	6.2388	26.00
981.2	1820.5	2801.7	2.5923	3.6321	6.2244	27.00
990.5	1811.5	2802.0	2.6105	3.6000	6.2105	28.00
999.5	1802.7	2802.2	2.6282	3.5687	6.1969	29.00
1008.3	1794.0	2802.3	2.6455	3.5383	6.1838	30.00
1017.0	1785.3	2802.3	2.6622	3.5088	6.1710	31.00
1025.4	1776.9	2802.3	2.6785	3.4800	6.1585	32.00
1033.7	1768.6	2802.3	2.6945	3.4518	6.1463	33.00
1041.8	1760.3	2802.1	2.7100	3.4245	6.1345	34.00
1049.7	1752.3	2802.0	2.7252	3.3977	6.1229	35.00
1057.5	1744.2	2801.7	2.7401	3.3714	6.1115	36.00
1065.2	1736.2	2801.4	2.7547	3.3457	6.1004	37.00
1072.7	1728.4	2801.1	2.7689	3.3207	6.0896	38.00
1080.1	1720.7	2800.8	2.7828	3.2961	6.0789	39.00
1087.4	1712.9	2800.3	2.7965	3.2720	6.0685	40.00
1101.6	1697.8	2799.4	2.8231	3.2251	6.0482	42.00
1115.4	1682.9	2798.3	2.8487	3.1799	6.0286	44.00
1128.8	1668.2	2797.0	2.8735	3.1362	6.0097	46.00
1141.8	1653.9	2795.7	2.8974	3.0939	5.9913	48.00
1154.5	1639.7	2794.2	2.9207	3.0528	5.9735	50.00
1166.9	1625.7	2792.6	2.9432	3.0129	5.9561	52.00
1179.0	1611.8	2790.8	2.9651	2.9741	5.9392	54.00
1190.8	1598.2	2789.0	2.9864	2.9363	5.9227	56.00
1202.4	1584.6	2787.0	3.0071	2.8994	5.9065	58.00
1213.7	1571.3	2785.0	3.0274	2.8632	5.8906	60.00

Saturated water and steam, to 221 bar

For definitions see p. 48

p	t_s	v_f	v_g	u_f	u_{fg}	u_g
62.00	277.71	1.3252	31.29	1216.7	1372.2	2588.9
64.00	279.80	1.3318	30.23	1227.3	1359.9	2587.2
66.00	281.85	1.3383	29.22	1237.7	1347.8	2585.5
68.00	283.85	1.3448	28.27	1248.0	1335.7	2583.7
70.00	285.80	1.3514	27.37	1258.0	1323.8	2581.8
72.00	287.71	1.3579	26.52	1267.9	1312.1	2580.0
74.00	289.59	1.3645	25.71	1277.7	1300.2	2577.9
76.00	291.42	1.3711	24.94	1287.3	1288.6	2575.9
78.00	293.22	1.3777	24.22	1296.8	1277.1	2573.8
80.00	294.98	1.3843	23.52	1306.1	1265.6	2571.7
82.00	296.71	1.3909	22.86	1315.3	1254.3	2569.5
84.00	298.40	1.3976	22.23	1324.4	1242.9	2567.3
86.00	300.07	1.4043	21.62	1333.3	1231.6	2564.9
88.00	301.71	1.4111	21.05	1342.3	1220.3	2562.6
90.00	303.31	1.4179	20.49	1351.0	1209.1	2560.2
92.00	304.89	1.4247	19.96	1359.7	1198.0	2557.6
94.00	306.45	1.4316	19.45	1368.2	1186.9	2555.1
96.00	307.98	1.4385	18.96	1376.8	1175.9	2552.6
98.00	309.48	1.4455	18.49	1385.2	1164.7	2550.0
100.00	310.96	1.4526	18.04	1393.6	1153.7	2547.3
102.00	312.42	1.4597	17.60	1401.8	1142.8	2544.6
104.00	313.86	1.4668	17.18	1409.9	1131.9	2541.9
106.00	315.27	1.4740	16.78	1418.1	1121.0	2539.1
108.00	316.67	1.4813	16.39	1426.2	1109.9	2536.1
110.00	318.04	1.4887	16.01	1434.2	1099.0	2533.2
112.00	319.40	1.4961	15.64	1442.1	1088.1	2530.2
114.00	320.73	1.5037	15.29	1450.1	1077.2	2527.2
116.00	322.05	1.5113	14.94	1457.9	1066.3	2524.2
118.00	323.35	1.5189	14.61	1465.7	1055.3	2521.0
120.00	324.64	1.5267	14.29	1473.4	1044.4	2517.8
122.00	325.90	1.5346	13.97	1481.1	1033.4	2514.5
124.00	327.15	1.5425	13.67	1488.8	1022.4	2511.1
126.00	328.39	1.5506	13.37	1496.4	1011.4	2507.7
128.00	329.61	1.5588	13.08	1503.9	1000.2	2504.2
130.00	330.81	1.5671	12.80	1511.5	989.1	2500.6
132.00	332.00	1.5755	12.53	1519.1	977.9	2497.0
134.00	333.18	1.5841	12.26	1526.6	966.6	2493.2
136.00	334.34	1.5927	12.00	1534.0	955.4	2489.4
138.00	335.49	1.6015	11.75	1541.5	944.0	2485.5
140.00	336.63	1.6105	11.50	1549.0	932.5	2481.4
142.00	337.75	1.6196	11.26	1556.4	920.9	2477.4
144.00	338.86	1.6289	11.02	1563.8	909.4	2473.2
146.00	339.96	1.6383	10.79	1571.3	897.6	2468.9
148.00	341.04	1.6480	10.56	1578.6	885.8	2464.5
150.00	342.12	1.6578	10.34	1586.0	873.9	2460.0
152.00	343.18	1.6678	10.13	1593.4	861.9	2455.4
154.00	344.23	1.6780	9.92	1600.9	849.8	2450.7
156.00	345.27	1.6885	9.71	1608.3	837.6	2445.8
158.00	346.30	1.6992	9.51	1615.7	825.3	2441.0
160.00	347.32	1.7102	9.31	1623.0	812.9	2435.9

h_f	h_{fg}	h_g	s_f	s_{fg}	s_g	p
1224·9	1558·0	2782·9	3·0472	2·8281	5·8753	62·00
1235·8	1544·8	2780·6	3·0665	2·7936	5·8601	64·00
1246·5	1531·8	2778·3	3·0854	2·7598	5·8452	66·00
1257·1	1518·8	2775·9	3·1039	2·7266	5·8305	68·00
1267·5	1505·9	2773·4	3·1220	2·6941	5·8161	70·00
1277·7	1493·2	2770·9	3·1398	2·6621	5·8019	72·00
1287·8	1480·4	2768·2	3·1572	2·6307	5·7879	74·00
1297·7	1467·8	2765·5	3·1743	2·5998	5·7741	76·00
1307·5	1455·2	2762·7	3·1912	2·5693	5·7605	78·00
1317·2	1442·7	2759·9	3·2077	2·5393	5·7470	80·00
1326·7	1430·3	2757·0	3·2240	2·5097	5·7337	82·00
1336·1	1417·9	2754·0	3·2400	2·4806	5·7206	84·00
1345·4	1405·5	2750·9	3·2558	2·4518	5·7076	86·00
1354·7	1393·1	2747·8	3·2714	2·4233	5·6947	88·00
1363·8	1380·8	2744·6	3·2867	2·3953	5·6820	90·00
1372·8	1368·5	2741·3	3·3019	2·3674	5·6693	92·00
1381·7	1356·3	2738·0	3·3168	2·3400	5·6568	94·00
1390·6	1344·1	2734·7	3·3316	2·3128	5·6444	96·00
1399·4	1331·8	2731·2	3·3462	2·2858	5·6320	98·00
1408·1	1319·6	2727·7	3·3606	2·2592	5·6198	100·00
1416·7	1307·5	2724·2	3·3748	2·2328	5·6076	102·00
1425·2	1295·4	2720·6	3·3889	2·2066	5·5955	104·00
1433·7	1283·2	2716·9	3·4029	2·1806	5·5835	106·00
1442·2	1270·9	2713·1	3·4167	2·1548	5·5715	108·00
1450·6	1258·7	2709·3	3·4304	2·1292	5·5596	110·00
1458·9	1246·5	2705·4	3·4439	2·1038	5·5477	112·00
1467·2	1234·3	2701·5	3·4574	2·0784	5·5358	114·00
1475·4	1222·1	2697·5	3·4707	2·0533	5·5240	116·00
1483·6	1209·8	2693·4	3·4840	2·0281	5·5121	118·00
1491·7	1197·5	2689·2	3·4971	2·0032	5·5003	120·00
1499·8	1185·1	2684·9	3·5101	1·9784	5·4885	122·00
1507·9	1172·7	2680·6	3·5231	1·9535	5·4766	124·00
1515·9	1160·3	2676·2	3·5359	1·9289	5·4648	126·00
1523·9	1147·7	2671·6	3·5487	1·9042	5·4529	128·00
1531·9	1135·1	2667·0	3·5614	1·8795	5·4409	130·00
1539·9	1122·4	2662·3	3·5741	1·8549	5·4290	132·00
1547·8	1109·7	2657·5	3·5867	1·8302	5·4169	134·00
1555·7	1096·9	2652·6	3·5992	1·8056	5·4048	136·00
1563·6	1084·0	2647·6	3·6116	1·7811	5·3927	138·00
1571·5	1070·9	2642·4	3·6241	1·7563	5·3804	140·00
1579·4	1057·8	2637·2	3·6365	1·7316	5·3681	142·00
1587·3	1044·6	2631·9	3·6488	1·7069	5·3557	144·00
1595·2	1031·2	2626·4	3·6611	1·6821	5·3432	146·00
1603·0	1017·8	2620·8	3·6734	1·6573	5·3307	148·00
1610·9	1004·2	2615·1	3·6857	1·6323	5·3180	150·00
1618·8	990·5	2609·3	3·6979	1·6074	5·3053	152·00
1626·7	976·7	2603·4	3·7102	1·5822	5·2924	154·00
1634·6	962·7	2597·3	3·7224	1·5571	5·2795	156·00
1642·5	948·7	2591·2	3·7347	1·5317	5·2664	158·00
1650·4	934·5	2584·9	3·7470	1·5063	5·2533	160·00

Saturated water and steam, to 221 bar

For definitions see p. 48

p	t_s	v_f	v_g	u_f	u_{fg}	u_g
162.00	348.32	1.7214	9.12	1630.5	800.4	2430.9
164.00	349.32	1.7330	8.93	1638.0	787.7	2425.7
166.00	350.31	1.7446	8.74	1645.4	775.1	2420.5
168.00	351.29	1.7569	8.55	1653.4	761.6	2415.0
170.00	352.26	1.7695	8.37	1661.5	747.8	2409.3
172.00	353.22	1.7825	8.19	1669.6	733.9	2403.5
174.00	354.16	1.7961	8.01	1677.7	719.9	2397.6
176.00	355.11	1.8101	7.84	1685.7	705.8	2391.5
178.00	356.04	1.8247	7.67	1693.7	691.6	2385.3
180.00	356.96	1.8399	7.50	1701.7	677.3	2378.9
182.00	357.87	1.8557	7.33	1709.7	662.6	2372.3
184.00	358.78	1.8722	7.16	1717.8	647.7	2365.5
186.00	359.67	1.8894	7.00	1725.8	632.7	2358.5
188.00	360.56	1.9074	6.84	1733.9	617.3	2351.3
190.00	361.44	1.9262	6.68	1742.1	601.6	2343.7
192.00	362.31	1.9460	6.52	1750.5	585.3	2335.8
194.00	363.17	1.9669	6.36	1758.9	568.8	2327.7
196.00	364.03	1.9889	6.20	1767.7	551.4	2319.2
198.00	364.87	2.0124	6.04	1776.7	533.5	2310.2
200.00	365.71	2.0374	5.87	1785.9	514.9	2300.7
202.00	366.54	2.0643	5.71	1795.5	495.2	2290.7
204.00	367.37	2.0935	5.55	1805.6	474.4	2280.0
206.00	368.18	2.1256	5.38	1816.2	452.2	2268.4
208.00	368.99	2.1612	5.20	1827.6	428.2	2255.9
210.00	369.79	2.2018	5.02	1840.1	402.0	2242.0
212.00	370.58	2.2489	4.83	1853.9	372.6	2226.5
214.00	371.37	2.3059	4.62	1869.7	338.8	2208.4
216.00	372.14	2.3788	4.39	1888.4	298.4	2186.8
218.00	372.92	2.4819	4.12	1912.8	245.8	2158.6
220.00	373.68	2.6675	3.73	1951.6	162.8	2114.4
221.20	374.15	3.1700	3.17	2037.3	0.0	2037.3

h_f	h_{fg}	h_g	s_f	s_{fg}	s_g	p
1658.4	920.2	2578.6	3.7592	1.4809	5.2401	162.00
1666.4	905.7	2572.1	3.7716	1.4552	5.2268	164.00
1674.4	891.2	2565.6	3.7841	1.4292	5.2133	166.00
1682.9	875.8	2558.7	3.7973	1.4022	5.1995	168.00
1691.6	860.0	2551.6	3.8106	1.3750	5.1856	170.00
1700.3	844.1	2544.4	3.8239	1.3475	5.1714	172.00
1709.0	828.1	2537.1	3.8372	1.3199	5.1571	174.00
1717.6	811.9	2529.5	3.8502	1.2923	5.1425	176.00
1726.2	795.6	2521.8	3.8635	1.2643	5.1278	178.00
1734.8	779.1	2513.9	3.8766	1.2361	5.1127	180.00
1743.5	762.2	2505.7	3.8897	1.2078	5.0975	182.00
1752.2	745.1	2497.3	3.9029	1.1790	5.0819	184.00
1760.9	727.8	2488.7	3.9161	1.1499	5.0660	186.00
1769.8	710.0	2479.8	3.9295	1.1202	5.0497	188.00
1778.7	691.8	2470.5	3.9430	1.0900	5.0330	190.00
1787.9	673.0	2460.9	3.9568	1.0590	5.0158	192.00
1797.1	653.9	2451.0	3.9708	1.0273	4.9981	194.00
1806.7	633.9	2440.6	3.9851	0.9947	4.9798	196.00
1816.5	613.2	2429.7	3.9998	0.9610	4.9608	198.00
1826.6	591.6	2418.2	4.0151	0.9259	4.9410	200.00
1837.2	568.9	2406.1	4.0310	0.8891	4.9201	202.00
1848.3	544.8	2393.1	4.0476	0.8506	4.8982	204.00
1860.0	519.2	2379.2	4.0653	0.8095	4.8748	206.00
1872.6	491.5	2364.1	4.0843	0.7653	4.8496	208.00
1886.3	461.2	2347.5	4.1049	0.7173	4.8222	210.00
1901.6	427.3	2328.9	4.1279	0.6638	4.7917	212.00
1919.0	388.4	2307.4	4.1543	0.6027	4.7570	214.00
1939.8	341.9	2281.7	4.1858	0.5299	4.7157	216.00
1966.9	281.5	2248.4	4.2271	0.4357	4.6628	218.00
2010.3	186.3	2196.6	4.2934	0.2880	4.5814	220.00
2107.4	0.0	2107.4	4.4429	0.0000	4.4429	221.20

For definitions see p. 48

Superheated steam, to 220 bar and 800°C

p (t_s)	t:	t_s	50	100	150	200	250
0							
	u		2571.7	2642.5	2714.6	2787.8	2862.4
	h		2594.8	2688.7	2783.8	2880.1	2977.8
0.02	v	67012	74524	86080	97628	109171	12071
(17.50)	u	2399.6	2445.4	2516.3	2588.4	2661.7	2736.
	h	2533.6	2594.4	2688.5	2783.7	2880.0	2977.7
	s	8.7246	8.9226	9.1934	9.4327	9.6479	9.8441
0.06	v	23741	24812	28676	32532	36383	40232
(36.20)	u	2425.1	2444.6	2515.9	2588.2	2661.5	2736.2
	h	2567.5	2593.5	2688.0	2783.4	2879.8	2977.6
	s	8.3312	8.4135	8.6854	8.9251	9.1406	9.3369
0.10	v	14674	14869	17195	19512	21825	24136
(45.80)	u	2438.1	2444.0	2515.6	2588.0	2661.4	2736.0
	h	2584.8	2592.7	2687.5	2783.1	2879.6	2977.4
	s	8.1511	8.1757	8.4486	8.6888	8.9045	9.1010
0.50	v	3240.1		3418.1	3889.3	4356.0	4820.5
(81.30)	u	2484.0		2511.7	2585.6	2659.9	2735.1
	h	2646.0		2682.6	2780.1	2877.7	2976.1
	s	7.5947		7.6953	7.9406	8.1587	8.3564
1.00	v	1693.7		1695.5	1936.3	2172.3	2406.1
(99.60)	u	2506.0		2506.7	2582.7	2658.2	2733.9
	h	2675.4		2676.2	2776.3	2875.4	2974.5
	s	7.3598		7.3618	7.6137	7.8349	8.0342
1.50	v	1159.0			1285.2	1444.4	1601.3
(111.4)	u	2519.6			2579.7	2656.2	2732.7
	h	2693.4			2772.5	2872.9	2972.9
	s	7.2234			7.4194	7.6439	7.8447
2	v	885.40			959.54	1080.4	1198.9
(120.2)	u	2529.2			2576.6	2654.4	2731.4
	h	2706.3			2768.5	2870.5	2971.2
	s	7.1268			7.2794	7.5072	7.7096
3	v	605.53			633.74	716.35	796.44
(133.5)	u	2543.0			2570.3	2650.6	2729.0
	h	2724.7			2760.4	2865.5	2967.9
	s	6.9909			7.0771	7.3119	7.5176
4	v	462.20			470.66	534.26	595.19
(143.6)	u	2552.7			2563.7	2646.7	2726.4
	h	2737.6			2752.0	2860.4	2964.5
	s	6.8943			6.9285	7.1708	7.3800
5	v	374.66				424.96	474.43
(151.8)	u	2560.2				2642.6	2723.9
	h	2747.5				2855.1	2961.1
	s	6.8192				7.0592	7.2721
6	v	315.46				352.04	393.91
(158.8)	u	2566.2				2638.5	2721.3
	h	2755.5				2849.7	2957.6
	s	6.7575				6.9662	7.1829
7	v	272.68				299.92	336.37
(165.0)	u	2571.1				2634.3	2718.5
	h	2762.0				2844.2	2954.0
	s	6.7052				6.8859	7.1066

300	400	500	600	700	800	p (t_s)
						0
2938.4	3095.1	3258.4	3428.7	3605.8	3789.5	
3076.9	3279.7	3489.2	3705.6	3928.9	4158.7	
132251	155329	178405	201482	224558	247634	0.02
2812.3	2969.0	3132.4	3302.6	3479.7	3663.4	(17.50)
3076.8	3279.7	3489.2	3705.6	3928.8	4158.7	
10.025	10.351	10.641	10.904	11.146	11.371	
44079	51773	59467	67159	74852	82544	0.06
2812.2	2969.0	3132.4	3302.6	3479.7	3663.4	(36.20)
3076.7	3279.6	3489.2	3705.6	3928.8	4158.7	
9.5179	9.8441	10.134	10.397	10.639	10.864	
26445	31062	35679	40295	44910	49526	0.10
2812.2	2969.0	3132.3	3302.6	3479.7	3663.4	(45.80)
3076.6	3279.6	3489.1	3705.5	3928.8	4158.7	
9.2820	9.6083	9.8984	10.162	10.404	10.628	
5283.9	6209.1	7133.5	8057.4	8981.0	9904.4	0.50
2811.5	2968.5	3132.0	3302.3	3479.5	3663.3	(81.30)
3075.7	3279.0	3488.7	3705.2	3928.6	4158.5	
8.5380	8.8649	9.1552	9.4185	9.6606	9.8855	
2638.7	3102.5	3565.3	4027.7	4489.8	4951.7	1.00
2810.6	2968.0	3131.6	3302.0	3479.2	3663.1	(99.60)
3074.5	3278.2	3488.1	3704.8	3928.2	4158.3	
8.2166	8.5442	8.8348	9.0982	9.3405	9.5654	
1757.0	2066.9	2375.9	2684.5	2992.7	3300.8	1.50
2809.8	2967.5	3131.2	3301.7	3479.0	3662.9	(111.4)
3073.3	3277.5	3487.6	3704.4	3927.9	4158.0	
8.0280	8.3562	8.6472	8.9108	9.1531	9.3781	
1316.2	1549.2	1781.2	2012.9	2244.2	2475.4	2
2808.9	2966.9	3130.8	3301.4	3478.8	3662.7	(120.2)
3072.1	3276.7	3487.0	3704.0	3927.6	4157.8	
7.8937	8.2226	8.5139	8.7776	9.0201	9.2452	
875.29	1031.4	1186.5	1341.2	1495.7	1649.9	3
2807.1	2965.8	3130.1	3300.8	3478.3	3662.3	(133.5)
3069.7	3275.2	3486.0	3703.2	3927.0	4157.3	
7.7034	8.0338	8.3257	8.5898	8.8325	9.0577	
654.85	772.50	889.19	1005.4	1121.4	1237.2	4
2805.3	2964.6	3129.2	3300.1	3477.8	3662.0	(143.6)
3067.2	3273.6	3484.9	3702.3	3926.4	4156.9	
7.5675	7.8994	8.1919	8.4563	8.6992	8.9246	
522.58	617.16	710.78	803.95	896.85	989.56	5
2803.5	2963.5	3128.4	3299.5	3477.4	3661.6	(151.8)
3064.8	3272.1	3483.8	3701.5	3925.8	4156.4	
7.4614	7.7948	8.0879	8.3526	8.5957	8.8213	
434.39	513.61	591.84	669.63	747.14	824.47	6
2801.7	2962.4	3127.6	3298.9	3476.8	3661.2	(158.8)
3062.3	3270.6	3482.7	3700.7	3925.1	4155.9	
7.3740	7.7090	8.0027	8.2678	8.5111	8.7368	
371.39	439.64	506.89	573.68	640.21	706.55	7
2799.8	2961.3	3126.8	3298.3	3476.4	3660.9	(165.0)
3059.8	3269.0	3481.6	3699.9	3924.5	4155.5	
7.2997	7.6362	7.9305	8.1959	8.4395	8.6653	

Superheated steam, to 220 bar and 800°C

For definitions see p. 48

p (t_s)	t:	t_s	50	100	150	200	250
8	v	240·26				260·79	293·21
(170·4)	u	2575·3				2630·0	2715·8
	h	2767·5				2838·6	2950·4
	s	6·6596				6·8148	7·0397
9	v	214·82				230·32	259·63
(175·4)	u	2578·8				2625·4	2713·1
	h	2772·1				2832·7	2946·8
	s	6·6192				6·7508	6·9800
10	v	194·30				205·92	232·75
(179·9)	u	2581·9				2620·9	2710·3
	h	2776·2				2826·8	2943·0
	s	6·5828				6·6922	6·9259
15	v	131·67				132·38	151·99
(198·3)	u	2592·4				2596·1	2695·5
	h	2789·9				2794·7	2923·5
	s	6·4406				6·4508	6·7099
20	v	99·549					111·45
(212·4)	u	2598·1					2679·5
	h	2797·2					2902·4
	s	6·3367					6·5454
30	v	66·632					70·551
(233·8)	u	2602·4					2643·1
	h	2802·3					2854·8
	s	6·1838					6·2857
40	v	49·749					
(250·3)	u	2601·3					
	h	2800·3					
	s	6·0685					
50	v	39·425					
(263·9)	u	2597·1					
	h	2794·2					
	s	5·9735					
60	v	32·433					
(275·6)	u	2590·4					
	h	2785·0					
	s	5·8907					
70	v	27·368					
(285·8)	u	2581·8					
	h	2773·4					
	s	5·8161					
80	v	23·521					
(295·0)	u	2571·7					
	h	2759·9					
	s	5·7470					
90	v	20·493					
(303·3)	u	2560·2					
	h	2744·6					
	s	5·6820					

300	400	500	600	700	800	p (t_s)
324.14	384.16	443.17	501.72	560.01	618.11	8
2798.0	2960.2	3126.0	3297.7	3475.9	3660.5	(170.4)
3057.3	3267.5	3480.5	3699.1	3923.9	4155.0	
7.2348	7.5729	7.8678	8.1336	8.3773	8.6033	
287.39	341.01	393.61	445.76	497.63	549.33	9
2796.0	2959.1	3125.2	3297.0	3475.4	3660.1	(175.4)
3054.7	3266.0	3479.4	3698.2	3923.3	4154.5	
7.1771	7.5169	7.8124	8.0785	8.3225	8.5486	
257.98	306.49	353.96	400.98	447.73	494.30	10
2794.1	2957.9	3124.3	3296.4	3475.0	3659.8	(179.9)
3052.1	3264.4	3478.3	3697.4	3922.7	4154.1	
7.1251	7.4665	7.7627	8.0292	8.2734	8.4997	
169.70	202.92	235.03	266.66	298.03	329.21	15
2784.4	2952.2	3120.3	3293.3	3472.6	3657.9	(198.3)
3038.9	3256.6	3472.8	3693.3	3919.6	4151.7	
6.9207	7.2709	7.5703	7.8385	8.0838	8.3108	
125.50	151.13	175.55	199.50	223.17	246.66	20
2774.0	2946.4	3116.2	3290.2	3470.2	3656.1	(212.4)
3025.0	3248.7	3467.3	3689.2	3916.5	4149.4	
6.7696	7.1296	7.4323	7.7022	7.9485	8.1763	
81.159	99.310	116.08	132.34	148.32	164.12	30
2751.6	2934.6	3108.0	3284.0	3465.3	3652.3	(233.8)
2995.1	3232.5	3456.2	3681.0	3910.3	4144.7	
6.5422	6.9246	7.2345	7.5079	7.7564	7.9857	
58.833	73.376	86.341	98.763	110.90	122.85	40
2726.7	2922.2	3099.6	3277.7	3460.5	3648.6	(250.3)
2962.0	3215.7	3445.0	3672.8	3904.1	4140.0	
6.3642	6.7733	7.0909	7.3680	7.6187	7.8495	
45.301	57.791	68.494	78.616	88.446	98.093	50
2699.0	2909.3	3091.2	3271.4	3455.7	3644.8	(263.9)
2925.5	3198.3	3433.7	3664.5	3897.9	4135.3	
6.2105	6.6508	6.9770	7.2578	7.5108	7.7431	
36.145	47.379	56.592	65.184	73.478	81.587	60
2668.1	2895.8	3082.6	3265.1	3450.8	3641.2	(275.6)
2885.0	3180.1	3422.2	3656.2	3891.7	4130.7	
6.0692	6.5462	6.8818	7.1664	7.4217	7.6554	
29.457	39.922	48.086	55.590	62.787	69.798	70
2633.2	2881.7	3074.0	3258.8	3445.9	3637.4	(285.8)
2839.4	3161.2	3410.6	3647.9	3885.4	4126.0	
5.9327	6.4536	6.7993	7.0880	7.3456	7.5808	
24.264	34.310	41.704	48.394	54.770	60.956	80
2592.7	2867.1	3065.2	3252.3	3441.0	3633.7	(295.0)
2786.8	3141.6	3398.8	3639.5	3879.2	4121.3	
5.7942	6.3694	6.7262	7.0191	7.2790	7.5158	
	29.929	36.737	42.798	48.534	54.080	90
	2851.8	3056.2	3245.9	3436.2	3630.0	(303.3)
	3121.2	3386.8	3631.1	3873.0	4116.7	
	6.2915	6.6600	6.9574	7.2196	7.4579	

For definitions see p. 48

Superheated steam, to 220 bar and 800°C

p (t_s)	t:	t_s	400	500	600	700	800
100	v	18.041	26.408	32.760	38.320	43.546	48.580
(311.0)	u	2547.3	2835.8	3047.0	3239.5	3431.3	3626.2
	h	2727.7	3099.9	3374.6	3622.7	3866.8	4112.0
	s	5.6198	6.2182	6.5994	6.9013	7.1660	7.4058
110	v	16.007	23.512	29.503	34.656	39.466	44.081
(318.0)	u	2533.2	2819.2	3037.7	3233.0	3426.4	3622.4
	h	2709.3	3077.8	3362.2	3614.2	3860.5	4107.3
	s	5.5596	6.1483	6.5432	6.8499	7.1170	7.3584
120	v	14.285	21.084	26.786	31.603	36.066	40.332
(324.6)	u	2517.8	2801.8	3028.2	3226.5	3421.5	3618.7
	h	2689.2	3054.8	3349.6	3605.7	3854.3	4102.7
	s	5.5003	6.0810	6.4906	6.8022	7.0718	7.3147
130	v	12.800	19.015	24.485	29.019	33.189	37.160
(330.8)	u	2500.6	2783.5	3018.5	3219.9	3416.5	3614.9
	h	2667.0	3030.7	3336.8	3597.1	3848.0	4098.0
	s	5.4409	6.0155	6.4409	6.7577	7.0298	7.2743
140	v	11.498	17.227	22.509	26.804	30.723	34.441
(336.6)	u	2481.4	2764.4	3008.7	3213.2	3411.6	3611.1
	h	2642.4	3005.6	3323.8	3588.5	3841.7	4093.3
	s	5.3804	5.9513	6.3937	6.7159	6.9906	7.2367
150	v	10.343	15.661	20.795	24.884	28.587	32.086
(342.1)	u	2460.0	2744.2	2998.7	3206.5	3406.6	3607.3
	h	2615.1	2979.1	3310.6	3579.8	3835.4	4088.6
	s	5.3180	5.8876	6.3487	6.6764	6.9536	7.2013
160	v	9.3099	14.275	19.293	23.204	26.717	30.025
(347.3)	u	2435.9	2722.9	2988.4	3199.7	3401.6	3603.6
	h	2584.9	2951.3	3297.1	3571.0	3829.1	4084.0
	s	5.2533	5.8240	6.3054	6.6389	6.9188	7.1681
170	v	8.3721	13.034	17.966	21.721	25.068	28.207
(352.3)	u	2409.3	2700.1	2978.1	3192.9	3396.6	3599.8
	h	2551.6	2921.7	3283.5	3562.2	3822.8	4079.3
	s	5.1856	5.7599	6.2636	6.6031	6.8857	7.1366
180	v	7.4973	11.913	16.785	20.403	23.603	26.591
(357.0)	u	2378.9	2675.9	2967.5	3186.1	3391.6	3596.0
	h	2513.9	2890.3	3269.6	3553.4	3816.5	4074.6
	s	5.1127	5.6947	6.2232	6.5688	6.8542	7.1067
190	v	6.6759	10.889	15.726	19.223	22.291	25.146
(361.4)	u	2343.7	2649.8	2956.6	3179.3	3386.7	3592.2
	h	2470.5	2856.7	3255.4	3544.5	3810.2	4070.0
	s	5.0330	5.6278	6.1839	6.5360	6.8241	7.0783
200	v	5.8745	9.9470	14.771	18.161	21.111	23.845
(365.7)	u	2300.7	2621.6	2945.7	3172.3	3381.6	3588.4
	h	2418.2	2820.5	3241.1	3535.5	3803.8	4065.3
	s	4.9410	5.5585	6.1456	6.5043	6.7953	7.0511
210	v	5.0225	9.0714	13.907	17.201	20.044	22.669
(369.8)	u	2242.0	2590.8	2934.5	3165.3	3376.6	3584.6
	h	2347.5	2781.3	3226.5	3526.5	3797.5	4060.6
	s	4.8222	5.4863	6.1082	6.4737	6.7677	7.0251
220	v	3.7347	8.2510	13.119	16.327	19.074	21.599
(373.7)	u	2114.4	2557.3	2923.1	3158.2	3371.5	3580.7
	h	2196.6	2738.8	3211.7	3517.4	3791.1	4055.9
	s	4.5814	5.4102	6.0716	6.4441	6.7410	7.0001

Supercritical steam, to 1000 bar and 800°C

p	t:	400	500	600	700	800
240	v	6·7392	11·737	14·798	17·377	19·729
	h	2641·2	3181·4	3499·1	3778·3	4046·6
	s	5·2430	6·0003	6·3876	6·6905	6·9529
260	v	5·2812	10·565	13·505	15·941	18·147
	h	2511·7	3150·2	3480·6	3765·5	4037·2
	s	5·0326	5·9311	6·3340	6·6431	6·9090
280	v	3·8219	9·5566	12·397	14·712	16·792
	h	2330·7	3118·1	3461·9	3752·6	4027·8
	s	4·7503	5·8636	6·2829	6·5984	6·8677
300	v	2·8306	8·6808	11·436	13·647	15·619
	h	2161·8	3085·0	3443·0	3739·7	4018·5
	s	4·4896	5·7972	6·2340	6·5560	6·8288
350	v	2·1108	6·9253	9·5194	11·520	13·275
	h	1993·1	2998·3	3395·1	3707·3	3995·1
	s	4·2214	5·6349	6·1194	6·4584	6·7400
400	v	1·9091	5·6156	8·0884	9·9302	11·521
	h	1934·1	2906·8	3346·4	3674·8	3971·7
	s	4·1190	5·4762	6·0135	6·3701	6·6606
450	v	1·8013	4·6249	6·9842	8·6988	10·160
	h	1900·6	2813·5	3297·4	3642·4	3948·4
	s	4·0554	5·3226	5·9143	6·2890	6·5885
500	v	1·7291	3·8822	6·1113	7·7197	9·0759
	h	1877·7	2723·0	3248·3	3610·2	3925·3
	s	4·0083	5·1782	5·8207	6·2138	6·5222
600	v	1·6324	2·9515	4·8350	6·2690	7·4603
	h	1847·3	2570·6	3151·6	3547·0	3879·6
	s	3·9383	4·9374	5·6477	6·0775	6·4031
700	v	1·5671	2·4668	3·9719	5·2566	6·3208
	h	1827·8	2467·1	3060·4	3486·3	3835·3
	s	3·8855	4·7688	5·4931	5·9562	6·2979
800	v	1·5180	2·1881	3·3792	4·5193	5·4805
	h	1814·2	2397·4	2980·3	3428·7	3792·8
	s	3·8425	4·6488	5·3595	5·8470	6·2034
900	v	1·4788	2·0129	2·9668	3·9642	4·8407
	h	1804·6	2349·9	2913·5	3374·6	3752·4
	s	3·8059	4·5602	5·2468	5·7479	6·1179
1000	v	1·4464	1·8934	2·6681	3·5356	4·3411
	h	1797·6	2316·1	2857·5	3324·4	3714·3
	s	3·7738	4·4913	5·1505	5·6579	6·0397

Saturated water and steam

v Specific volume ($m^3\,kg^{-1}$)
c_p Specific heat capacity at constant pressure ($kJ\,kg^{-1}\,K^{-1}$)
k Thermal conductivity ($W\,m^{-1}\,K^{-1}$)
μ Dynamic viscosity (m Pa s, or cP)
Pr Prandtl number, $c_p\mu/k$

For other definitions see p. 48

t_s	v_f	v_g	c_{pf}	c_{pg}	k_f	k_g	μ_f	μ_g	Pr_f	Pr_g	t_s
0·01	0·00100	206·2	4·217	1·854	0·569	0·0173	1·755	0·0088	13·02	0·942	0·01* Triple poin
10	0·00100	106·4	4·193	1·860	0·587	0·0185	1·301	0·0091	9·29	0·915	10
20	0·00100	57·8	4·182	1·866	0·603	0·0191	1·002	0·0094	6·95	0·918	20
30	0·00100	32·9	4·179	1·875	0·618	0·0198	0·797	0·0097	5·39	0·923	30
40	0·00101	19·5	4·179	1·885	0·632	0·0204	0·651	0·0101	4·31	0·930	40
50	0·00101	12·05	4·181	1·899	0·643	0·0210	0·544	0·0104	3·53	0·939	50
60	0·00102	7·68	4·185	1·915	0·653	0·0217	0·462	0·0107	2·96	0·947	60
70	0·00102	5·05	4·190	1·936	0·662	0·0224	0·400	0·0111	2·53	0·956	70
80	0·00103	3·41	4·197	1·962	0·670	0·0231	0·350	0·0114	2·19	0·966	80
90	0·00104	2·36	4·205	1·992	0·676	0·0240	0·311	0·0117	1·93	0·976	90
100	0·00104	1·673	4·216	2·028	0·681	0·0249	0·278	0·0121	1·723	0·986	100
125	0·00107	0·770	4·254	2·147	0·687	0·0272	0·219	0·0133	1·358	1·047	125
150	0·00109	0·392	4·310	2·314	0·687	0·0300	0·180	0·0144	1·133	1·110	150
175	0·00112	0·217	4·389	2·542	0·679	0·0334	0·153	0·0156	0·990	1·185	175
200	0·00116	0·127	4·497	2·843	0·665	0·0375	0·133	0·0167	0·902	1·270	200
225	0·00120	0·0783	4·648	3·238	0·644	0·0427	0·1182	0·0179	0·853	1·36	225
250	0·00125	0·0500	4·867	3·772	0·616	0·0495	0·1065	0·0191	0·841	1·45	250
275	0·00132	0·0327	5·202	4·561	0·582	0·0587	0·0972	0·0202	0·869	1·56	275
300	0·00140	0·0216	5·762	5·863	0·541	0·0719	0·0897	0·0214	0·955	1·74	300
325	0·00153	0·0142	6·861	8·440	0·493	0·0929	0·0790	0·0230	1·100	2·09	325
350	0·00174	0·00880	10·10	17·15	0·437	0·1343	0·0648	0·0258	1·50	3·29	350
360	0·00190	0·00694	14·6	25·1	0·400	0·168	0·0582	0·0275	2·11	3·89	360
374·15	0·00317	0·00317	∞	∞	0·24	0·24	0·045	0·045	∞	∞	374·15† Critical poi

* Triple point.
† Critical point.

Ammonia, NH_3 (p = pressure (MPa), v = specific volume ($m^3\,kg^{-1}$); for other definitions see p. 48)

		Saturated						Superheated By 50 K		Superheated By 100 K	
t_s	p	v_f	v_g	h_f	h_g	s_f	s_g	h	s	h	s
−40	0·0718	0·00145	1·552	zero	1390	zero	5·963	1499	6·387	1606	6·736
−35	0·0932	0·00146	1·216	22·3	1398	0·095	5·872	1508	6·292	1616	6·639
−30	0·1196	0·00148	0·963	44·7	1406	0·188	5·785	1517	6·203	1626	6·547
−25	0·1516	0·00149	0·772	67·2	1413	0·279	5·703	1526	6·119	1636	6·461
−20	0·190	0·00150	0·624	89·8	1420	0·368	5·624	1535	6·039	1646	6·379
−15	0·236	0·00152	0·509	112·3	1426	0·457	5·549	1543	5·963	1656	6·301
−10	0·291	0·00153	0·418	135·4	1433	0·544	5·477	1552	5·891	1665	6·227
−5	0·355	0·00155	0·347	158·2	1439	0·630	5·407	1560	5·822	1675	6·157
0	0·429	0·00157	0·289	181·2	1444	0·715	5·340	1568	5·756	1684	6·090
5	0·516	0·00158	0·243	204·5	1450	0·799	5·276	1576	5·694	1693	6·027
10	0·615	0·00160	0·206	227·7	1454	0·881	5·214	1583	5·634	1702	5·967
15	0·728	0·00162	0·175	251·4	1459	0·963	5·154	1590	5·576	1711	5·909
20	0·857	0·00164	0·149	275·2	1463	1·044	5·095	1597	5·521	1719	5·853
25	1·001	0·00166	0·128	298·9	1466	1·124	5·039	1604	5·468	1728	5·800
30	1·167	0·00168	0·111	323·1	1469	1·204	4·984	1610	5·417	1736	5·750
35	1·350	0·00170	0·096	347·5	1471	1·282	4·930	1616	5·368	1744	5·702
40	1·554	0·00173	0·083	371·5	1473	1·360	4·877	1622	5·321	1752	5·655
45	1·782	0·00175	0·073	396·8	1474	1·437	4·825	1628	5·275	1760	5·610
50	2·033	0·00178	0·063	421·9	1475	1·515	4·773	1633	5·230	1767	5·567

Freezing point at 1 atm = − 77·7 °C
Critical point: Celsius temp. = 132·4 °C
pressure = 11·30 MPa (113·0 bar)

Dichlorofluoromethane, CCl_2F_2 (Refrigerant-12, Arcton-12, or Freon-12)

(p = pressure (MPa), v = specific volume ($m^3\,kg^{-1}$); for other definitions see p. 48)

		Saturated						Superheated By 20 K		Superheated By 40 K	
t_s	p	v_f	v_g	h_f	h_g	s_f	s_g	h	s	h	s
−40	0·0641	0·00066	0·2421	zero	169·6	zero	0·7274	180·8	0·7737	192·4	0·8178
−35	0·0806	0·00067	0·1955	4·4	171·9	0·0187	0·7220	183·3	0·7681	195·1	0·8120
−30	0·1003	0·00067	0·1595	8·9	174·2	0·0371	0·7171	185·8	0·7631	197·8	0·8068
−25	0·1236	0·00068	0·1313	13·3	176·5	0·0552	0·7127	188·3	0·7586	200·4	0·8021
−20	0·1508	0·00069	0·1089	17·8	178·7	0·0731	0·7088	190·8	0·7546	203·1	0·7979
−15	0·1825	0·00069	0·0911	22·3	181·0	0·0906	0·7052	193·2	0·7510	205·7	0·7942
−10	0·219	0·00070	0·0767	26·9	183·2	0·1080	0·7020	195·7	0·7477	208·3	0·7909
−5	0·261	0·00071	0·0650	31·4	185·4	0·1251	0·6991	198·1	0·7449	210·9	0·7879
0	0·308	0·00072	0·0554	36·1	187·5	0·1420	0·6966	200·5	0·7423	213·5	0·7853
5	0·362	0·00072	0·0475	40·7	189·7	0·1587	0·6942	202·9	0·7401	216·5	0·7830
10	0·423	0·00073	0·0409	45·4	191·7	0·1752	0·6921	205·2	0·7381	218·6	0·7810
15	0·491	0·00074	0·0354	50·1	193·8	0·1915	0·6902	207·5	0·7363	221·2	0·7792
20	0·567	0·00075	0·0308	54·9	195·8	0·2078	0·6885	209·8	0·7348	223·7	0·7777
25	0·651	0·00076	0·0269	59·7	197·7	0·2239	0·6869	212·1	0·7334	226·1	0·7763
30	0·745	0·00077	0·0235	64·6	199·6	0·2399	0·6854	214·3	0·7321	228·6	0·7751
35	0·847	0·00079	0·0206	69·5	201·5	0·2559	0·6839	216·4	0·7310	231·0	0·7741
40	0·960	0·00080	0·0182	74·6	203·2	0·2718	0·6825	218·5	0·7300	233·4	0·7732
45	1·084	0·00081	0·0160	79·7	204·9	0·2877	0·6812	220·6	0·7291	235·7	0·7724
50	1·219	0·00083	0·0142	84·9	206·5	0·3037	0·6797	222·6	0·7282	238·0	0·7718

Triple point temp· ≈ − 158 °C
Freezing point at 1 atm = − 155·0 °C
Critical point: Celsius temp. = 112·0 °C
pressure = 4·115 MPa (41·15 bar)

Carbon dioxide, CO_2 (p = pressure (MPa), v = specific volume ($m^3\,kg^{-1}$); for other definitions see p. 48)

		Saturated						Superheated By 30 K		Superheated By 60 K	
t_s	p	v_f	v_g	h_f	h_g	s_f	s_g	h	s	h	s
−40	1·005	0·00090	0·0382	zero	321·1	zero	1·377	355·4	1·507	383·0	1·611
−35	1·20	0·00091	0·0320	9·7	322·2	0·039	1·352	356·9	1·485	385·6	1·588
−30	1·43	0·00093	0·0270	19·5	323·1	0·079	1·328	358·7	1·464	388·0	1·566
−25	1·68	0·00095	0·0229	29·5	323·7	0·119	1·304	360·4	1·442	390·3	1·545
−20	1·97	0·00097	0·0195	39·7	323·7	0·158	1·280	361·8	1·421	392·5	1·525
−15	2·29	0·00099	0·0166	50·2	323·2	0·198	1·256	363·0	1·401	394·5	1·505
−10	2·65	0·00102	0·0142	60·9	322·3	0·238	1·231	363·9	1·381	396·2	1·486
−5	3·04	0·00105	0·0122	72·0	320·5	0·278	1·205	364·6	1·361	397·8	1·467
0	3·48	0·00108	0·0104	83·7	318·1	0·320	1·178	364·9	1·342	399·3	1·449
5	3·97	0·00111	0·00879	96·0	312·9	0·364	1·143	364·9	1·322	400·4	1·431
10	4·50	0·00116	0·00743	109·1	307·2	0·407	1·107	364·7	1·302	401·4	1·414
15	5·08	0·00121	0·00623	123·3	301·0	0·454	1·071	364·0	1·282	402·2	1·396
20	5·73	0·00129	0·00516	139·1	292·3	0·506	1·028	362·9	1·261	402·7	1·379
25	6·44	0·00140	0·00413	159·7	279·9	0·573	0·976	361·5	1·241	403·0	1·362
30	7·21	0·00169	0·00294	191·2	253·1	0·682	0·886	359·6	1·220	402·9	1·345
†31·05	7·38	0·00214	0·00214	223·0	223·0	0·780	0·780	359·1	1·216	402·9	1·341 Critical point

Triple point: temperature = − 56·6 °C, pressure = 0·518 MN/m^2
Sublimation point at 1 atm = − 78·5 °C

Air at atmospheric pressure

t Temperature (° C)
v Specific volume ($m^3\,kg^{-1}$)
c_p Specific heat capacity at constant pressure ($kJ\,kg^{-1}\,K^{-1}$)
k Thermal conductivity ($W\,m^{-1}\,K^{-1}$)
μ Dynamic viscosity (m Pa s, or cP)
Pr Prandtl number, $c_p\mu/k$

t	v	c_p	k	μ	Pr	t
−100	0.488	1.01	0.016	0.012	0.75	−100
0	0.773	1.01	0.024	0.017	0.72	0
100	1.057	1.02	0.032	0.022	0.70	100
200	1.341	1.03	0.039	0.026	0.69	200
300	1.624	1.05	0.045	0.030	0.69	300
400	1.908	1.07	0.051	0.033	0.70	400
500	2.191	1.10	0.056	0.036	0.70	500
600	2.473	1.12	0.061	0.039	0.71	600
700	2.756	1.14	0.066	0.042	0.72	700
800	3.039	1.16	0.071	0.044	0.73	800

This Table may be used with reasonable accuracy for values of c_p, k, μ and Pr of N_2, O_2 and CO.

International Standard Atmosphere

z Altitude above sea level (km)
p Absolute pressure (bar)
T Absolute temperature (K)
ρ Mass density ($kg\,m^{-3}$) (at sea level $\rho = \rho_0 = 1{\cdot}225$)
ν Kinematic viscosity ($mm^2\,s^{-1}$, or cSt)
k Thermal conductivity ($W\,m^{-1}\,K^{-1}$)
λ Mean free path (μm)
a Speed of sound ($m\,s^{-1}$)
For other definitions see p. 48

z	p	T	ρ/ρ_0	ν	k	λ	a
−2·5	1·352	304·4	1·263	12·07	0·0266	0·053	349·8
−2·0	1·278	301·2	1·207	12·53	0·0264	0·055	347·9
−1·5	1·207	297·9	1·152	13·01	0·0261	0·058	346·0
−1·0	1·139	294·7	1·100	13·52	0·0259	0·060	344·1
−0·5	1·075	291·4	1·049	14·05	0·0256	0·063	342·2
0	1·013	288·2	1·000	14·61	0·0253	0·066	340·3
0·5	0·955	284·9	0·953	15·20	0·0251	0·070	338·4
1·0	0·899	281·7	0·907	15·81	0·0248	0·073	336·4
1·5	0·846	278·4	0·864	16·46	0·0246	0·077	334·5
2·0	0·795	275·2	0·822	17·15	0·0243	0·081	332·5
2·5	0·747	271·9	0·781	17·87	0·0241	0·085	330·6
3·0	0·701	268·7	0·742	18·63	0·0238	0·089	328·6
3·5	0·658	265·4	0·705	19·43	0·0235	0·094	326·6
4·0	0·617	262·2	0·669	20·28	0·0233	0·099	324·6
4·5	0·578	258·9	0·634	21·17	0·0230	0·105	322·6
5·0	0·540	255·7	0·601	22·11	0·0228	0·110	320·5
6·0	0·472	249·2	0·539	24·16	0·0222	0·123	316·5
7·0	0·411	242·7	0·482	26·46	0·0217	0·138	312·3
8·0	0·357	236·2	0·429	29·04	0·0212	0·155	308·1
9·0	0·308	229·7	0·381	31·96	0·0206	0·174	303·8
10·0	0·265	223·3	0·338	35·25	0·0201	0·196	299·5

z	p	T	ρ/ρ_0	ν	k	λ	a
11	0·227	216·8	0·298	38·99	0·0195	0·223	295·2
12	0·194	216·7	0·255	45·57	0·0195	0·260	295·1
13	0·166	216·7	0·218	53·33	0·0195	0·305	295·1
14	0·142	216·7	0·186	62·39	0·0195	0·357	295·1
15	0·121	216·7	0·159	73·00	0·0195	0·417	295·1
16	0·104	216·7	0·136	85·40	0·0195	0·488	295·1
17	0·088	216·7	0·116	99·90	0·0195	0·571	295·1
18	0·076	216·7	0·099	116·9	0·0195	0·668	295·1
19	0·065	216·7	0·085	136·7	0·0195	0·781	295·1
20	0·055	216·7	0·073	159·9	0·0195	0·914	295·1
21	0·047	217·6	0·062	188·4	0·0196	1·073	295·7
22	0·040	218·6	0·053	222·0	0·0197	1·260	296·4
23	0·035	219·6	0·045	261·4	0·0198	1·477	297·0
24	0·030	220·6	0·038	307·4	0·0199	1·731	297·7
25	0·025	221·6	0·033	361·4	0·0199	2·027	298·4
26	0·022	222·5	0·028	424·4	0·0200	2·372	299·1
27	0·019	223·5	0·024	498·1	0·0201	2·773	299·7
28	0·016	224·5	0·020	584·1	0·0202	3·240	300·4
29	0·014	225·5	0·018	684·4	0·0203	3·783	301·0
30	0·012	226·5	0·015	801·3	0·0203	4·413	301·7

At sea level the normal composition of clean dry air, by volume and (weight) percent, is:

Nitrogen	78·08 (75·52)	Carbon dioxide	0·03 (0·05)	Methane	0·0002 (0·0001)	Nitrous oxide	5×10^{-5} (8 × 10⁻
Oxygen	20·95 (23·14)	Neon	0·002 (0·001)	Krypton	0·0001 (0·0003)	Xenon	9×10^{-6} (4 × 10⁻
Argon	0·93 (1.28)	Helium	0·0005 (7×10^{-5})	Hydrogen	5×10^{-5} (3×10^{-6})		

The approximate composition may be taken to be: nitrogen/oxygen 79/21 (77/23).

Properties of gases

MW	Molecular weight or molar mass (a.m.u. or kg kmol^{-1})
ρ	Mass density at s.t.p.* (kg m^{-3})
R	Gas constant (kJ kg^{-1} K^{-1})
t_f, t_b	Freezing, boiling points at 1 atm (° C)
t_c, p_c, ρ_c	Critical temperature, pressure, density (° C, bar, kg m^{-3})
k	Thermal conductivity at s.t.p* (m W m^{-1} K^{-1})
c_p	Specific heat capacity at constant pressure, for s.t.p* (kJ kg^{-1} K^{-1})
γ	Ratio of specific heat capacities c_p/c_v at s.t.p*
μ	Dynamic viscosity at 1 atm, 20° C (μPa s)
λ	Mean free path at s.t.p* (nm)
ϵ_r	Dielectric constant at s.t.p* and 10 GHz

*Standard temperature and pressure: 0° C, 1 bar; but the figures are for the former standard pressure of 1 atm (1·013 bar).

	MW	ρ	R	t_f	t_b	t_c	p_c	ρ_c	k	c_p	γ	μ	λ	ϵ_r
Air	28·96	1·293	0·287			−141	37·7		24·1	1·004	1·40	18·1		1·000 58
Oxygen	32·00	1·429	0·260	−219	−183	−119	50·4	430	24·4	0·915	1·40	20·0	63	1·000 53
Nitrogen	28·01	1·250	0·297	−210	−196	−147	33·9	311	24·3	1·039	1·40	17·4	59	1·000 59
Hydrogen	2·02	0·090	4·124	−259	−253	−240	13·0	31	168·4	14·2	1·41	8·8	111	1·000 26
Carbon monoxide	28·01	1·250	0·297	−205	−192	−140	35·0	300	22·9	1·040	1·40	17·4	59	1·000 70
Carbon dioxide	44·01	1·977	0·189	−57		31	74·0	460	14·5	0·819	1·30	14·6	40	1·000 99
Helium	4·00	0·178	2·077		−269	−268	2·3	69	141·5	5·19	1·63	19·4	174	1·000 70
Neon	20·18	0·900	0·412	−249	−246	−229	27·2	484	46·5		1·64	31·0	124	1·001 27
Argon	39·95	1·784	0·208	−189	−186	−122	48·6	531	16·2	0·520	1·67	22·2	63	1·000 56

Variation of specific heat c_p with temperature

These values apply to a wide range of pressure

t(° C)	100	200	300	400	500	1000	1500	2000	2500	3000
Air	1·010	1·025	1·045	1·069	1·092	1·184	1·235	1·265	1·287	1·302
Oxygen	0·934	0·963	0·995	1·024	1·048	1·123	1·164	1·200	1·234	1·260
Nitrogen	1·042	1·052	1·069	1·092	1·115	1·215	1·269	1·298	1·316	1·331
Hydrogen	14·45	14·50	14·53	14·58	14·66	15·52	16·56	17·39	18·01	18·55
Carbon monoxide	1·045	1·058	1·080	1·106	1·132	1·231	1·280	1·307	1·323	1·336
Carbon dioxide	0·914	0·993	1·057	1·110	1·155	1·290	1·350	1·378	1·388	1·394

Thermochemical data for equilibrium reactions

Stoichiometric equations

$\sum \nu_i A_i = 0$, where ν_i is the *stoichiometric coefficient* of the substance whose *chemical symbol* is A_i.

(1) $-2H + H_2 = 0$
(2) $-2N + N_2 = 0$
(3) $-2O + O_2 = 0$
(4) $-2NO + N_2 + O_2 = 0$
(5) $-H_2 - \frac{1}{2}O_2 + H_2O = 0$
(6) $-\frac{1}{2}H_2 - OH + H_2O = 0$
(7) $-CO - \frac{1}{2}O_2 + CO_2 = 0$
(8) $-CO - H_2O + CO_2 + H_2 = 0$
(9) $-\frac{1}{2}N_2 - \frac{3}{2}H_2 + NH_3 = 0$

Standard enthalpy of reaction

$\Delta H_T^\circ = \sum_i \nu_i [\bar{h}_i]_T^\circ$ where $[\bar{h}_i]_T^\circ$ = enthalpy per kmol of substance A_i, at 1 bar pressure and absolute temperature T.

Warning: This table lists *absolute* temperatures.

Temp. K	Reaction number 1 (ΔH_T°/MJ)	2	3	4	5	6	7	8	9	Temp. K
200	−434.7	−944.1	−496.9	−180.4	−240.9	−280.2	−282.1	−41.21	−43.71	200
298	−436.0	−945.3	−498.4	−180.6	−241.8	−281.3	−283.0	−41.17	−45.90	298
400	−437.3	−946.6	−499.8	−180.7	−242.8	−282.4	−283.5	−40.63	−48.04	400
600	−439.7	−948.9	−502.1	−180.7	−244.8	−284.1	−283.6	−38.88	−51.39	600
800	−442.1	−951.1	−503.9	−180.8	−246.5	−285.5	−283.3	−36.82	−53.66	800
1000	−444.5	−953.0	−505.4	−180.9	−247.9	−286.6	−282.6	−34.74	−55.07	1000
1200	−446.7	−954.7	−506.7	−180.9	−249.0	−287.4	−281.8	−32.79	−55.83	1200
1400	−448.7	−956.1	−507.8	−181.0	−249.9	−287.9	−280.9	−30.98	−56.07	1400
1600	−450.6	−957.5	−508.9	−181.0	−250.6	−288.4	−279.9	−29.29	−55.99	1600
1800	−452.3	−958.7	−509.8	−181.0	−251.2	−288.6	−278.9	−27.71	−55.66	1800
2000	−453.8	−959.9	−510.6	−181.0	−251.7	−288.8	−277.9	−26.22	−55.19	2000
2200	−455.2	−961.0	−511.4	−180.8	−252.1	−288.9	−276.8	−24.79	−54.61	2200
2400	−456.4	−962.1	−512.0	−180.7	−252.4	−289.0	−275.8	−23.41	−53.92	2400
2600	−457.6	−963.1	−512.5	−180.4	−252.7	−289.0	−274.8	−22.07	−53.12	2600
2800	−458.6	−964.1	−513.0	−180.1	−253.0	−288.9	−273.7	−20.77	−52.22	2800
3000	−459.6	−965.0	−513.4	−179.7	−253.3	−288.9	−272.7	−19.49	−51.20	3000
3200	−460.4	−966.0	−513.8	−179.3	−253.5	−288.8	−271.7	−18.19	−50.10	3200
3400	−461.2	−967.0	−514.1	−178.7	−253.8	−288.7	−270.7	−16.91	−48.94	3400
3600	−461.9	−968.1	−514.4	−178.2	−254.1	−288.6	−269.8	−15.62	−47.75	3600
3800	−462.5	−969.2	−514.6	−177.6	−254.5	−288.5	−268.8	−14.33	−46.49	3800
4000	−463.0	−970.4	−514.8	−176.9	−254.8	−288.4	−267.8	−13.00	−45.19	4000
4500	−464.0	−973.8	−515.3	−175.2	−255.9	−288.1	−265.5	−9.57	−41.68	4500
5000	−464.6	−977.9	−515.9	−173.2	−257.2	−288.0	−263.1	−5.95	−37.79	5000
5500	−464.8	−982.9	−516.5	−171.1	−258.6	−287.9	−260.7	−2.10	−33.56	5500
6000	−464.7	−989.0	−517.2	−169.0	−260.3	−287.9	−258.2	2.01	−28.98	6000

Equilibrium constants

$\ln K_p = \sum_i \nu_i \ln p_i^*$ where the dimensionless quantity p_i^* is numerically equal to the partial pressure of substance A_i, in bars.

Warning: This table lists *absolute* temperatures.

Reaction number; values are $\ln K_p$

Temp. K	1	2	3	4	5	6	7	8	9	Temp. K
200	250.149	554.472	285.471	105.592	139.972	161.789	159.692	19.719	15.433	200
298	163.986	367.479	186.975	69.865	92.207	106.228	103.762	11.554	6.593	298
400	119.150	270.329	135.715	51.311	67.321	77.284	74.669	7.348	1.778	400
600	75.217	175.356	85.523	33.203	42.897	48.905	46.245	3.348	−3.191	600
800	53.126	127.753	60.319	24.145	30.592	34.634	32.036	1.444	−5.822	800
1000	39.803	99.127	45.150	18.706	23.162	26.033	23.528	0.366	−7.457	1000
1200	30.874	80.011	35.005	15.082	18.182	20.281	17.871	−0.311	−8.570	1200
1400	24.463	66.329	27.742	12.489	14.608	16.160	13.841	−0.767	−9.371	1400
1600	19.632	56.055	22.285	10.546	11.921	13.065	10.829	−1.091	−9.972	1600
1800	15.865	48.051	18.030	9.035	9.825	10.657	8.497	−1.329	−10.439	1800
2000	12.835	41.645	14.622	7.824	8.145	8.727	6.634	−1.510	−10.810	2000
2200	10.353	36.391	11.827	6.834	6.768	7.148	5.119	−1.649	−11.109	2200
2400	8.276	32.011	9.497	6.010	5.619	5.831	3.859	−1.759	−11.358	2400
2600	6.512	28.304	7.521	5.314	4.647	4.718	2.800	−1.847	−11.563	2600
2800	5.002	25.117	5.286	4.720	3.811	3.763	1.893	−1.918	−11.738	2800
3000	3.685	22.359	4.357	4.205	3.086	2.936	1.110	−1.976	−11.885	3000
3200	2.533	19.936	3.072	3.753	2.450	2.211	0.429	−2.022	−12.012	3200
3400	1.516	17.800	1.935	3.357	1.891	1.575	−0.170	−2.061	−12.122	3400
3600	0.609	15.898	0.926	3.007	1.391	1.007	−0.702	−2.093	−12.217	3600
3800	−0.207	14.198	0.019	2.694	0.944	0.500	−1.176	−2.121	−12.300	3800
4000	−0.939	12.660	−0.796	2.413	0.541	0.044	−1.600	−2.141	−12.373	4000
4500	−2.486	9.414	−2.514	1.828	−0.313	−0.921	−2.491	−2.178	−12.519	4500
5000	−3.725	6.807	−3.895	1.363	−0.997	−1.690	−3.198	−2.201	−12.624	5000
5500	−4.743	4.666	−5.024	0.986	−1.561	−2.318	−3.771	−2.210	−12.703	5500
6000	−5.590	2.865	−5.963	0.677	−2.033	−2.843	−4.246	−2.213	−12.760	6000

Standard free enthalpy of reaction

At a given temperature, the standard free enthalpy of reaction ΔG_T° (or *standard Gibbs function change*) may be calculated from the listed value of $\ln K_p$ by the following equation:

$$\Delta G_T^\circ = -R_0 T \ln K_p$$
$$= -8.3145 T \ln K_p \text{ kJ kmol}^{-1}$$

Thermodynamics and fluid mechanics

In the following, T is absolute temperature, Q heat *input* per unit mass, W work *output* per unit mass, V velocity μ dynamic viscosity and v kinematic viscosity other symbols are as defined on page 48 or below.

$Q = UA\Delta T_{(LM)}$

$T_A - T_1 = \frac{Q}{2\pi r_1 h_1}$; $T_1 - T_2 = \frac{Q}{2\pi k_w} \ln\left(\frac{r_2}{r_1}\right)$

$\frac{1}{U_1} = \frac{1}{h_1} + f_1 + \frac{r_1 \ln\left(\frac{r_2}{r_1}\right)}{k_w} + \frac{r_1}{r_2 \ldots}$

Thermodynamic relations

Basic relations

Enthalpy: $h = u + pv$

Helmholtz function: $f = u - Ts$

Gibbs function: $g = h - Ts$

$du = T\,ds - p\,dv$

$dh = T\,ds + v\,dp$

$df = -p\,dv - s\,dT$

$dg = v\,dp - s\,dT$

$Nu = \frac{h_1 d}{k_{fluid}}$

$Re = \frac{\rho U_m d}{\mu}$

$Pr = \frac{\mu c_p}{k_{fluid}}$

$St = \frac{Nu}{Re\,Pr}$

Maxwell's relations

$$\left(\frac{\partial T}{\partial v}\right)_s = -\left(\frac{\partial p}{\partial s}\right)_v$$

$$\left(\frac{\partial T}{\partial p}\right)_s = \left(\frac{\partial v}{\partial s}\right)_p$$

$$\left(\frac{\partial p}{\partial T}\right)_v = \left(\frac{\partial s}{\partial v}\right)_T$$

$$\left(\frac{\partial v}{\partial T}\right)_p = -\left(\frac{\partial s}{\partial p}\right)_T$$

Specific heats

$$c_p = \left(\frac{\partial h}{\partial T}\right)_p ; c_v = \left(\frac{\partial u}{\partial T}\right)_v$$

$$c_p - c_v = \beta^2 \frac{Tv}{\kappa}$$

(for a perfect gas $c_p - c_v = R$).

Coefficients

Volume expansion $\beta = \frac{1}{v}\left(\frac{\partial v}{\partial T}\right)_p$

Compressibility $\kappa = -\frac{1}{v}\left(\frac{\partial v}{\partial p}\right)_T$

Joule-Thomson $\mu = \left(\frac{\partial T}{\partial p}\right)_h$

(for a perfect gas $\mu = 0$)

Equations of state

For unit mass:

Perfect gas $\quad pv = RT = \frac{R_0 T}{M}$

for a gas of molecular weight M with gas constant R.

Van der Waals' gas $\quad (p + a/v^2)(v - b) = RT$

where a, b are constants of the gas.

Process relations

Reversible polytropic, closed system (adiabatic)

$pv^n = \text{constant}$

$$W = \frac{p_2 v_2 - p_1 v_1}{1 - n} \quad (n \neq 1)$$

For a perfect gas,

$$W = \frac{R}{1-n}(T_2 - T_1)$$

$$Q = \left(c_v + \frac{R}{1-n}\right)(T_2 - T_1)$$

$$\frac{T_2}{T_1} = \left(\frac{p_2}{p_1}\right)^{(n-1)/n} = \left(\frac{V_1}{V_2}\right)^{(\gamma-1)}$$

and: if adiabatic ($Q = 0$), $n = \gamma$, the ratio of specific heats c_p/c_v if isothermal, $n = 1$ and

$$Q = W = RT\ln(v_2/v_1)$$

Steady flow

In terms of enthalpy, the energy equation relating any two sections of a steady flow at which properties are uniform is

$$Q - W_s = h_2 - h_1 + \tfrac{1}{2}(V_2^2 - V_1^2) + g(z_2 - z_1)$$

where g is the acceleration of gravity, W_s is shaft work and z is height above a horizontal datum.

$\Delta T_{LM} = \frac{\Delta T_A - \Delta T_B}{\ln\left(\frac{\Delta T_A}{\Delta T_B}\right)}$; $T_A = T_{HA} - T_{CA}$

Nozzle flow

For an expansion in which pv^n is constant the critical pressure ratio is

$$\frac{p_2}{p_1} = \left(\frac{2}{n+1}\right)^{n/(n-1)}$$

which gives the mass flux

$$\frac{\dot{m}}{A_2} = \sqrt{\left\{n\left(\frac{2}{n+1}\right)^{(n+1)/(n-1)}\frac{p_1}{v_1}\right\}}.$$

Pressure and area are related by

$$\frac{dA}{dp} = vA\left(\frac{1}{V^2} - \frac{1}{a^2}\right)$$

where a is the sonic velocity, given by

$$a^2 = -v^2\left(\frac{\partial p}{\partial v}\right)_s = \left(\frac{\partial p}{\partial \rho}\right)_s$$

For *any process* the specific entropy change of a perfect gas is

$$s_2 - s_1 = c_v \ln\frac{p_2}{p_1} + c_p \ln\frac{v_2}{v_1}$$

$$= c_v \ln\frac{T_2}{T_1} + R\ln\frac{v_2}{v_1}$$

$$= c_p \ln\frac{T_2}{T_1} - R\ln\frac{p_2}{p_1}$$

Equations for fluid flow

Continuity

$$\frac{\partial \rho}{\partial t} = -\operatorname{div}(\rho\mathbf{V})$$

(for an incompressible fluid, $\operatorname{div}\mathbf{V} = 0$)

Momentum

$$\mathbf{F} = \frac{\partial}{\partial t}\int_\tau \rho\mathbf{V}\,d\tau + \int_A \rho\mathbf{V}(\mathbf{V}.d\mathbf{A})$$

where $\mathbf{F}$ is the net force on a control volume τ bounded by a surface A. For steady flow

$$\mathbf{F} = \int_A \rho\mathbf{V}(\mathbf{V}.d\mathbf{A})$$

Energy

$$\dot{q} + \operatorname{div}(k \operatorname{grad} T) = \rho\left(\frac{\partial u}{\partial t} + \mathbf{V}.\operatorname{grad} u\right) + p\operatorname{div}\mathbf{V}$$

at any point; terms for dissipation and radiation may be included in $\dot{q}$, the rate of heat production per unit volume. The integrated form for a control volume τ bounded by a surface A is

$$\dot{q} - \dot{w} = \frac{\partial}{\partial t}\int_\tau (u + \tfrac{1}{2}V^2 + gz)\rho\,d\tau + \int_A \rho(u + \tfrac{1}{2}V^2 + gz + \frac{p}{\rho})\mathbf{V}.d\mathbf{A}$$

where $\dot{q}$, $\dot{w}$ are now the total rates of heat input and work output for the volume. For steady flow,

$$\dot{q} - \dot{w} = \int \rho\left(u + \tfrac{1}{2}V^2 + gz + \frac{p}{\rho}\right)\mathbf{V}.d\mathbf{A}$$

The Navier-Stokes equations

In Cartesian coordinates

$$\rho\left(\frac{\partial V_x}{\partial t} + V_x\frac{\partial V_x}{\partial x} + V_y\frac{\partial V_x}{\partial y} + V_z\frac{\partial V_x}{\partial z}\right)$$

$$= F_x + \frac{\partial \sigma_x}{\partial x} + \frac{\partial \tau_{yx}}{\partial y} + \frac{\partial \tau_{zx}}{\partial z}$$

etc., where F_x is a body force per unit volume and

$$\sigma_x = -p + 2\mu\frac{\partial V_x}{\partial x} - \tfrac{2}{3}\mu\left(\frac{\partial V_x}{\partial x} + \frac{\partial V_y}{\partial y} + \frac{\partial V_z}{\partial z}\right)$$

etc.

$$\tau_{xy} = \tau_{yx} = \mu\left(\frac{\partial V_x}{\partial y} + \frac{\partial V_y}{\partial x}\right)$$

etc. These give, for constant viscosity,

$$\frac{DV_x}{Dt} = -\frac{1}{\rho}\frac{\partial p}{\partial x} + \nu\left(\frac{\partial^2 V_x}{\partial x^2} + \frac{\partial^2 V_x}{\partial y^2} + \frac{\partial^2 V_x}{\partial z^2}\right) + \tfrac{1}{3}\nu\frac{\partial \theta}{\partial x} + \frac{F_x}{\rho}$$

etc. or

$$\frac{D\mathbf{V}}{Dt} = -\frac{1}{\rho}\nabla p + \nu\nabla^2\mathbf{V} + \tfrac{1}{3}\nu\nabla\theta + \frac{\mathbf{F}}{\rho}$$

where

$$\frac{D}{Dt} \equiv \frac{\partial}{\partial t} + \mathbf{V}.\nabla$$

and

$$\theta = \frac{\partial V_x}{\partial x} + \frac{\partial V_y}{\partial y} + \frac{\partial V_z}{\partial z} = \nabla.\mathbf{V}$$

Stream function and velocity potential

For two-dimensional incompressible flow the stream function is a scalar field ψ such that, in Cartesian coordinates,

$$V_x = -\frac{\partial \psi}{\partial y},\ V_y = \frac{\partial \psi}{\partial x}.$$

If in addition the flow is irrotational, the velocity potential is a scalar field ϕ such that

$$V_x = -\frac{\partial \phi}{\partial x},\ V_y = -\frac{\partial \phi}{\partial y}.$$

ϕ and ψ are both harmonic functions satisfying Laplace's equation; the complex potential $\phi + j\psi$ satisfies the Cauchy-Riemann conditions. It may also be so defined as to reverse the signs in all four of the above relations.

$Ra = Gr\,Pr$

$Fo = \frac{\alpha t}{L^2}$

Dimensionless groups

General

h heat transfer coefficient (rate of heat flow per unit area and per degree of temperature difference ΔT)

L characteristic dimension (length or diameter)

Froude number $\quad Fr = \frac{V}{\sqrt{Lg}}$ or $\frac{V^2}{Lg}$

Grashof number $\quad Gr = \frac{\beta g \rho^2 L^3 \Delta T}{\mu^2}$ $\quad \beta = \frac{1}{T}$

Mach number $\quad M = \frac{V}{a}$

Nusselt number $\quad Nu = \frac{hL}{k}$ fluid

$\bar{h} = \frac{Nu\, k_f}{D} = \frac{1}{L}\int_0^L h(x)\,dx.$

Prandtl number $\quad Pr = \frac{c_p \mu}{k}$

Reynolds number $\quad Re = \frac{VL\rho}{\mu}$

Stanton number $\quad St = \frac{h}{\rho V c_p}$

$\alpha = \frac{k}{\rho c_p}$

Hydraulic machines

P power
N speed
D diameter
Q discharge
H head

Power number $\quad \frac{P}{\rho N^3 D^5}$

Discharge number $\quad \frac{Q}{ND^3}$

Head number $\quad \frac{gH}{N^2 D^2}$

Reynolds number $\quad \frac{\rho N D^2}{\mu}$

Addison's shape parameter:

Turbine $\quad \frac{N\sqrt{(P/\rho)}}{(gH)^{5/4}}$

Pump $\quad \frac{N\sqrt{Q}}{(gH)^{3/4}}$

$Q = -kA\,dT \qquad = \varepsilon\sigma(T_1^4 - T_2^4)A.$

$\int_{D/2}^{\infty} \frac{dr}{r^2} = -\frac{4\pi k}{Q}\int_{T_s}^{T_\infty} dT.$

$\varphi = \theta - \frac{G}{km^2} \qquad m = \sqrt{\frac{2h}{kR}}$

$\theta = C_1 \sinh mx + C_2 \cosh mx.$

$$Q = \frac{\Delta T\, 2\pi L}{\frac{1}{r_1 h_1} + \frac{\ln\left(\frac{r_2}{r_1}\right)}{k_1} + \frac{1}{r_2 h_2}}$$

Nusselt numbers for convective heat transfer

In calculating heat transfer the temperature difference ΔT to be used is that between the wall temperature T_w and either the undisturbed free-stream temperature T_f (for external convection) or the average temperature T_b of the bulk flow (for convection in pipes). The fluid properties are to be evaluated at the mean film temperature $\frac{1}{2}(T_w + T_f)$, for external convection, or at T_b for pipe flow.

Unless otherwise indicated, the dimension L specified is that to be used in the groups Gr, Nu and Re (see *Dimensionless groups* above) and the expressions given are for constant wall temperature T_w.

Natural convection

Over the ranges indicated, Nu can be expressed as $C(\mathrm{GrPr})^n$ where C and n are as shown.

	GrPr	C	n
Vertical plates and cylinders (height L)			
Laminar flow	10^4–10^9	0·56	$\frac{1}{4}$
Turbulent flow	10^9–10^{13}	0·13	$\frac{1}{3}$
Horizontal cylinders (diameter L)			
Laminar flow	10^4–10^9	0·53	$\frac{1}{4}$
Turbulent flow	10^9–10^{12}	0·126	$\frac{1}{3}$
Horizontal plates			
Heated facing up, or cooled facing down	10^5–10^7	0·54	$\frac{1}{4}$
	10^7–10^{11}	0·14	$\frac{1}{3}$
Heated facing down, or cooled facing up	10^5–10^{11}	0·27	$\frac{1}{4}$

Forced convection

Flat plates (streamwise length L).

Laminar flow, $\mathrm{Re} < 5 \times 10^5$: $\mathrm{Nu} = 0{\cdot}664\,\mathrm{Re}^{1/2}\,\mathrm{Pr}^{1/3}$

For laminar flow with constant heat flux, the *local* Nusselt number at any distance L from the leading edge is: $\mathrm{Nu} = 0{\cdot}453\,\mathrm{Re}^{1/2}\,\mathrm{Pr}^{1/3}$

Turbulent flow, $\mathrm{Re} > 5 \times 10^5$: $\mathrm{Nu} = 0{\cdot}036\,\mathrm{Re}^{0{\cdot}8}\,\mathrm{Pr}^{1/3}$

Flow in circular pipes (length L, diameter D, Re and Nu based on D)

Laminar flow, $\mathrm{Re} < 2000$: $\mathrm{Nu} = 1{\cdot}615\,(\mathrm{Re\,Pr}\,D/L)^{1/3}$

For laminar flow with constant heat flux: $\mathrm{Nu} = 4{\cdot}36$

Turbulent flow, fully developed, $\mathrm{Re} > 2000$: $\mathrm{Nu} = 0{\cdot}023\,\mathrm{Re}^{0{\cdot}8}\,\mathrm{Pr}^{0{\cdot}4}$

Cylinder in cross flow (diameter L)

For air the Nusselt number can be approximated as

$$\mathrm{Nu} = C\,\mathrm{Re}^n$$

where C and n depend on Re as follows.

Re	C	n
0·0001–0·004	0·437	0·0895
0·004 –0·09	0·565	0·136
0·09 –1·0	0·800	0·280
1–35	0·795	0·384
35–5 000	0·583	0·471
5 000–50 000	0·148	0·633
50 000–200 000	0·0208	0·814

For other fluids the following approximations may be used:

$\mathrm{Nu} = 0{\cdot}91\,\mathrm{Re}^{0{\cdot}385}\,\mathrm{Pr}^{0{\cdot}31}$ $(0{\cdot}1 < \mathrm{Re} < 50)$

$\mathrm{Nu} = 0{\cdot}60\,\mathrm{Re}^{0{\cdot}5}\,\mathrm{Pr}^{0{\cdot}31}$ $(50 < \mathrm{Re} < 10\,000)$

Friction in pipes

Moody diagram

The head loss for an average flow velocity V in a pipe of diameter d and length L is given by Darcy's equation:

$$h_L = f \cdot \frac{4L}{d} \cdot \frac{V^2}{2g}$$

in which f is the friction coefficient given by $\tau / \frac{1}{2}\rho V^2$ where τ is the shear stress at the wall. The curves below show f as a function of Re for various values of k/d, where k is the effective surface roughness and Re is based on the diameter d.

Coefficients of loss for pipe fittings

The loss of head incurred by fittings, valves or sudden contractions of area is given by the loss coefficient K_L according to the relation

$$h_L = K_L(V^2/2g)$$

where V is the average flow velocity. Values of K_L for fittings, valves and contractions of area ratio A_2/A_1 are given below

	K_L
Glove valve, fully open	10·0
Angle valve, fully open	5·0

	K_L
Swing check valve, fully open	2·5
Gate valve, fully open	0·19
Three-quarters open	1·15
One-half open	5·6
One-quarter open	24·0
Close return bend	2·2
Standard tee	1·8
Standard 90° elbow	0·9
Medium sweep 90° elbow	0·75
Long sweep 90° elbow	0·60
45° elbow	0·42
Rounded inlet	0·04
Re-entrant inlet	0·8
Sharp-edged inlet	0·5
Contraction, A_2/A_1 = 0·1	0·37
= 0·2	0·35
= 0·4	0·27
= 0·6	0·17
= 0·8	0·06
= 0·9	0·02

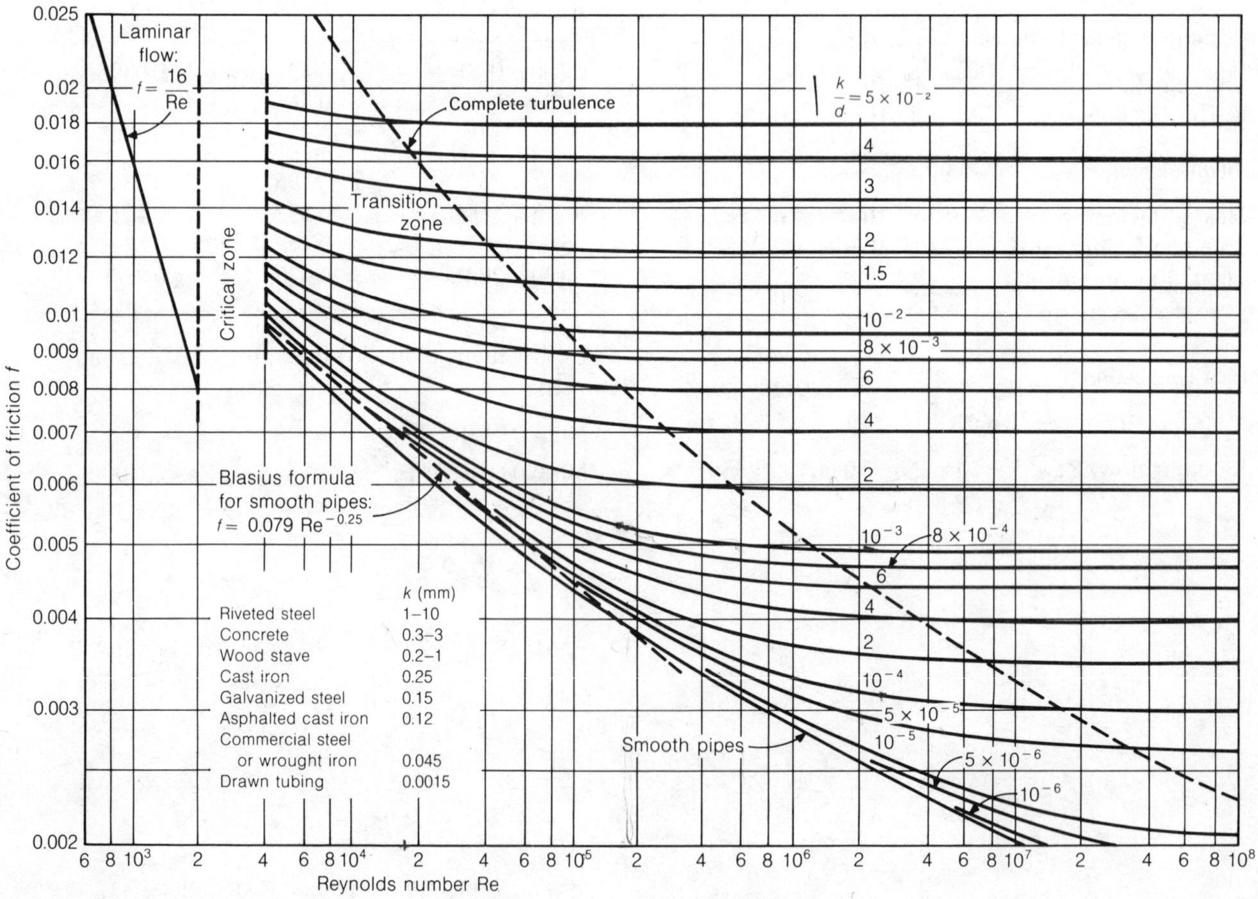

Boundary-layer friction and drag

In a two-dimensional constant-pressure boundary layer the local skin-friction coefficient f at distance x from the leading edge, and the friction drag coefficient C_F for unit width of a plate of length l (one surface) are as follows.

Laminar, $0 < \mathrm{Re} \lesssim 10^5$ (Blasius solution)

$$f = \tau/\tfrac{1}{2}\rho V^2 = 0{\cdot}664\left(\frac{Vx}{\nu}\right)^{-1/2}$$

$$C_F = D/\tfrac{1}{2}\rho V^2 l = 1{\cdot}328\left(\frac{Vl}{\nu}\right)^{-1/2}$$

Turbulent, $\mathrm{Re} \gtrsim 10^{\circ}$ ($\frac{1}{7}$th root velocity profile)

$$f = 0{\cdot}0576\left(\frac{Vx}{\nu}\right)^{-1/5}$$

$$C_F = 0{\cdot}072\left(\frac{Vl}{\nu}\right)^{-1/5}$$

or

$$C_F = 0{\cdot}455\left\{\log_{10}\left(\frac{Vl}{\nu}\right)\right\}^{-2{\cdot}58}$$

which is the *Prandtl-Schlichting* formula.

In the above τ is the shear stress at the wall, V the velocity outside the boundary layer and D the drag.

Open-channel flow

The velocity V of uniform flow in an open channel of slope S may be estimated from the Chézy formula

$$V = C\sqrt{(RS)}$$

or from the Manning formula

$$V = \frac{0{\cdot}82}{n} R^{2/3} S^{1/2}$$

in which C is the Chézy coefficient given by $\sqrt{(8g/f)}$ or $\sqrt{(2g/C_f)}$, and R is the hydraulic radius (ratio of flow section to wetted perimeter) of the channel. Values of n are given below.

	n (m$^{1/6}$)
Smooth surface	0·008
Neat cement surface	0·009
Finished concrete, planed wood, or steel surface	0·010
Mortar, clay, or glazed brick surface	0·011
Vitrified clay surface	0·012
Brick surface lined with cement mortar	0·012
Unfinished cement surface	0·014
Rubble masonry or corrugated metal surface	0·016
Earth channel with gravel bottom	0·020
Earth channel with dense weed	0·030
Natural channel with clean bottom, brush on sides	0·040
Flood plain with dense brush	0·080

Black-body radiation

The power radiated by a black body in all directions over a solid angle 2π, per unit surface area and per unit frequency interval is

$$E_\nu = \frac{2\pi h\nu^3/c^2}{e^{h\nu/kT} - 1}$$

in the region of the frequency ν, where h is the Planck constant, k the Boltzmann constant, and c the velocity of light; alternatively, the power per unit wavelength interval in the region of the wavelength λ is

$$E_\lambda = \frac{2\pi hc^2/\lambda^5}{e^{hc/\lambda kT} - 1}$$

The total power per unit area is

$$E = \int_0^\infty E_\nu \, d\nu = \int_0^\infty E_\lambda \, d\lambda = \sigma T^4$$

in which σ is the Stefan-Boltzmann constant. The wavelength λ_m at which E_λ is a maximum is given by the *Wien displacement law:*

$$\lambda_m T = 0{\cdot}0029 \text{ m K}$$

Radiation exchange

The net power exchanged between two black (or diffusely radiating) surface elements dA_1, and dA_2 distance r apart is

$$\dot{q} = (E_1 - E_2)\cos\phi_1 \cos\phi_2 \, dA_1 \, dA_2$$

where ϕ_1, ϕ_2 are the angles between r and each normal.

Generalized compressibility chart

The compressibility factor Z as a function of reduced pressure and temperature p_R and T_R is to a good approximation the same for all gases; the function is plotted here with p_R as independent variable and T_R as parameter. Reduced pressure and temperature are the ratios of actual values to the critical values p_c, T_c.

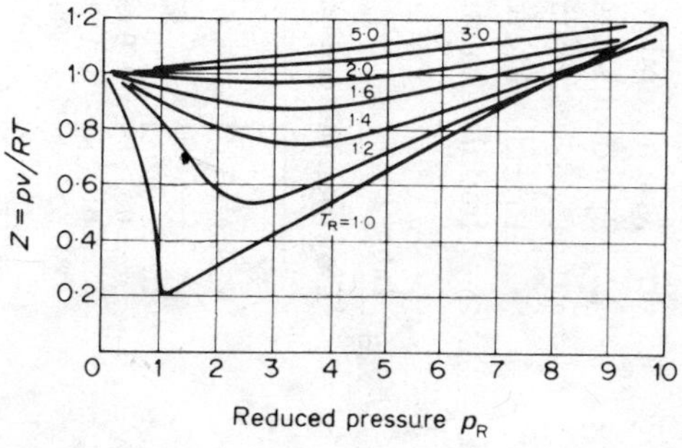

Tables for compressible flow of a perfect gas

M Mach number
a Speed of sound
A Cross-sectional area of duct
θ Prandtl–Meyer angle
I Impulse function

L_{max} Maximum length for choked flow in a duct of diameter D with friction coefficient f
Asterisk (*) denotes reference value when $M = 1$
Subscript or superscript 0 denotes stagnation condition; subscripts 1, 2 denote conditions upstream, downstream of shock; subscript n denotes normal to shock

Table 1

Flow parameters versus M for subsonic isentropic flow, $\gamma = 1{\cdot}4$

M	p/p_0	ρ/ρ_0	T/T_0	a/a_0	A^*/A
.00	1.0000	1.0000	1.0000	1.0000	.00000
.01	.9999	1.0000	1.0000	1.0000	.01728
.02	.9997	.9998	.9999	1.0000	.03455
.03	.9994	.9996	.9998	.9999	.05181
.04	.9989	.9992	.9997	.9998	.06905
.05	.9983	.9988	.9995	.9998	.08627
.06	.9975	.9982	.9993	.9996	.1035
.07	.9966	.9976	.9990	.9995	.1206
.08	.9955	.9968	.9987	.9994	.1377
.09	.9944	.9960	.9984	.9992	.1548
.10	.9930	.9950	.9980	.9990	.1718
.11	.9916	.9940	.9976	.9988	.1887
.12	.9900	.9928	.9971	.9986	.2056
.13	.9883	.9916	.9966	.9983	.2224
.14	.9864	.9903	.9961	.9980	.2391
.15	.9844	.9888	.9955	.9978	.2557
.16	.9823	.9873	.9949	.9974	.2723
.17	.9800	.9857	.9943	.9971	.2887
.18	.9776	.9840	.9936	.9968	.3051
.19	.9751	.9822	.9928	.9964	.3213
.20	.9725	.9803	.9921	.9960	.3374
.21	.9697	.9783	.9913	.9956	.3534
.22	.9668	.9762	.9904	.9952	.3693
.23	.9638	.9740	.9895	.9948	.3851
.24	.9607	.9718	.9886	.9943	.4007
.25	.9575	.9694	.9877	.9938	.4162
.26	.9541	.9670	.9867	.9933	.4315
.27	.9506	.9645	.9856	.9928	.4467
.28	.9470	.9619	.9846	.9923	.4618
.29	.9433	.9592	.9835	.9917	.4767
.30	.9395	.9564	.9823	.9911	.4914
.31	.9355	.9535	.9811	.9905	.5059
.32	.9315	.9506	.9799	.9899	.5203
.33	.9274	.9476	.9787	.9893	.5345
.34	.9231	.9445	.9774	.9886	.5486
.35	.9188	.9413	.9761	.9880	.5624
.36	.9143	.9380	.9747	.9873	.5761
.37	.9098	.9347	.9733	.9866	.5896
.38	.9052	.9313	.9719	.9859	.6029
.39	.9004	.9278	.9705	.9851	.6160
.40	.8956	.9243	.9690	.9844	.6289
.41	.8907	.9207	.9675	.9836	.6416
.42	.8857	.9170	.9659	.9828	.6541
.43	.8807	.9132	.9643	.9820	.6663
.44	.8755	.9094	.9627	.9812	.6784
.45	.8703	.9055	.9611	.9803	.6903
.46	.8650	.9016	.9594	.9795	.7019
.47	.8596	.8976	.9577	.9786	.7134
.48	.8541	.8935	.9560	.9777	.7246
.49	.8486	.8894	.9542	.9768	.7356
.50	.8430	.8852	.9524	.9759	.7464
.51	.8374	.8809	.9506	.9750	.7569
.52	.8317	.8766	.9487	.9740	.7672
.53	.8259	.8723	.9468	.9730	.7773
.54	.8201	.8679	.9449	.9721	.7872
.55	.8142	.8634	.9430	.9711	.7968
.56	.8082	.8589	.9410	.9701	.8063
.57	.8022	.8544	.9390	.9690	.8155
.58	.7962	.8498	.9370	.9680	.8244
.59	.7901	.8451	.9349	.9669	.8331
.60	.7840	.8405	.9328	.9658	.8416
.61	.7778	.8357	.9307	.9647	.8499
.62	.7716	.8310	.9286	.9636	.8579

Table 1 (Continued)

M	p/p_0	ρ/ρ_0	T/T_0	a/a_0	A^*/A
.63	.7654	.8262	.9265	.9625	.8657
.64	.7591	.8213	.9243	.9614	.8732
65	.7528	.8164	.9221	.9603	.8806
.66	.7465	.8115	.9199	.9591	.8877
.67	.7401	.8066	.9176	.9579	.8945
.68	.7338	.8016	.9153	.9567	.9012
.69	.7274	.7966	.9131	.9555	.9076
.70	.7209	.7916	.9107	.9543	.9138
.71	.7145	.7865	.9084	.9531	.9197
.72	.7080	.7814	.9061	.9519	.9254
.73	.7016	.7763	.9037	.9506	.9309
.74	.6951	.7712	.9013	.9494	.9362
.75	.6886	.7660	.8989	.9481	.9412
.76	.6821	.7609	.8964	.9468	.9461
.77	.6756	.7557	.8940	.9455	.9507
.78	.6690	.7505	.8915	.9442	.9551
.79	.6625	.7452	.8890	.9429	.9592
.80	.6560	.7400	.8865	.9416	.9632
.81	.6495	.7347	.8840	.9402	.9669
.82	.6430	.7295	.8815	.9389	.9704
.83	.6365	.7242	.8789	.9375	.9737
.84	.6300	.7189	.8763	.9361	.9769
.85	.6235	.7136	.8737	.9347	.9797
.86	.6170	.7083	.8711	.9333	.9824
.87	.6106	.7030	.8685	.9319	.9849
.88	.6041	.6977	.8659	.9305	.9872
.89	.5977	.6924	.8632	.9291	.9893
.90	.5913	.6870	.8606	.9277	.9912
.91	.5849	.6817	.8579	.9262	.9929
.92	.5785	.6764	.8552	.9248	.9944
.93	.5721	.6711	.8525	.9233	.9958
.94	.5658	.6658	.8498	.9218	.9969
.95	.5595	.6604	.8471	.9204	.9979
.96	.5532	.6551	.8444	.9189	.9986
.97	.5469	.6498	.8416	.9174	.9992
.98	.5407	.6445	.8389	.9159	.9997
.99	.5345	.6392	.8361	.9144	.9999
1.00	.5283	.6339	.8333	.9129	1.0000

Table 2

Flow parameters versus M for supersonic isentropic flow, $\gamma = 1{\cdot}4$

M	$\frac{p}{p_0}$	$\frac{\rho}{\rho_0}$	$\frac{T}{T_0}$	$\frac{a}{a_0}$	$\frac{A^*}{A}$	$\frac{\frac{\rho}{2}V^2}{p_0}$	θ
1.00	.5283	.6339	.8333	.9129	1.0000	.3698	0
1.01	.5221	.6287	.8306	.9113	.9999	.3728	.04473
1.02	.5160	.6234	.8278	.9098	.9997	.3758	.1257
1.03	.5099	.6181	.8250	.9083	.9993	.3787	.2294
1.04	.5039	.6129	.8222	.9067	.9987	.3815	.3510
1.05	.4979	.6077	.8193	.9052	.9980	.3842	.4874
1.06	.4919	.6024	.8165	.9036	.9971	.3869	.6367
1.07	.4860	.5972	.8137	.9020	.9961	.3895	.7973
1.08	.4800	.5920	.8108	.9005	.9949	.3919	.9680
1.09	.4742	.5869	.8080	.8989	.9936	.3944	1.148
1.10	.4684	.5817	.8052	.8973	.9921	.3967	1.336
1.11	.4626	.5766	.8023	.8957	.9905	.3990	1.532
1.12	.4568	.5714	.7994	.8941	.9888	.4011	1.735
1.13	.4511	.5663	.7966	.8925	.9870	.4032	1.944
1.14	.4455	.5612	.7937	.8909	.9850	.4052	2.160
1.15	.4398	.5562	.7908	.8893	.9828	.4072	2.381
1.16	.4343	.5511	.7879	.8877	.9806	.4090	2.607
1.17	.4287	.5461	.7851	.8860	.9782	.4108	2.839
1.18	.4232	.5411	.7822	.8844	.9758	.4125	3.074
1.19	.4178	.5361	.7793	.8828	.9732	.4141	3.314
1.20	.4124	.5311	.7764	.8811	.9705	.4157	3.558
1.21	.4070	.5262	.7735	.8795	.9676	.4171	3.806
1.22	.4017	.5213	.7706	.8778	.9647	.4185	4.057
1.23	.3964	.5164	.7677	.8762	.9617	.4198	4.312
1.24	.3912	.5115	.7648	.8745	.9586	.4211	4.569
1.25	.3861	.5067	.7619	.8729	.9553	.4223	4.830
1.26	.3809	.5019	.7590	.8712	.9520	.4233	5.093
1.27	.3759	.4971	.7561	.8695	.9486	.4244	5.359
1.28	.3708	.4923	.7532	.8679	.9451	.4253	5.627
1.29	.3658	.4876	.7503	.8662	.9415	.4262	5.898
1.30	.3609	.4829	.7474	.8645	.9378	.4270	6.170
1.31	.3560	.4782	.7445	.8628	.9341	.4277	6.445
1.32	.3512	.4736	.7416	.8611	.9302	.4283	6.721
1.33	.3464	.4690	.7387	.8595	.9263	.4289	7.000
1.34	.3417	.4644	.7358	.8578	.9223	.4294	7.279

Table 2 (Continued)

M	$\frac{p}{p_0}$	$\frac{\rho}{\rho_0}$	$\frac{T}{T_0}$	$\frac{a}{a_0}$	$\frac{A^*}{A}$	$\frac{\frac{\rho}{2}V^2}{p_0}$	θ
1.35	.3370	.4598	.7329	.8561	.9182	.4299	7.561
1.36	.3323	.4553	.7300	.8544	.9141	.4303	7.844
1.37	.3277	.4508	.7271	.8527	.9099	.4306	8.128
1.38	.3232	.4463	.7242	.8510	.9056	.4308	8.413
1.39	.3187	.4418	.7213	.8493	.9013	.4310	8.699
1.40	.3142	.4374	.7184	.8476	.8969	.4311	8.987
1.41	.3098	.4330	.7155	.8459	.8925	.4312	9.276
1.42	.3055	.4287	.7126	.8442	.8880	.4312	9.565
1.43	.3012	.4244	.7097	.8425	.8834	.4311	9.855
1.44	.2969	.4201	.7069	.8407	.8788	.4310	10.15
1.45	.2927	.4158	.7040	.8390	.8742	.4308	10.44
1.46	.2886	.4116	.7011	.8373	.8695	.4306	10.73
1.47	.2845	.4074	.6982	.8356	.8647	.4303	11.02
1.48	.2804	.4032	.6954	.8339	.8599	.4299	11.32
1.49	.2764	.3991	.6925	.8322	.8551	.4295	11.61
1.50	.2724	.3950	.6897	.8305	.8502	.4290	11.91
1.51	.2685	.3909	.6868	.8287	.8453	.4285	12.20
1.52	.2646	.3869	.6840	.8270	.8404	.4279	12.49
1.53	.2608	.3829	.6811	.8253	.8354	.4273	12.79
1.54	.2570	.3789	.6783	.8236	.8304	.4266	13.09
1.55	.2533	.3750	.6754	.8219	.8254	.4259	13.38
1.56	.2496	.3710	.6726	.8201	.8203	.4252	13.68
1.57	.2459	.3672	.6698	.8184	.8152	.4243	13.97
1.58	.2423	.3633	.6670	.8167	.8101	.4235	14.27
1.59	.2388	.3595	.6642	.8150	.8050	.4226	14.56
1.60	.2353	.3557	.6614	.8133	.7998	.4216	14.86
1.61	.2318	.3520	.6586	.8115	.7947	.4206	15.16
1.62	.2284	.3483	.6558	.8098	.7895	.4196	15.45
1.63	.2250	.3446	.6530	.8081	.7843	.4185	15.75
1.64	.2217	.3409	.6502	.8064	.7791	.4174	16.04
1.65	.2184	.3373	.6475	.8046	.7739	.4162	16.34
1.66	.2151	.3337	.6447	.8029	.7686	.4150	16.63
1.67	.2119	.3302	.6419	.8012	.7634	.4138	16.93
1.68	.2088	.3266	.6392	.7995	.7581	.4125	17.22
1.69	.2057	.3232	.6364	.7978	.7529	.4112	17.52
1.70	.2026	.3197	.6337	.7961	.7476	.4098	17.81
1.71	.1996	.3163	.6310	.7943	.7423	.4086	18.10
1.72	.1966	.3129	.6283	.7926	.7371	.4071	18.40
1.73	.1936	.3095	.6256	.7909	.7318	.4056	18.69
1.74	.1907	.3062	.6229	.7892	.7265	.4041	18.98
1.75	.1878	.3029	.6202	.7875	.7212	.4026	19.27
1.76	.1850	.2996	.6175	.7858	.7160	.4011	19.56
1.77	.1822	.2964	.6148	.7841	.7107	.3996	19.86
1.78	.1794	.2932	.6121	.7824	.7054	.3980	20.15
1.79	.1767	.2900	.6095	.7807	.7002	.3964	20.44
1.80	.1740	.2868	.6068	.7790	.6949	.3947	20.73
1.81	.1714	.2837	.6041	.7773	.6897	.3931	21.01
1.82	.1688	.2806	.6015	.7756	.6845	.3914	21.30
1.83	.1662	.2776	.5989	.7739	.6792	.3897	21.59
1.84	.1637	.2745	.5963	.7722	.6740	.3879	21.88
1.85	.1612	.2715	.5936	.7705	.6688	.3862	22.16
1.86	.1587	.2686	.5910	.7688	.6636	.3844	22.45
1.87	.1563	.2656	.5884	.7671	.6584	.3826	22.73
1.88	.1539	.2627	.5859	.7654	.6533	.3808	23.02
1.89	.1516	.2598	.5833	.7637	.6481	.3790	23.30
1.90	.1492	.2570	.5807	.7620	.6430	.3771	23.59
1.91	.1470	.2542	.5782	.7604	.6379	.3753	23.87
1.92	.1447	.2514	.5756	.7587	.6328	.3734	24.15
1.93	.1425	.2486	.5731	.7570	.6277	.3715	24.43
1.94	.1403	.2459	.5705	.7553	.6226	.3696	24.71
1.95	.1381	.2432	.5680	.7537	.6175	.3677	24.99
1.96	.1360	.2405	.5655	.7520	.6125	.3657	25.27
1.97	.1339	.2378	.5630	.7503	.6075	.3638	25.55
1.98	.1318	.2352	.5605	.7487	.6025	.3618	25.83
1.99	.1298	.2326	.5580	.7470	.5975	.3598	26.10
2.00	.1278	.2300	.5556	.7454	.5926	.3579	26.38
2.01	.1258	.2275	.5531	.7437	.5877	.3559	26.66
2.02	.1239	.2250	.5506	.7420	.5828	.3539	26.93
2.03	.1220	.2225	.5482	.7404	.5779	.3518	27.20
2.04	.1201	.2200	.5458	.7388	.5730	.3498	27.48

Table 2 (Continued)

M	$\frac{p}{p_0}$	$\frac{\rho}{\rho_0}$	$\frac{T}{T_0}$	$\frac{a}{a_0}$	$\frac{A^*}{A}$	$\frac{\frac{\rho}{2}V^2}{p_0}$	θ
2.05	.1182	.2176	.5433	.7371	.5682	.3478	27.75
2.06	.1164	.2152	.5409	.7355	.5634	.3458	28.02
2.07	.1146	.2128	.5385	.7338	.5586	.3437	28.29
2.08	.1128	.2104	.5361	.7322	.5538	.3417	28.56
2.09	.1111	.2081	.5337	.7306	.5491	.3396	28.83
2.10	.1094	.2058	.5313	.7289	.5444	.3376	29.10
2.11	.1077	.2035	.5290	.7273	.5397	.3355	29.36
2.12	.1060	.2013	.5266	.7257	.5350	.3334	29.63
2.13	.1043	.1990	.5243	.7241	.5304	.3314	29.90
2.14	.1027	.1968	.5219	.7225	.5258	.3293	30.16
2.15	.1011	.1946	.5196	.7208	.5212	.3272	30.43
2.16	.09956	.1925	.5173	.7192	.5167	.3252	30.69
2.17	.09802	.1903	.5150	.7176	.5122	.3231	30.95
2.18	.09650	.1882	.5127	.7160	.5077	.3210	31.21
2.19	.09500	.1861	.5104	.7144	.5032	.3189	31.47
2.20	.09352	.1841	.5081	.7128	.4988	.3169	31.73
2.21	.09207	.1820	.5059	.7112	.4944	.3148	31.99
2.22	.09064	.1800	.5036	.7097	.4900	.3127	32.25
2.23	.08923	.1780	.5014	.7081	.4856	.3106	32.51
2.24	.08785	.1760	.4991	.7065	.4813	.3085	32.76
2.25	.08648	.1740	.4969	.7049	.4770	.3065	33.02
2.26	.08514	.1721	.4947	.7033	.4727	.3044	33.27
2.27	.08382	.1702	.4925	.7018	.4685	.3023	33.53
2.28	.08252	.1683	.4903	.7002	.4643	.3003	33.78
2.29	.08123	.1664	.4881	.6986	.4601	.2982	34.03
2.30	.07997	.1646	.4859	.6971	.4560	.2961	34.28
2.31	.07873	.1628	.4837	.6955	.4519	.2941	34.53
2.32	.07751	.1609	.4816	.6940	.4478	.2920	34.78
2.33	.07631	.1592	.4794	.6924	.4437	.2900	35.03
2.34	.07512	.1574	.4773	.6909	.4397	.2879	35.28
2.35	.07396	.1556	.4752	.6893	.4357	.2859	35.53
2.36	.07281	.1539	.4731	.6878	.4317	.2839	35.77
2.37	.07168	.1522	.4709	.6863	.4278	.2818	36.02
2.38	.07057	.1505	.4688	.6847	.4239	.2798	36.26
2.39	.06948	.1488	.4668	.6832	.4200	.2778	36.50

M	$\frac{p}{p_0}$	$\frac{\rho}{\rho_0}$	$\frac{T}{T_0}$	$\frac{a}{a_0}$	$\frac{A^*}{A}$	$\frac{\frac{\rho}{2}V^2}{p_0}$	θ
2.40	.06840	.1472	.4647	.6817	.4161	.2758	36.75
2.41	.06734	.1456	.4626	.6802	.4123	.2738	36.99
2.42	.06630	.1439	.4606	.6786	.4085	.2718	37.23
2.43	.06527	.1424	.4585	.6771	.4048	.2698	37.47
2.44	.06426	.1408	.4565	.6756	.4010	.2678	37.71
2.45	.06327	.1392	.4544	.6741	.3973	.2658	37.95
2.46	.06229	.1377	.4524	.6726	.3937	.2639	38.18
2.47	.06133	.1362	.4504	.6711	.3900	.2619	38.42
2.48	.06038	.1347	.4484	.6696	.3864	.2599	38.66
2.49	.05945	.1332	.4464	.6681	.3828	.2580	38.89
2.50	.05853	.1317	.4444	.6667	.3793	.2561	39.12
2.51	.05762	.1302	.4425	.6652	.3757	.2541	39.36
2.52	.05674	.1288	.4405	.6637	.3722	.2522	39.59
2.53	.05586	.1274	.4386	.6622	.3688	.2503	39.82
2.54	.05500	.1260	.4366	.6608	.3653	.2484	40.05
2.55	.05415	.1246	.4347	.6593	.3619	.2465	40.28
2.56	.05332	.1232	.4328	.6579	.3585	.2446	40.51
2.57	.05250	.1218	.4309	.6564	.3552	.2427	40.75
2.58	.05169	.1205	.4289	.6549	.3519	.2409	40.96
2.59	.05090	.1192	.4271	.6535	.3486	.2390	41.19
2.60	.05012	.1179	.4252	.6521	.3453	.2371	41.41
2.61	.04935	.1166	.4233	.6506	.3421	.2353	41.64
2.62	.04859	.1153	.4214	.6492	.3389	.2335	41.86
2.63	.04784	.1140	.4196	.6477	.3357	.2317	42.09
2.64	.04711	.1128	.4177	.6463	.3325	.2298	42.31
2.65	.04639	.1115	.4159	.6449	.3294	.2280	42.53
2.66	.04568	.1103	.4141	.6435	.3263	.2262	42.75
2.67	.04498	.1091	.4122	.6421	.3232	.2245	42.97
2.68	.04429	.1079	.4104	.6406	.3202	.2227	43.19
2.69	.04362	.1067	.4086	.6392	.3172	.2209	43.40
2.70	.04295	.1056	.4068	.6378	.3142	.2192	43.62
2.71	.04229	.1044	.4051	.6364	.3112	.2174	43.84
2.72	.04165	.1033	.4033	.6350	.3083	.2157	44.05
2.73	.04102	.1022	.4015	.6337	.3054	.2140	44.27
2.74	.04039	.1010	.3998	.6323	.3025	.2123	44.48

Table 2 (Continued)

M	$\frac{p}{p_0}$	$\frac{\rho}{\rho_0}$	$\frac{T}{T_0}$	$\frac{a}{a_0}$	$\frac{A^*}{A}$	$\frac{\frac{\rho}{2}V^2}{p_0}$	θ
2.75	.03978	.09994	.3980	.6309	.2996	.2106	44.69
2.76	.03917	.09885	.3963	.6295	.2968	.2089	44.91
2.77	.03858	.09778	.3945	.6281	.2940	.2072	45.12
2.78	.03799	.09671	.3928	.6268	.2912	.2055	45.33
2.79	.03742	.09566	.3911	.6254	.2884	.2039	45.54
2.80	.03685	.09463	.3894	.6240	.2857	.2022	45.75
2.81	.03629	.09360	.3877	.6227	.2830	.2006	45.95
2.82	.03574	.09259	.3860	.6213	.2803	.1990	46.16
2.83	.03520	.09158	.3844	.6200	.2777	.1973	46.37
2.84	.03467	.09059	.3827	.6186	.2750	.1957	46.57
2.85	.03415	.08962	.3810	.6173	.2724	.1941	46.78
2.86	.03363	.08865	.3794	.6159	.2698	.1926	46.98
2.87	.03312	.08769	.3777	.6146	.2673	.1910	47.19
2.88	.03263	.08675	.3761	.6133	.2648	.1894	47.39
2.89	.03213	.08581	.3745	.6119	.2622	.1879	47.59
2.90	.03165	.08489	.3729	.6106	.2598	.1863	47.79
2.91	.03118	.08398	.3712	.6093	.2573	.1848	47.99
2.92	.03071	.08307	.3696	.6080	.2549	.1833	48.19
2.93	.03025	.08218	.3681	.6067	.2524	.1818	48.39
2.94	.02980	.08130	.3665	.6054	.2500	.1803	48.59
2.95	.02935	.08043	.3649	.6041	.2477	.1788	48.78
2.96	.02891	.07957	.3633	.6028	.2453	.1773	48.98
2.97	.02848	.07872	.3618	.6015	.2430	.1758	49.18
2.98	.02805	.07788	.3602	.6002	.2407	.1744	49.37
2.99	.02764	.97705	.3587	.5989	.2384	.1729	49.56
3.00	.02722	.07623	.3571	.5976	.2362	.1715	49.76
3.01	.02682	.07541	.3556	.5963	.2339	.1701	49.95
3.02	.02642	.07461	.3541	.5951	.2317	.1687	50.14
3.03	.02603	.07382	.3526	.5938	.2295	.1673	50.33
3.04	.02564	.07303	.3511	.5925	.2273	.1659	50.52
3.05	.02526	.07226	.3496	.5913	.2252	.1645	50.71
3.06	.02489	.07149	.3481	.5900	.2230	.1631	50.90
3.07	.02452	.07074	.3466	.5887	.2209	.1618	51.09
3.08	.02416	.06999	.3452	.5875	.2188	.1604	51.28
3.09	.02380	.06925	.3437	.5862	.2168	.1591	51.46

M	$\frac{p}{p_0}$	$\frac{\rho}{\rho_0}$	$\frac{T}{T_0}$	$\frac{a}{a_0}$	$\frac{A^*}{A}$	$\frac{\frac{\rho}{2}V^2}{p_0}$	θ
3.10	.02345	.06852	.3422	.5850	.2147	.1577	51.65
3.11	.02310	.06779	.3408	.5838	.2127	.1564	51.84
3.12	.02276	.06708	.3393	.5825	.2107	.1551	52.02
3.13	.02243	.06637	.3379	.5813	.2087	.1538	52.20
3.14	.02210	.06568	.3365	.5801	.2067	.1525	52.39
3.15	.02177	.06499	.3351	.5788	.2048	.1512	52.57
3.16	.02146	.06430	.3337	.5776	.2028	.1500	52.75
3.17	.02114	.06363	.3323	.5764	.2009	.1487	52.93
3.18	.02083	.06296	.3309	.5752	.1990	.1475	53.11
3.19	.02053	.06231	.3295	.5740	.1971	.1462	53.29
3.20	.02023	.06165	.3281	.5728	.1953	.1450	53.47
3.21	.01993	.06101	.3267	.5716	.1934	.1438	53.65
3.22	.01964	.06037	.3253	.5704	.1916	.1426	53.83
3.23	.01936	.05975	.3240	.5692	.1898	.1414	54.00
3.24	.01908	.05912	.3226	.5680	.1880	.1402	54.18
3.25	.01880	.05851	.3213	.5668	.1863	.1390	54.35
3.26	.01853	.05790	.3199	.5656	.1845	.1378	54.53
3.27	.01826	.05730	.3186	.5645	.1828	.1367	54.71
3.28	.01799	.05671	.3173	.5633	.1810	.1355	54.88
3.29	.01773	.05612	.3160	.5621	.1793	.1344	55.05
3.30	.01748	.05554	.3147	.5609	.1777	.1332	55.22
3.31	.01722	.05497	.3134	.5598	.1760	.1321	55.39
3.32	.01698	.05440	.3121	.5586	.1743	.1310	55.56
3.33	.01673	.05384	.3108	.5575	.1727	.1299	55.73
3.34	.01649	.05329	.3095	.5563	.1711	.1288	55.90
3.35	.01625	.05274	.3082	.5552	.1695	.1277	56.07
3.36	.01602	.05220	.3069	.5540	.1679	.1266	56.24
3.37	.01579	.05166	.3057	.5529	.1663	.1255	56.41
3.38	.01557	.05113	.3044	.5517	.1648	.1245	56.58
3.39	.01534	.05061	.3032	.5506	.1632	.1234	56.75
3.40	.01513	.05009	.3019	.5495	.1617	.1224	56.91
3.41	.01491	.04958	.3007	.5484	.1602	.1214	57.07
3.42	.01470	.04908	.2995	.5472	.1587	.1203	57.24
3.43	.01449	.04858	.2982	.5461	.1572	.1193	57.40
3.44	.01428	.04808	.2970	.5450	.1558	.1183	57.56

Table 2 (Continued)

M	$\frac{p}{p_0}$	$\frac{\rho}{\rho_0}$	$\frac{T}{T_0}$	$\frac{a}{a_0}$	$\frac{A^*}{A}$	$\frac{\frac{\rho}{2}V^2}{p_0}$	θ
3.45	.01408	.04759	.2958	.5439	.1543	.1173	57.73
3.46	.01388	.04711	.2946	.5428	.1529	.1163	57.89
3.47	.01368	.04663	.2934	.5417	.1515	.1153	58.05
3.48	.01349	.04616	.2922	.5406	.1501	.1144	58.21
3.49	.01330	.04569	.2910	.5395	.1487	.1134	58.37
3.50	.01311	.04523	.2899	.5384	.1473	.1124	58.53
3.60	.01138	.04089	.2784	.5276	.1342	.1033	60.09
3.70	9.903×10^{-3}	.03702	.2675	.5172	.1224	.09490	61.60
3.80	8.629×10^{-3}	.03355	.2572	.5072	.1117	.08722	63.04
3.90	7.532×10^{-3}	.03044	.2474	.4974	.1021	.08019	64.44
4.00	6.586×10^{-3}	.02766	.2381	.4880	.09329	.07376	65.78
4.10	5.769×10^{-3}	.02516	.2293	.4788	.08536	.06788	67.08
4.20	5.062×10^{-3}	.02292	.2208	.4699	.07818	.06251	68.33
4.30	4.449×10^{-3}	.02090	.2129	.4614	.07166	.05759	69.54
4.40	3.918×10^{-3}	.01909	.2053	.4531	.06575	.05309	70.71
4.50	3.455×10^{-3}	.01745	.1980	.4450	.06038	.04898	71.83
4.60	3.053×10^{-3}	.01597	.1911	.4372	.05550	.04521	72.92
4.70	2.701×10^{-3}	.01464	.1846	.4296	.05107	.04177	73.97
4.80	2.394×10^{-3}	.01343	.1783	.4223	.04703	.03861	74.99
4.90	2.126×10^{-3}	.01233	.1724	.4152	.04335	.03572	75.97
5.00	1.890×10^{-3}	.01134	.1667	.4082	.04000	.03308	76.92
6.00	6.334×10^{-4}	5.194×10^{-3}	.1220	.3492	.01880	.01596	84.96
7.00	2.416×10^{-4}	2.609×10^{-3}	.09259	.3043	9.602×10^{-3}	8.285×10^{-3}	90.97
8.00	1.024×10^{-4}	1.414×10^{-3}	.07246	.2692	5.260×10^{-3}	4.589×10^{-3}	95.62
9.00	4.739×10^{-5}	8.150×10^{-4}	.05814	.2411	3.056×10^{-3}	2.687×10^{-3}	99.32
10.00	2.356×10^{-5}	4.948×10^{-4}	.04762	.2182	1.866×10^{-3}	1.649×10^{-3}	102.3
100.00	2.790×10^{-12}	5.583×10^{-9}	4.998×10^{-4}	.02236	2.157×10^{-8}	1.953×10^{-8}	127.6
∞	0	0	0	0	0	0	130.5

Table 3

Parameters for shock flow, $\gamma = 1 \cdot 4$

M_{1n}	p_2/p_1	ρ_2/ρ_1	T_2/T_1	a_2/a_1	p_2^0/p_1^0	M_2 for Normal Shocks Only
1.00	1.000	1.000	1.000	1.000	1.0000	1.0000
1.01	1.023	1.017	1.007	1.003	1.0000	.9901
1.02	1.047	1.033	1.013	1.007	1.0000	.9805
1.03	1.071	1.050	1.020	1.010	1.0000	.9712
1.04	1.095	1.067	1.026	1.013	.9999	.9620
1.05	1.120	1.084	1.033	1.016	.9999	.9531
1.06	1.144	1.101	1.039	1.019	.9998	.9444
1.07	1.169	1.118	1.046	1.023	.9996	.9360
1.08	1.194	1.135	1.052	1.026	.9994	.9277
1.09	1.219	1.152	1.059	1.029	.9992	.9196
1.10	1.245	1.169	1.065	1.032	.9989	.9118
1.11	1.271	1.186	1.071	1.035	.9986	.9041
1.12	1.297	1.203	1.078	1.038	.9982	.8966
1.13	1.323	1.221	1.084	1.041	.9978	.8892
1.14	1.350	1.238	1.090	1.044	.9973	.8820
1.15	1.376	1.255	1.097	1.047	.9967	.8750
1.16	1.403	1.272	1.103	1.050	.9961	.8682
1.17	1.430	1.290	1.109	1.053	.9953	.8615
1.18	1.458	1.307	1.115	1.056	.9946	.8549
1.19	1.485	1.324	1.122	1.059	.9937	.8485
1.20	1.513	1.342	1.128	1.062	.9928	.8422
1.21	1.541	1.359	1.134	1.065	.9918	.8360
1.22	1.570	1.376	1.141	1.068	.9907	.8300
1.23	1.598	1.394	1.147	1.071	.9896	.8241
1.24	1.627	1.411	1.153	1.074	.9884	.8183
1.25	1.656	1.429	1.159	1.077	.9871	.8126
1.26	1.686	1.446	1.166	1.080	.9857	.8071
1.27	1.715	1.463	1.172	1.083	.9842	.8016
1.28	1.745	1.481	1.178	1.085	.9827	.7963
1.29	1.775	1.498	1.185	1.088	.9811	.7911
1.30	1.805	1.516	1.191	1.091	.9794	.7860
1.31	1.835	1.533	1.197	1.094	.9776	.7809
1.32	1.866	1.551	1.204	1.097	.9758	.7760
1.33	1.897	1.568	1.210	1.100	.9738	.7712
1.34	1.928	1.585	1.216	1.103	.9718	.7664
1.35	1.960	1.603	1.223	1.106	.9697	.7618
1.36	1.991	1.620	1.229	1.109	.9676	.7572
1.37	2.023	1.638	1.235	1.111	.9653	.7527
1.38	2.055	1.655	1.242	1.114	.9630	.7483
1.39	2.087	1.672	1.248	1.117	.9606	.7440
1.40	2.120	1.690	1.255	1.120	.9582	.7397
1.41	2.153	1.707	1.261	1.123	.9557	.7355
1.42	2.186	1.724	1.268	1.126	.9531	.7314
1.43	2.219	1.742	1.274	1.129	.9504	.7274
1.44	2.253	1.759	1.281	1.132	.9476	.7235
1.45	2.286	1.776	1.287	1.135	.9448	.7196
1.46	2.320	1.793	1.294	1.137	.9420	.7157
1.47	2.354	1.811	1.300	1.140	.9390	.7120
1.48	2.389	1.828	1.307	1.143	.9360	.7083
1.49	2.423	1.845	1.314	1.146	.9329	.7047
1.50	2.458	1.862	1.320	1.149	.9298	.7011
1.51	2.493	1.879	1.327	1.152	.9266	.6976
1.52	2.529	1.896	1.334	1.155	.9233	.6941
1.53	2.564	1.913	1.340	1.158	.9200	.6907
1.54	2.600	1.930	1.347	1.161	.9166	.6874
1.55	2.636	1.947	1.354	1.164	.9132	.6841
1.56	2.673	1.964	1.361	1.166	.9097	.6809
1.57	2.709	1.981	1.367	1.169	.9061	.6777
1.58	2.746	1.998	1.374	1.172	.9026	.6746
1.59	2.783	2.015	1.381	1.175	.8989	.6715
1.60	2.820	2.032	1.388	1.178	.8952	.6684
1.61	2.857	2.049	1.395	1.181	.8914	.6655
1.62	2.895	2.065	1.402	1.184	.8877	.6625
1.63	2.933	2.082	1.409	1.187	.8838	.6596
1.64	2.971	2.099	1.416	1.190	.8799	.6568
1.65	3.010	2.115	1.423	1.193	.8760	.6540
1.66	3.048	2.132	1.430	1.196	.8720	.6512
1.67	3.087	2.148	1.437	1.199	.8680	.6485
1.68	3.126	2.165	1.444	1.202	.8640	.6458
1.69	3.165	2.181	1.451	1.205	.8599	.6431

Table 3 (Continued)

M_{1n}	p_2/p_1	ρ_2/ρ_1	T_2/T_1	a_2/a_1	p_2^0/p_1^0	M_2 for Normal Shocks Only
1.70	3.205	2.198	1.458	1.208	.8557	.6405
1.71	3.245	2.214	1.466	1.211	.8516	.6380
1.72	3.285	2.230	1.473	1.214	.8474	.6355
1.73	3.325	2.247	1.480	1.217	.8431	.6330
1.74	3.366	2.263	1.487	1.220	.8389	.6305
1.75	3.406	2.279	1.495	1.223	.8346	.6281
1.76	3.447	2.295	1.502	1.226	.8302	.6257
1.77	3.488	2.311	1.509	1.229	.8259	.6234
1.78	3.530	2.327	1.517	1.232	.8215	.6210
1.79	3.571	2.343	1.524	1.235	.8171	.6188
1.80	3.613	2.359	1.532	1.238	.8127	.6165
1.81	3.655	2.375	1.539	1.241	.8082	.6143
1.82	3.698	2.391	1.547	1.244	.8038	.6121
1.83	3.740	2.407	1.554	1.247	.7993	.6099
1.84	3.783	2.422	1.562	1.250	.7948	.6078
1.85	3.826	2.438	1.569	1.253	.7902	.6057
1.86	3.870	2.454	1.577	1.256	.7857	.6036
1.87	3.913	2.469	1.585	1.259	.7811	.6016
1.88	3.957	2.485	1.592	1.262	.7765	.5996
1.89	4.001	2.500	1.600	1.265	.7720	.5976
1.90	4.045	2.516	1.608	1.268	.7674	.5956
1.91	4.089	2.531	1.616	1.271	.7628	.5937
1.92	4.134	2.546	1.624	1.274	.7581	.5918
1.93	4.179	2.562	1.631	1.277	.7535	.5899
1.94	4.224	2.577	1.639	1.280	.7488	.5880
1.95	4.270	2.592	1.647	1.283	.7442	.5862
1.96	4.315	2.607	1.655	1.287	.7395	.5844
1.97	4.361	2.622	1.663	1.290	.7349	.5826
1.98	4.407	2.637	1.671	1.293	.7302	.5808
1.99	4.453	2.652	1.679	1.296	.7255	.5791
2.00	4.500	2.667	1.688	1.299	.7209	.5773
2.01	4.547	2.681	1.696	1.302	.7162	.5757
2.02	4.594	2.696	1.704	1.305	.7115	.5740
2.03	4.641	2.711	1.712	1.308	.7069	.5723
2.04	4.689	2.725	1.720	1.312	.7022	.5707
2.05	4.736	2.740	1.729	1.315	.6975	.5691
2.06	4.784	2.755	1.737	1.318	.6928	.5675
2.07	4.832	2.769	1.745	1.321	.6882	.5659
2.08	4.881	2.783	1.754	1.324	.6835	.5643
2.09	4.929	2.798	1.762	1.327	.6789	.5628
2.10	4.978	2.812	1.770	1.331	.6742	.5613
2.11	5.027	2.826	1.779	1.334	.6696	.5598
2.12	5.077	2.840	1.787	1.337	.6649	.5583
2.13	5.126	2.854	1.796	1.340	.6603	.5568
2.14	5.176	2.868	1.805	1.343	.6557	.5554
2.15	5.226	2.882	1.813	1.347	.6511	.5540
2.16	5.277	2.896	1.822	1.350	.6464	.5525
2.17	5.327	2.910	1.831	1.353	.6419	.5511
2.18	5.378	2.924	1.839	1.356	.6373	.5498
2.19	5.429	2.938	1.848	1.359	.6327	.5484
2.20	5.480	2.951	1.857	1.363	.6281	.5471
2.21	5.531	2.965	1.866	1.366	.6236	.5457
2.22	5.583	2.978	1.875	1.369	.6191	.5444
2.23	5.635	2.992	1.883	1.372	.6145	.5431
2.24	5.687	3.005	1.892	1.376	.6100	.5418
2.25	5.740	3.019	1.901	1.379	.6055	.5406
2.26	5.792	3.032	1.910	1.382	.6011	.5393
2.27	5.845	3.045	1.919	1.385	.5966	.5381
2.28	5.898	3.058	1.929	1.389	.5921	.5368
2.29	5.951	3.071	1.938	1.392	.5877	.5356
2.30	6.005	3.085	1.947	1.395	.5833	.5344
2.31	6.059	3.098	1.956	1.399	.5789	.5332
2.32	6.113	3.110	1.965	1.402	.5745	.5321
2.33	6.167	3.123	1.974	1.405	.5702	.5309
2.34	6.222	3.136	1.984	1.408	.5658	.5297
2.35	6.276	3.149	1.993	1.412	.5615	.5286
2.36	6.331	3.162	2.002	1.415	.5572	.5275
2.37	6.386	3.174	2.012	1.418	.5529	.5264
2.38	6.442	3.187	2.021	1.422	.5486	.5253
2.39	6.497	3.199	2.031	1.425	.5444	.5242

Table 3 (Continued)

M_{1n}	p_2/p_1	ρ_2/ρ_1	T_2/T_1	a_2/a_1	p_2^0/p_1^0	M_2 for Normal Shocks Only
2.40	6.553	3.212	2.040	1.428	.5401	.5231
2.41	6.609	3.224	2.050	1.432	.5359	.5221
2.42	6.666	3.237	2.059	1.435	.5317	.5210
2.43	6.722	3.249	2.069	1.438	.5276	.5200
2.44	6.779	3.261	2.079	1.442	.5234	.5189
2.45	6.836	3.273	2.088	1.445	.5193	.5179
2.46	6.894	3.285	2.098	1.449	.5152	.5169
2.47	6.951	3.298	2.108	1.452	.5111	.5159
2.48	7.009	3.310	2.118	1.455	.5071	.5149
2.49	7.067	3.321	2.128	1.459	.5030	.5140
2.50	7.125	3.333	2.138	1.462	.4990	.5130
2.51	7.183	3.345	2.147	1.465	.4950	.5120
2.52	7.242	3.357	2.157	1.469	.4911	.5111
2.53	7.301	3.369	2.167	1.472	.4871	.5102
2.54	7.360	3.380	2.177	1.476	.4832	.5092
2.55	7.420	3.392	2.187	1.479	.4793	.5083
2.56	7.479	3.403	2.198	1.482	.4754	.5074
2.57	7.539	3.415	2.208	1.486	.4715	.5065
2.58	7.599	3.426	2.218	1.489	.4677	.5056
2.59	7.659	3.438	2.228	1.493	.4639	.5047
2.60	7.720	3.449	2.238	1.496	.4601	.5039
2.61	7.781	3.460	2.249	1.500	.4564	.5030
2.62	7.842	3.471	2.259	1.503	.4526	.5022
2.63	7.903	3.483	2.269	1.506	.4489	.5013
2.64	7.965	3.494	2.280	1.510	.4452	.5005
2.65	8.026	3.505	2.290	1.513	.4416	.4996
2.66	8.088	3.516	2.301	1.517	.4379	.4988
2.67	8.150	3.527	2.311	1.520	.4343	.4980
2.68	8.213	3.537	2.322	1.524	.4307	.4972
2.69	8.275	3.548	2.332	1.527	.4271	.4964
2.70	8.338	3.559	2.343	1.531	.4236	.4956
2.71	8.401	3.570	2.354	1.534	.4201	.4949
2.72	8.465	3.580	2.364	1.538	.4166	.4941
2.73	8.528	3.591	2.375	1.541	.4131	.4933
2.74	8.592	3.601	2.386	1.545	.4097	.4926
2.75	8.656	3.612	2.397	1.548	.4062	.4918
2.76	8.721	3.622	2.407	1.552	.4028	.4911
2.77	8.785	3.633	2.418	1.555	.3994	.4903
2.78	8.850	3.643	2.429	1.559	.3961	.4896
2.79	8.915	3.653	2.440	1.562	.3928	.4889
2.80	8.980	3.664	2.451	1.566	.3895	.4882
2.81	9.045	3.674	2.462	1.569	.3862	.4875
2.82	9.111	3.684	2.473	1.573	.3829	.4868
2.83	9.177	3.694	2.484	1.576	.3797	.4861
2.84	9.243	3.704	2.496	1.580	.3765	.4854
2.85	9.310	3.714	2.507	1.583	.3733	.4847
2.86	9.376	3.724	2.518	1.587	.3701	.4840
2.87	9.443	3.734	2.529	1.590	.3670	.4833
2.88	9.510	3.743	2.540	1.594	.3639	.4827
2.89	9.577	3.753	2.552	1.597	.3608	.4820
2.90	9.645	3.763	2.563	1.601	.3577	.4814
2.91	9.713	3.773	2.575	1.605	.3547	.4807
2.92	9.781	3.782	2.586	1.608	.3517	.4801
2.93	9.849	3.792	2.598	1.612	.3487	.4795
2.94	9.918	3.801	2.609	1.615	.3457	.4788
2.95	9.986	3.811	2.621	1.619	.3428	.4782
2.96	10.06	3.820	2.632	1.622	.3398	.4776
2.97	10.12	3.829	2.644	1.626	.3369	.4770
2.98	10.19	3.839	2.656	1.630	.3340	.4764
2.99	10.26	3.848	2.667	1.633	.3312	.4758
3.00	10.33	3.857	2.679	1.637	.3283	.4752
3.10	11.05	3.947	2.799	1.673	.3012	.4695
3.20	11.78	4.031	2.922	1.709	.2762	.4643
3.30	12.54	4.112	3.049	1.746	.2533	.4596
3.40	13.32	4.188	3.180	1.783	.2322	.4552
3.50	14.13	4.261	3.315	1.821	.2129	.4512
3.60	14.95	4.330	3.454	1.858	.1953	.4474
3.70	15.80	4.395	3.596	1.896	.1792	.4439
3.80	16.68	4.457	3.743	1.935	.1645	.4407
3.90	17.58	4.516	3.893	1.973	.1510	.4377

Table 3 (Continued)

M_{1n}	p_2/p_1	ρ_2/ρ_1	T_2/T_1	a_2/a_1	p_2^0/p_1^0	M_2 for Normal Shocks Only
4.00	18.50	4.571	4.047	2.012	.1388	.4350
5.00	29.00	5.000	5.800	2.408	.06172	.4152
6.00	41.83	5.268	7.941	2.818	.02965	.4042
7.00	57.00	5.444	10.47	3.236	.01535	.3974
8.00	74.50	5.565	13.39	3.659	8.488×10^{-3}	.3929
9.00	94.33	5.651	16.69	4.086	4.964×10^{-3}	.3898
10.00	116.5	5.714	20.39	4.515	3.045×10^{-3}	.3876
100.00	11,666.5	5.997	1945.4	44.11	3.593×10^{-8}	.3781
∞	∞	6	∞	∞	0	.3780

Table 4

Fanno line–one-dimensional, adiabatic, constant-area flow of a perfect gas.
(Constant specific heat and molecular weight)
$\gamma = 1{\cdot}4$

M	$\frac{T}{T^*}$	$\frac{p}{p^*}$	$\frac{p_0}{p_0^*}$	$\frac{V}{V^*}$	$\frac{I}{I^*}$	$\frac{fL_{max}}{D}$
0	1.2000	∞	∞	0	∞	∞
0.01	1.2000	109.544	57.874	.01095	45.650	7134.40
.02	1.1999	54.770	28.942	.02191	22.834	1778.45
.03	1.1998	36.511	19.300	.03286	15.232	787.08
.04	1.1996	27.382	14.482	.04381	11.435	440.35
.05	1.1994	21.903	11.5914	.05476	9.1584	280.02
.06	1.1991	18.251	9.6659	.06570	7.6428	193.03
.07	1.1988	15.642	8.2915	.07664	6.5620	140.66
.08	1.1985	13.684	7.2616	.08758	5.7529	106.72
.09	1.1981	12.162	6.4614	.09851	5.1249	83.496
.10	1.1976	10.9435	5.8218	.10943	4.6236	66.922
.11	1.1971	9.9465	5.2992	.12035	4.2146	54.688
.12	1.1966	9.1156	4.8643	.13126	3.8747	45.408
.13	1.1960	8.4123	4.4968	.14216	3.5880	38.207
.14	1.1953	7.8093	4.1824	.15306	3.3432	32.511
.15	1.1946	7.2866	3.9103	.16395	3.1317	27.932
.16	1.1939	6.8291	3.6727	.17482	2.9474	24.198
.17	1.1931	6.4252	3.4635	.18568	2.7855	21.115
.18	1.1923	6.0662	3.2779	.19654	2.6422	18.543
.19	1.1914	5.7448	3.1123	.20739	2.5146	16.375
.20	1.1905	5.4555	2.9635	.21822	2.4004	14.533
.21	1.1895	5.1936	2.8293	.22904	2.2976	12.956
.22	1.1885	4.9554	2.7076	.23984	2.2046	11.596
.23	1.1874	4.7378	2.5968	.25063	2.1203	10.416
.24	1.1863	4.5383	2.4956	.26141	2.0434	9.3865
.25	1.1852	4.3546	2.4027	.27217	1.9732	8.4834
.26	1.1840	4.1850	2.3173	.28291	1.9088	7.6876
.27	1.1828	4.0280	2.2385	.29364	1.8496	6.9832
.28	1.1815	3.8820	2.1656	.30435	1.7950	6.3572
.29	1.1802	3.7460	2.0979	.31504	1.7446	5.7989
.30	1.1788	3.6190	2.0351	.32572	1.6979	5.2992
.31	1.1774	3.5002	1.9765	.33637	1.6546	4.8507
.32	1.1759	3.3888	1.9219	.34700	1.6144	4.4468
.33	1.1744	3.2840	1.8708	.35762	1.5769	4.0821
.34	1.1729	3.1853	1.8229	.36822	1.5420	3.7520
.35	1.1713	3.0922	1.7780	.37880	1.5094	3.4525
.36	1.1697	3.0042	1.7358	.38935	1.4789	3.1801
.37	1.1680	2.9209	1.6961	.39988	1.4503	2.9320
.38	1.1663	2.8420	1.6587	.41039	1.4236	2.7055
.39	1.1646	2.7671	1.6234	.42087	1.3985	2.4983

Table 4 (Continued)

$\gamma = 1{\cdot}4$

M	$\frac{T}{T^*}$	$\frac{p}{p^*}$	$\frac{p_0}{p_0^*}$	$\frac{V}{V^*}$	$\frac{I}{I^*}$	$\frac{fL_{max}}{D}$
0.40	1.1628	2.6958	1.5901	.43133	1.3749	2.3085
.41	1.1610	2.6280	1.5587	.44177	1.3527	2.1344
.42	1.1591	2.5634	1.5289	.45218	1.3318	1.9744
.43	1.1572	2.5017	1.5007	.46257	1.3122	1.8272
.44	1.1553	2.4428	1.4739	.47293	1.2937	1.6915
.45	1.1533	2.3865	1.4486	.48326	1.2763	1.5664
.46	1.1513	2.3326	1.4246	.49357	1.2598	1.4509
.47	1.1492	2.2809	1.4018	.50385	1.2443	1.3442
.48	1.1471	2.2314	1.3801	.51410	1.2296	1.2453
.49	1.1450	2.1838	1.3595	.52433	1.2158	1.1539
.50	1.1429	2.1381	1.3399	.53453	1.2027	1.06908
.51	1.1407	2.0942	1.3212	.54469	1.1903	.99042
.52	1.1384	2.0519	1.3034	.55482	1.1786	.91741
.53	1.1362	2.0112	1.2864	.56493	1.1675	.84963
.54	1.1339	1.9719	1.2702	.57501	1.1571	.78662
.55	1.1315	1.9341	1.2549	.58506	1.1472	.72805
.56	1.1292	1.8976	1.2403	.59507	1.1378	.67357
.57	1.1268	1.8623	1.2263	.60505	1.1289	.62286
.58	1.1244	1.8282	1.2130	.61500	1.1205	.57568
.59	1.1219	1.7952	1.2003	.62492	1.1126	.53174
.60	1.1194	1.7634	1.1882	.63481	1.10504	.49081
.61	1.1169	1.7325	1.1766	.64467	1.09793	.45270
.62	1.1144	1.7026	1.1656	.65449	1.09120	.41720
.63	1.1118	1.6737	1.1551	.66427	1.08485	.38411
.64	1.1091	1.6456	1.1451	.67402	1.07883	.35330
.65	1.10650	1.6183	1.1356	.68374	1.07314	.32460
.66	1.10383	1.5919	1.1265	.69342	1.06777	.29785
.67	1.10114	1.5662	1.1179	.70306	1.06271	.27295
.68	1.09842	1.5413	1.1097	.71267	1.05792	.24978
.69	1.09567	1.5170	1.1018	.72225	1.05340	.22821
.70	1.09290	1.4934	1.09436	.73179	1.04915	.20814
.71	1.09010	1.4705	1.08729	.74129	1.04514	.18949
.72	1.08727	1.4482	1.08057	.75076	1.04137	.17215
.73	1.08442	1.4265	1.07419	.76019	1.03783	.15606
.74	1.08155	1.4054	1.06815	.76958	1.03450	.14113
.75	1.07865	1.3848	1.06242	.77893	1.03137	.12728
.76	1.07573	1.3647	1.05700	.78825	1.02844	.11446
.77	1.07279	1.3451	1.05188	.79753	1.02570	.10262
.78	1.06982	1.3260	1.04705	.80677	1.02314	.09167
.79	1.06684	1.3074	1.04250	.81598	1.02075	.08159

$\gamma = 1{\cdot}4$

M	$\frac{T}{T^*}$	$\frac{p}{p^*}$	$\frac{p_0}{p_0^*}$	$\frac{V}{V^*}$	$\frac{I}{I^*}$	$\frac{fL_{max}}{D}$
0.80	1.06383	1.2892	1.03823	.82514	1.01853	.07229
.81	1.06080	1.2715	1.03422	.83426	1.01646	.06375
.82	1.05775	1.2542	1.03047	.84334	1.01455	.05593
.83	1.05468	1.2373	1.02696	.85239	1.01278	.04878
.84	1.05160	1.2208	1.02370	.86140	1.01115	.04226
.85	1.04849	1.2047	1.02067	.87037	1.00966	.03632
.86	1.04537	1.1889	1.01787	.87929	1.00829	.03097
.87	1.04223	1.1735	1.01529	.88818	1.00704	.02613
.88	1.03907	1.1584	1.01294	.89703	1.00591	.02180
.89	1.03589	1.1436	1.01080	.90583	1.00490	.01793
.90	1.03270	1.12913	1.00887	.91459	1.00399	.014513
.91	1.02950	1.11500	1.00714	.92332	1.00318	.011519
.92	1.02627	1.10114	1.00560	.93201	1.00248	.008916
.93	1.02304	1.08758	1.00426	.94065	1.00188	.006694
.94	1.01978	1.07430	1.00311	.94925	1.00136	.004815
.95	1.01652	1.06129	1.00215	.95782	1.00093	.003280
.96	1.01324	1.04854	1.00137	.96634	1.00059	.002056
.97	1.00995	1.03605	1.00076	.97481	1.00033	.001135
.98	1.00664	1.02379	1.00033	.98324	1.00014	.000493
.99	1.00333	1.01178	1.00008	.99164	1.00003	.000120
1.00	1.00000	1.00000	1.00000	1.00000	1.00000	0
1.01	.99666	.98844	1.00008	1.00831	1.00003	.000114
1.02	.99331	.97711	1.00033	1.01658	1.00013	.000458
1.03	.98995	.96598	1.00073	1.02481	1.00030	.001013
1.04	.98658	.95506	1.00130	1.03300	1.00053	.001771
1.05	.98320	.94435	1.00203	1.04115	1.00082	.002712
1.06	.97982	.93383	1.00291	1.04925	1.00116	.003837
1.07	.97642	.92350	1.00394	1.05731	1.00155	.005129
1.08	.97302	.91335	1.00512	1.06533	1.00200	.006582
1.09	.96960	.90338	1.00645	1.07331	1.00250	.008185
1.10	.96618	.89359	1.00793	1.08124	1.00305	.009933
1.11	.96276	.88397	1.00955	1.08913	1.00365	.011813
1.12	.95933	.87451	1.01131	1.09698	1.00429	.013824
1.13	.95589	.86522	1.01322	1.10479	1.00497	.015949
1.14	.95244	.85608	1.01527	1.11256	1.00569	.018187
1.15	.94899	.84710	1.01746	1.1203	1.00646	.02053
1.16	.94554	.83827	1.01978	1.1280	1.00726	.02298
1.17	.94208	.82958	1.02224	1.1356	1.00810	.02552
1.18	.93862	.82104	1.02484	1.1432	1.00897	.02814
1.19	.93515	.81263	1.02757	1.1508	1.00988	.03085

Table 4 (Continued)

$\gamma = 1{\cdot}4$

M	T/T^*	p/p^*	p_0/p_0^*	V/V^*	I/I^*	fL_{max}/D
1.20	.93168	.80436	1.03044	1.1583	1.01082	.03364
1.21	.92820	.79623	1.03344	1.1658	1.01178	.03650
1.22	.92473	.78822	1.03657	1.1732	1.01278	.03942
1.23	.92125	.78034	1.03983	1.1806	1.01381	.04241
1.24	.91777	.77258	1.04323	1.1879	1.01486	.04547
1.25	.91429	.76495	1.04676	1.1952	1.01594	.04858
1.26	.91080	.75743	1.05041	1.2025	1.01705	.05174
1.27	.90732	.75003	1.05419	1.2097	1.01818	.05494
1.28	.90383	.74274	1.05809	1.2169	1.01933	.05820
1.29	.90035	.73556	1.06213	1.2240	1.02050	.06150
1.30	.89686	.72848	1.06630	1.2311	1.02169	.06483
1.31	.89338	.72152	1.07060	1.2382	1.02291	.06820
1.32	.88989	.71465	1.07502	1.2452	1.02415	.07161
1.33	.88641	.70789	1.07957	1.2522	1.02540	.07504
1.34	.88292	.70123	1.08424	1.2591	1.02666	.07850
1.35	.87944	.69466	1.08904	1.2660	1.02794	.08199
1.36	.87596	.68818	1.09397	1.2729	1.02924	.08550
1.37	.87249	.68180	1.09902	1.2797	1.03056	.08904
1.38	.86901	.67551	1.10419	1.2864	1.03189	.09259
1.39	.86554	.66931	1.10948	1.2932	1.03323	.09616
1.40	.86207	.66320	1.1149	1.2999	1.03458	.09974
1.41	.85860	.65717	1.1205	1.3065	1.03595	.10333
1.42	.85514	.65122	1.1262	1.3131	1.03733	.10694
1.43	.85168	.64536	1.1320	1.3197	1.03872	.11056
1.44	.84822	.63958	1.1379	1.3262	1.04012	.11419
1.45	.84477	.63387	1.1440	1.3327	1.04153	.11782
1.46	.84133	.62824	1.1502	1.3392	1.04295	.12146
1.47	.83788	.62269	1.1565	1.3456	1.04438	.12510
1.48	.83445	.61722	1.1629	1.3520	1.04581	.12875
1.49	.83101	.61181	1.1695	1.3583	1.04725	.13240
1.50	.82759	.60648	1.1762	1.3646	1.04870	.13605
1.51	.82416	.60122	1.1830	1.3708	1.05016	.13970
1.52	.82075	.59602	1.1899	1.3770	1.05162	.14335
1.53	.81734	.59089	1.1970	1.3832	1.05309	.14699
1.54	.81394	.58583	1.2043	1.3894	1.05456	.15063
1.55	.81054	.58084	1.2116	1.3955	1.05604	.15427
1.56	.80715	.57591	1.2190	1.4015	1.05752	.15790
1.57	.80376	.57104	1.2266	1.4075	1.05900	.16152
1.58	.80038	.56623	1.2343	1.4135	1.06049	.16514
1.59	.79701	.56148	1.2422	1.4195	1.06198	.16876

$\gamma = 1{\cdot}4$

M	T/T^*	p/p^*	p_0/p_0^*	V/V^*	I/I^*	fL_{max}/D
1.60	.79365	.55679	1.2502	1.4254	1.06348	.17236
1.61	.79030	.55216	1.2583	1.4313	1.06498	.17595
1.62	.78695	.54759	1.2666	1.4371	1.06648	.17953
1.63	.78361	.54308	1.2750	1.4429	1.06798	.18311
1.64	.78028	.53862	1.2835	1.4487	1.06948	.18667
1.65	.77695	.53421	1.2922	1.4544	1.07098	.19022
1.66	.77363	.52986	1.3010	1.4601	1.07249	.19376
1.67	.77033	.52556	1.3099	1.4657	1.07399	.19729
1.68	.76703	.52131	1.3190	1.4713	1.07550	.20081
1.69	.76374	.51711	1.3282	1.4769	1.07701	.20431
1.70	.76046	.51297	1.3376	1.4825	1.07851	.20780
1.71	.75718	.50887	1.3471	1.4880	1.08002	.21128
1.72	.75392	.50482	1.3567	1.4935	1.08152	.21474
1.73	.75067	.50082	1.3665	1.4989	1.08302	.21819
1.74	.74742	.49686	1.3764	1.5043	1.08453	.22162
1.75	.74419	.49295	1.3865	1.5097	1.08603	.22504
1.76	.74096	.48909	1.3967	1.5150	1.08753	.22844
1.77	.73774	.48527	1.4070	1.5203	1.08903	.23183
1.78	.73453	.48149	1.4175	1.5256	1.09053	.23520
1.79	.73134	.47776	1.4282	1.5308	1.09202	.23855
1.80	.72816	.47407	1.4390	1.5360	1.09352	.24189
1.81	.72498	.47042	1.4499	1.5412	1.09500	.24521
1.82	.72181	.46681	1.4610	1.5463	1.09649	.24851
1.83	.71865	.46324	1.4723	1.5514	1.09798	.25180
1.84	.71551	.45972	1.4837	1.5564	1.09946	.25507
1.85	.71238	.45623	1.4952	1.5614	1.1009	.25832
1.86	.70925	.45278	1.5069	1.5664	1.1024	.26156
1.87	.70614	.44937	1.5188	1.5714	1.1039	.26478
1.88	.70304	.44600	1.5308	1.5763	1.1054	.26798
1.89	.69995	.44266	1.5429	1.5812	1.1068	.27116
1.90	.69686	.43936	1.5552	1.5861	1.1083	.27433
1.91	.69379	.43610	1.5677	1.5909	1.1097	.27748
1.92	.69074	.43287	1.5804	1.5957	1.1112	.28061
1.93	.68769	.42967	1.5932	1.6005	1.1126	.28372
1.94	.68465	.42651	1.6062	1.6052	1.1141	.28681
1.95	.68162	.42339	1.6193	1.6099	1.1155	.28989
1.96	.67861	.42030	1.6326	1.6146	1.1170	.29295
1.97	.67561	.41724	1.6461	1.6193	1.1184	.29599
1.98	.67262	.41421	1.6597	1.6239	1.1198	.29901
1.99	.66964	.41121	1.6735	1.6284	1.1213	.30201

Table 4 (Continued)

$\gamma = 1{\cdot}4$

M	$\frac{T}{T^*}$	$\frac{p}{p^*}$	$\frac{p_0}{p_0^*}$	$\frac{V}{V^*}$	$\frac{I}{I^*}$	$\frac{fL_{max}}{D}$
2.00	.66667	.40825	1.6875	1.6330	1.1227	.30499
2.01	.66371	.40532	1.7017	1.6375	1.1241	.30796
2.02	.66076	.40241	1.7160	1.6420	1.1255	.31091
2.03	.65783	.39954	1.7305	1.6465	1.1269	.31384
2.04	.65491	.39670	1.7452	1.6509	1.1283	.31675
2.05	.65200	.39389	1.7600	1.6553	1.1297	.31965
2.06	.64910	.39110	1.7750	1.6597	1.1311	.32253
2.07	.64621	.38834	1.7902	1.6640	1.1325	.32538
2.08	.64333	.38562	1.8056	1.6683	1.1339	.32822
2.09	.64047	.38292	1.8212	1.6726	1.1352	.33104
2.10	.63762	.38024	1.8369	1.6769	1.1366	.33385
2.11	.63478	.37760	1.8528	1.6811	1.1380	.33664
2.12	.63195	.37498	1.8690	1.6853	1.1393	.33940
2.13	.62914	.37239	1.8853	1.6895	1.1407	.34215
2.14	.62633	.36982	1.9018	1.6936	1.1420	.34488
2.15	.62354	.36728	1.9185	1.6977	1.1434	.34760
2.16	.62076	.36476	1.9354	1.7018	1.1447	.35030
2.17	.61799	.36227	1.9525	1.7059	1.1460	.35298
2.18	.61523	.35980	1.9698	1.7099	1.1474	.35564
2.19	.61249	.35736	1.9873	1.7139	1.1487	.35828
2.20	.60976	.35494	2.0050	1.7179	1.1500	.36091
2.21	.60704	.35254	2.0228	1.7219	1.1513	.36352
2.22	.60433	.35017	2.0409	1.7258	1.1526	.36611
2.23	.60163	.34782	2.0592	1.7297	1.1539	.36868
2.24	.59895	.34550	2.0777	1.7336	1.1552	.37124
2.25	.59627	.34319	2.0964	1.7374	1.1565	.37378
2.26	.59361	.34091	2.1154	1.7412	1.1578	.37630
2.27	.59096	.33865	2.1345	1.7450	1.1590	.37881
2.28	.58833	.33641	2.1538	1.7488	1.1603	.38130
2.29	.58570	.33420	2.1733	1.7526	1.1616	.38377
2.30	.58309	.33200	2.1931	1.7563	1.1629	.38623
2.31	.58049	.32983	2.2131	1.7600	1.1641	.38867
2.32	.57790	.32767	2.2333	1.7637	1.1653	.39109
2.33	.57532	.32554	2.2537	1.7673	1.1666	.39350
2.34	.57276	.32342	2.2744	1.7709	1.1678	.39589
2.35	.57021	.32133	2.2953	1.7745	1.1690	.39826
2.36	.56767	.31925	2.3164	1.7781	1.1703	.40062
2.37	.56514	.31720	2.3377	1.7817	1.1715	.40296
2.38	.56262	.31516	2.3593	1.7852	1.1727	.40528
2.39	.56011	.31314	2.3811	1.7887	1.1739	.40760

$\gamma = 1{\cdot}4$

M	$\frac{T}{T^*}$	$\frac{p}{p^*}$	$\frac{p_0}{p_0^*}$	$\frac{V}{V^*}$	$\frac{I}{I^*}$	$\frac{fL_{max}}{D}$
2.40	.55762	.31114	2.4031	1.7922	1.1751	.40989
2.41	.55514	.30916	2.4254	1.7956	1.1763	.41216
2.42	.55267	.30720	2.4479	1.7991	1.1775	.41442
2.43	.55021	.30525	2.4706	1.8025	1.1786	.41667
2.44	.54776	.30332	2.4936	1.8059	1.1798	.41891
2.45	.54533	.30141	2.5168	1.8092	1.1810	.42113
2.46	.54291	.29952	2.5403	1.8126	1.1821	.42333
2.47	.54050	.29765	2.5640	1.8159	1.1833	.42551
2.48	.53810	.29579	2.5880	1.8192	1.1844	.42768
2.49	.53571	.29395	2.6122	1.8225	1.1856	.42983
2.50	.53333	.29212	2.6367	1.8257	1.1867	.43197
2.51	.53097	.29031	2.6615	1.8290	1.1879	.43410
2.52	.52862	.28852	2.6865	1.8322	1.1890	.43621
2.53	.52627	.28674	2.7117	1.8354	1.1901	.43831
2.54	.52394	.28498	2.7372	1.8386	1.1912	.44040
2.55	.52163	.28323	2.7630	1.8417	1.1923	.44247
2.56	.51932	.28150	2.7891	1.8448	1.1934	.44452
2.57	.51702	.27978	2.8154	1.8479	1.1945	.44655
2.58	.51474	.27808	2.8420	1.8510	1.1956	.44857
2.59	.51247	.27640	2.8689	1.8541	1.1967	.45059
2.60	.51020	.27473	2.8960	1.8571	1.1978	.45259
2.61	.50795	.27307	2.9234	1.8602	1.1989	.45457
2.62	.50571	.27143	2.9511	1.8632	1.2000	.45654
2.63	.50349	.26980	2.9791	1.8662	1.2011	.45850
2.64	.50127	.26818	3.0074	1.8691	1.2021	.46044
2.65	.49906	.26658	3.0359	1.8721	1.2031	.46237
2.66	.49687	.26499	3.0647	1.8750	1.2042	.46429
2.67	.49469	.26342	3.0938	1.8779	1.2052	.46619
2.68	.49251	.26186	3.1234	1.8808	1.2062	.46807
2.69	.49035	.26032	3.1530	1.8837	1.2073	.46996
2.70	.48820	.25878	3.1830	1.8865	1.2083	.47182
2.71	.48606	.25726	3.2133	1.8894	1.2093	.47367
2.72	.48393	.25575	3.2440	1.8922	1.2103	.47551
2.73	.48182	.25426	3.2749	1.8950	1.2113	.47734
2.74	.47971	.25278	3.3061	1.8978	1.2123	.47915
2.75	.47761	.25131	3.3376	1.9005	1.2133	.48095
2.76	.47553	.24985	3.3695	1.9032	1.2143	.48274
2.77	.47346	.24840	3.4017	1.9060	1.2153	.48452
2.78	.47139	.24697	3.4342	1.9087	1.2163	.48628
2.79	.46933	.24555	3.4670	1.9114	1.2173	.48803

Table 4 (Continued)

$\gamma = 1{\cdot}4$

M	$\frac{T}{T^*}$	$\frac{p}{p^*}$	$\frac{p_0}{p_0^*}$	$\frac{V}{V^*}$	$\frac{I}{I^*}$	$\frac{fL_{max}}{D}$
2.80	.46729	.24414	3.5001	1.9140	1.2182	.48976
2.81	.46526	.24274	3.5336	1.9167	1.2192	.49148
2.82	.46324	.24135	3.5674	1.9193	1.2202	.49321
2.83	.46122	.23997	3.6015	1.9220	1.2211	.49491
2.84	.45922	.23861	3.6359	1.9246	1.2221	.49660
2.85	.45723	.23726	3.6707	1.9271	1.2230	.49828
2.86	.45525	.23592	3.7058	1.9297	1.2240	.49995
2.87	.45328	.23458	3.7413	1.9322	1.2249	.50161
2.88	.45132	.23326	3.7771	1.9348	1.2258	.50326
2.89	.44937	.23196	3.8133	1.9373	1.2268	.50489
2.90	.44743	.23066	3.8498	1.9398	1.2277	.50651
2.91	.44550	.22937	3.8866	1.9423	1.2286	.50812
2.92	.44358	.22809	3.9238	1.9448	1.2295	.50973
2.93	.44167	.22682	3.9614	1.9472	1.2304	.51133
2.94	.43977	.22556	3.9993	1.9497	1.2313	.51291
2.95	.43788	.22431	4.0376	1.9521	1.2322	.51447
2.96	.43600	.22307	4.0763	1.9545	1.2331	.51603
2.97	.43413	.22185	4.1153	1.9569	1.2340	.51758
2.98	.43226	.22063	4.1547	1.9592	1.2348	.51912
2.99	.43041	.21942	4.1944	1.9616	1.2357	.52064
3.0	.42857	.21822	4.2346	1.9640	1.2366	.52216
3.5	.34783	.16850	6.7896	2.0642	1.2743	.58643
4.0	.28571	.13363	10.719	2.1381	1.3029	.63306
4.5	.23762	.10833	16.562	2.1936	1.3247	.66764
5.0	.20000	.08944	25.000	2.2361	1.3416	.69381
6.0	.14634	.06376	53.180	2.2953	1.3655	.72987
7.0	.11111	.04762	104.14	2.3333	1.3810	.75281
8.0	.08696	.03686	190.11	2.3591	1.3915	.76820
9.0	.06977	.02935	327.19	2.3772	1.3989	.77898
10.0	.05714	.02390	535.94	2.3905	1.4044	.78683
∞	0	0	∞	2.4495	1.4289	.82153

$\gamma = 1{\cdot}0$

M	$\frac{T}{T^*}$	$\frac{p}{p^*}$	$\frac{p_0}{p_0^*}$	$\frac{V}{V^*}$	$\frac{I}{I^*}$	$\frac{fL_{max}}{D}$
0	1.000	∞	∞	0	∞	∞
0.05		20.000	12.146	.0500	10.025	393.01
.10		10.000	6.096	.1000	5.050	94.39
.15		6.667	4.089	.1500	3.408	39.65
.20		5.000	3.094	.2000	2.600	20.78
.25		4.000	2.503	.2500	2.125	12.227
.30		3.333	2.115	.3000	1.817	7.703
.35		2.857	1.842	.3500	1.604	5.064
.40		2.500	1.643	.4000	1.450	3.417
.45		2.222	1.492	.4500	1.336	2.341
.50		2.000	1.375	.5000	1.250	1.614
.55		1.818	1.283	.5500	1.184	1.110
.60		1.667	1.210	.6000	1.133	.7561
.65		1.539	1.153	.6500	1.0942	.5053
.70		1.429	1.107	.7000	1.0643	.3275
.75		1.333	1.0714	.7500	1.0417	.2024
.80		1.250	1.0441	.8000	1.0250	.1162
.85		1.176	1.0240	.8500	1.0132	.05904
.90		1.111	1.0104	.9000	1.0056	.02385
.95		1.0526	1.0026	.9500	1.0013	.00545
1.00		1.0000	1.0000	1.0000	1.0000	0
1.05		.9524	1.0025	1.0500	1.0012	.00461
1.10		.9091	1.0097	1.100	1.0045	.01707
1.15		.8695	1.0217	1.150	1.0098	.03567
1.20		.8333	1.0384	1.200	1.0167	.05909
1.25		.8000	1.0598	1.250	1.0250	.08629
1.30		.7692	1.0862	1.300	1.0346	.1164
1.35		.7407	1.118	1.350	1.0453	.1489
1.40		.7143	1.154	1.400	1.0571	.1831
1.45		.6897	1.196	1.450	1.0698	.2188
1.50		.6667	1.245	1.500	1.0833	.2554
1.55		.6452	1.300	1.550	1.0976	.2927
1.60		.6250	1.363	1.600	1.112	.3306
1.65		.6061	1.434	1.650	1.128	.3689
1.70		.5882	1.514	1.700	1.144	.4073

Table 4 (Continued)

$\gamma = 1{\cdot}0$

M	$\frac{T}{T^*}$	$\frac{p}{p^*}$	$\frac{p_0}{p_0^*}$	$\frac{V}{V^*}$	$\frac{I}{I^*}$	$\frac{fL_{max}}{D}$
1.75	1.000	.5714	1.603	1.750	1.161	.4458
1.80		.5556	1.703	1.800	1.178	.4842
1.85		.5406	1.815	1.850	1.195	.5225
1.90		.5263	1.941	1.900	1.213	.5607
1.95		.5128	2.082	1.950	1.231	.5986
2.00		.5000	2.241	2.000	1.250	.6363
2.05		.4878	2.419	2.050	1.269	.6736
2.10		.4762	2.620	2.100	1.288	.7106
2.15		.4651	2.846	2.150	1.308	.7472
2.20		.4545	3.100	2.200	1.327	7835
2.25		.4444	3.388	2.250	1.347	.8194
2.30		.4348	3.714	2.300	1.367	.8549
2.35		.4256	4.083	2.350	1.388	.8900
2.40		.4167	4.502	2.400	1.408	.9246
2.45		.4082	4.979	2.450	1.429	.9588
2.50		.4000	5.522	2.500	1.450	.9926
2.55		.3922	6.142	2.550	1.471	1.0260
2.60		.3847	6.852	2.600	1.492	1.0590
2.65		.3774	7.665	2.650	1.514	1.0916
2.70		.3704	8.600	2.700	1.535	1.1237
2.75		.3636	9.676	2.750	1.557	1.155
2.80		.3571	10.92	2.800	1.579	1.187
2.85		.3509	12.35	2.850	1.600	1.218
2.90		.3449	14.02	2.900	1.622	1.248
2.95		.3390	15.95	2.950	1.644	1.279
3.00		.3333	18.20	3.000	1.667	1.308
3.50		.2857	79.22	3.500	1.893	1.587
4.00		.2500	452.01	4.000	2.125	1.835
4.50		.2222	3364	4.500	2.361	2.058
5.00		.2000	32550	5.000	2.600	2.259
6.00		.1667	$664(10)^4$	6.000	3.083	2.611
7.00		.1429	$378(10)^7$	7.000	3.571	2.912
8.00		.1250	$599(10)^{10}$	8.000	4.062	3.174
9.00		.1111	$262(10)^{14}$	9.000	4.556	3.407
10.00		.1000	$314(10)^{18}$	10.000	5.050	3.615
∞	1.000	0	∞	∞	∞	∞

$\gamma = 1{\cdot}1$

M	$\frac{T}{T^*}$	$\frac{p}{p^*}$	$\frac{p_0}{p_0^*}$	$\frac{V}{V^*}$	$\frac{I}{I^*}$	$\frac{fL_{max}}{D}$
0	1.0500	∞	∞	0	∞	∞
0.05	1.0499	20.493	11.999	.05123	9.785	357.05
.10	1.0495	10.244	6.023	.1024	4.932	85.65
.15	1.0488	6.828	4.042	.1536	3.332	35.92
.20	1.0479	5.118	3.059	.2047	2.545	18.79
.25	1.0467	4.092	2.476	.2558	2.083	11.03
.30	1.0453	3.408	2.094	.3067	1.784	6.936
.35	1.0436	2.919	1.825	.3575	1.577	4.549
.40	1.0417	2.552	1.628	.4082	1.429	3.062
.45	1.0395	2.266	1.480	.4588	1.319	2.093
.50	1.0370	2.037	1.365	.5092	1.237	1.439
.55	1.0343	1.849	1.275	.5594	1.174	.9871
.60	1.0314	1.693	1.204	.6094	1.125	.6705
.65	1.0283	1.560	1.148	.6591	1.0882	.4468
.70	1.0249	1.446	1.104	.7086	1.0599	.2887
.75	1.0213	1.347	1.0689	.7579	1.0386	.1780
.80	1.0174	1.261	1.0425	.8069	1.0231	.1019
.85	1.0133	1.184	1.0231	.8557	1.0122	.05160
.90	1.0091	1.116	1.0100	.9041	1.0051	.02078
.95	1.0047	1.0551	1.0024	.9522	1.0012	.00472
1.00	1.0000	1.0000	1.0000	1.0000	1.0000	0
1.05	.9951	.9501	1.0023	1.0474	1.0011	.00398
1.10	.9901	.9046	1.0092	1.0945	1.0041	.01468
1.15	.9849	.8630	1.0204	1.1412	1.0087	.03058
1.20	.9795	.8247	1.0360	1.1876	1.0148	.05050
1.25	.9739	.7895	1.0559	1.234	1.0221	.07350
1.30	.9682	.7569	1.0801	1.279	1.0304	.09885
1.35	.9623	.7266	1.109	1.324	1.0397	.1260
1.40	.9563	.6985	1.142	1.369	1.0498	.1544
1.45	.9501	.6722	1.180	1.413	1.0605	.1838
1.50	.9438	.6476	1.223	1.457	1.0717	.2138
1.55	.9374	.6246	1.272	1.501	1.0835	.2443
1.60	.9309	.6030	1.326	1.544	1.0958	.2749
1.65	.9242	.5826	1.387	1.586	1.108	.3056
1.70	.9174	.5634	1.454	1.628	1.121	.3362

Table 4 (Continued)

$\gamma = 1{\cdot}1$

M	$\frac{T}{T^*}$	$\frac{p}{p^*}$	$\frac{p_0}{p_0^*}$	$\frac{V}{V^*}$	$\frac{I}{I^*}$	$\frac{fL_{max}}{D}$
1.75	.9105	.5453	1.528	1.670	1.134	.3667
1.80	.9036	.5281	1.610	1.711	1.148	.3969
1.85	.8966	.5118	1.701	1.752	1.161	.4268
1.90	.8895	.4964	1.801	1.792	1.175	.4563
1.95	.8823	.4817	1.911	1.832	1.189	.4854
2.00	.8750	.4677	2.032	1.871	1.203	.5140
2.05	.8677	.4544	2.165	1.910	1.217	.5422
2.10	.8603	.4417	2.312	1.948	1.231	.5698
2.15	.8529	.4295	2.473	1.986	1.245	.5970
2.20	.8454	.4179	2.651	2.023	1.259	.6237
2.25	.8379	.4068	2.846	2.060	1.273	.6498
2.30	.8304	.3962	3.061	2.096	1.286	.6754
2.35	.8228	.3860	3.299	2.132	1.300	.7005
2.40	.8152	.3762	3.560	2.167	1.314	.7251
2.45	.8076	.3668	3.848	2.202	1.328	.7491
2.50	.8000	.3578	4.165	2.236	1.342	.7726
2.55	.7924	.3491	4.515	2.270	1.355	.7957
2.60	.7848	.3407	4.902	2.303	1.369	.8182
2.65	.7771	.3327	5.328	2.336	1.382	.8402
2.70	.7695	.3249	5.799	2.368	1.395	.8617
2.75	.7619	.3174	6.320	2.400	1.409	.8828
2.80	.7543	.3102	6.895	2.432	1.422	.9034
2.85	.7467	.3032	7.532	2.463	1.434	.9235
2.90	.7392	.2965	8.237	2.493	1.447	.9432
2.95	.7316	.2900	9.016	2.523	1.460	.9624
3.00	.7241	.2837	9.880	2.553	1.472	.9812
3.50	.6512	.2305	25.83	2.824	1.589	1.147
4.00	.5833	.1909	71.74	3.055	1.691	1.280
4.50	.5217	.1605	205.7	3.250	1.779	1.386
5.00	.4667	.1366	597.3	3.416	1.854	1.472
6.00	.3750	.1021	4911	3.674	1.973	1.601
7.00	.3043	.07881	37919	3.862	2.060	1.689
8.00	.2500	.06250	$263(10)^3$	4.000	2.125	1.752
9.00	.2079	.05067	$161(10)^4$	4.104	2.174	1.798
10.00	.1750	.04183	$889(10)^4$	4.183	2.211	1.832
∞	0	0	∞	4.583	2.400	1.997

$\gamma = 1{\cdot}2$

M	$\frac{T}{T^*}$	$\frac{p}{p^*}$	$\frac{p_0}{p_0^*}$	$\frac{V}{V^*}$	$\frac{I}{I^*}$	$\frac{fL_{max}}{D}$
0	1.1000	∞	∞	0	∞	∞
0.05	1.0997	20.974	11.857	.05243	9.562	327.09
.10	1.0989	10.483	5.953	.1048	4.822	78.36
.15	1.0975	6.984	3.996	.1571	3.260	32.81
.20	1.0956	5.234	3.026	.2093	2.493	17.13
.25	1.0932	4.182	2.451	.2614	2.044	10.04
.30	1.0902	3.480	2.073	.3133	1.753	6.298
.35	1.0867	2.978	1.809	.3649	1.553	4.121
.40	1.0827	2.601	1.615	.4162	1.409	2.768
.45	1.0782	2.307	1.469	.4672	1.304	1.887
.50	1.0732	2.072	1.356	.5179	1.224	1.294
.55	1.0677	1.879	1.268	.5683	1.164	.8855
.60	1.0618	1.717	1.199	.6183	1.118	.5999
.65	1.0554	1.581	1.144	.6678	1.0826	.3987
.70	1.0486	1.463	1.100	.7168	1.0561	.2570
.75	1.0414	1.361	1.0666	.7654	1.0360	.1579
.80	1.0338	1.271	1.0410	.8134	1.0214	.09016
.85	1.0259	1.192	1.0222	.8609	1.0112	.04554
.90	1.0176	1.121	1.0096	.9078	1.0047	.01829
.95	1.0089	1.0573	1.0023	.9542	1.0011	.00414
1.00	1.0000	1.0000	1.0000	1.0000	1.0000	0
1.05	.9908	.9480	1.0022	1.0451	1.0010	.00347
1.10	.9813	.9005	1.0087	1.0896	1.0037	.01277
1.15	.9715	.8571	1.0194	1.134	1.0079	.02657
1.20	.9615	.8172	1.0340	1.177	1.0134	.04368
1.25	.9514	.7803	1.0525	1.219	1.0197	.06338
1.30	.9410	.7462	1.0749	1.261	1.0270	.08500
1.35	.9304	.7145	1.101	1.302	1.0351	.1080
1.40	.9197	.6850	1.132	1.342	1.0437	.1320
1.45	.9089	.6575	1.166	1.382	1.0529	.1567
1.50	.8980	.6317	1.205	1.421	1.0625	.1817
1.55	.8869	.6076	1.248	1.459	1.0724	.2069
1.60	.8758	.5849	1.296	1.497	1.0826	.2323
1.65	.8646	.5635	1.349	1.534	1.0930	.2575
1.70	.8534	.5434	1.407	1.570	1.1036	.2825

Table 4 (Continued)

$\gamma = 1{\cdot}2$

M	$\frac{T}{T^*}$	$\frac{p}{p^*}$	$\frac{p_0}{p_0^*}$	$\frac{V}{V^*}$	$\frac{I}{I^*}$	$\frac{fL_{max}}{D}$
1.75	.8421	.5244	1.471	1.606	1.114	.3072
1.80	.8308	.5064	1.540	1.641	1.125	.3316
1.85	.8195	.4894	1.615	1.675	1.136	.3556
1.90	.8082	.4732	1.697	1.708	1.147	.3791
1.95	.7970	.4578	1.787	1.741	1.158	.4021
2.00	.7857	.4432	1.884	1.773	1.168	.4247
2.05	.7745	.4293	1.989	1.804	1.179	.4468
2.10	.7634	.4160	2.103	1.835	1.190	.4684
2.15	.7523	.4034	2.226	1.865	1.201	.4894
2.20	.7413	.3913	2.359	1.894	1.211	.5099
2.25	.7303	.3798	2.504	1.923	1.221	.5299
2.30	.7194	.3688	2.660	1.951	1.232	.5493
2.35	.7086	.3582	2.829	1.978	1.242	.5683
2.40	.6980	.3481	3.011	2.005	1.252	.5868
2.45	.6874	.3384	3.208	2.031	1.262	.6047
2.50	.6769	.3291	3.420	2.057	1.272	.6222
2.55	.6665	.3202	3.650	2.082	1.281	.6392
2.60	.6563	.3116	3.898	2.106	1.291	.6557
2.65	.6462	.3033	4.166	2.130	1.300	.6718
2.70	.6362	.2954	4.455	2.154	1.309	.6874
2.75	.6263	.2878	4.767	2.176	1.318	.7026
2.80	.6166	.2804	5.103	2.199	1.327	.7173
2.85	.6070	.2733	5.466	2.220	1.335	.7316
2.90	.5975	.2665	5.858	2.242	1.344	.7456
2.95	.5882	.2600	6.280	2.263	1.352	.7592
3.00	.5789	.2536	6.735	2.283	1.360	.7724
3.50	.4944	.2009	13.76	2.461	1.434	.8857
4.00	.4231	.1626	28.35	2.602	1.493	.9718
4.50	.3636	.1340	57.96	2.714	1.541	1.0380
5.00	.3143	.1121	116.31	2.803	1.580	1.0896
6.00	.2391	.08150	434.7	2.934	1.637	1.163
7.00	.1864	.06168	1458	3.023	1.677	1.212
8.00	.1486	.04819	4353	3.084	1.704	1.245
9.00	.1209	.03863	13156	3.129	1.724	1.268
10.00	.1000	.03162	29601	3.162	1.739	1.286
∞	0	0	∞	3.317	1.809	1.365

$\gamma = 1{\cdot}3$

M	$\frac{T}{T^*}$	$\frac{p}{p^*}$	$\frac{p_0}{p_0^*}$	$\frac{V}{V^*}$	$\frac{I}{I^*}$	$\frac{fL_{max}}{D}$
0	1.150	∞	∞	0	∞	∞
0.05	1.149	21.444	11.721	.05361	9.354	301.74
.10	1.148	10.716	5.885	.1072	4.720	72.20
.15	1.146	7.137	3.952	.1606	3.194	30.18
.20	1.143	5.346	2.994	.2138	2.445	15.73
.25	1.139	4.270	2.426	.2668	2.007	9.201
.30	1.134	3.551	2.054	.3195	1.724	5.759
.35	1.129	3.036	1.793	.3719	1.530	3.760
.40	1.123	2.649	1.602	.4239	1.391	2.520
.45	1.116	2.348	1.459	.4754	1.289	1.714
.50	1.1084	2.106	1.348	.5264	1.213	1.172
.55	1.1001	1.907	1.261	.5769	1.155	.8004
.60	1.0911	1.741	1.193	.6267	1.111	.5409
.65	1.0815	1.600	1.140	.6759	1.0777	.3586
.70	1.0713	1.479	1.0972	.7245	1.0524	.2305
.75	1.0605	1.373	1.0644	.7724	1.0336	.14131
.80	1.0493	1.280	1.0395	.8195	1.0199	.08044
.85	1.0376	1.198	1.0214	.8658	1.0104	.04053
.90	1.0254	1.125	1.0092	.9113	1.0043	.01623
.95	1.0129	1.0594	1.0022	.9561	1.0010	.00367
1.00	1.0000	1.0000	1.0000	1.0000	1.0000	0
1.05	.9868	.9461	1.0021	1.0430	1.0009	.00305
1.10	.9733	.8969	1.0083	1.0852	1.0033	.01122
1.15	.9596	.8518	1.0183	1.1266	1.0071	.02324
1.20	.9457	.8104	1.0321	1.1670	1.0120	.03820
1.25	.9316	.7722	1.0495	1.206	1.0177	.05524
1.30	.9174	.7368	1.0704	1.245	1.0241	.07388
1.35	.9031	.7039	1.0948	1.283	1.0312	.09365
1.40	.8887	.6734	1.1227	1.320	1.0388	.11417
1.45	.8743	.6448	1.1543	1.356	1.0467	.13513
1.50	.8598	.6182	1.189	1.391	1.0549	.1564
1.55	.8454	.5932	1.228	1.425	1.0634	.1777
1.60	.8309	.5697	1.271	1.458	1.0721	.1989
1.65	.8165	.5477	1.318	1.491	1.0808	.2200
1.70	.8022	.5269	1.369	1.523	1.0897	.2408

Table 4 (Continued)

$\gamma = 1{\cdot}3$

M	$\frac{T}{T^*}$	$\frac{p}{p^*}$	$\frac{p_0}{p_0^*}$	$\frac{V}{V^*}$	$\frac{I}{I^*}$	$\frac{fL_{max}}{D}$
1.75	.7880	.5073	1.424	1.554	1.0986	.2613
1.80	.7739	.4887	1.484	1.584	1.108	.2814
1.85	.7599	.4712	1.549	1.613	1.116	.3010
1.90	.7460	.4546	1.618	1.641	1.125	.3202
1.95	.7323	.4388	1.693	1.669	1.134	.3390
2.00	.7188	.4239	1.773	1.696	1.143	.3573
2.05	.7054	.4097	1.859	1.722	1.151	.3751
2.10	.6922	.3962	1.951	1.747	1.160	.3924
2.15	.6791	.3833	2.050	1.772	1.168	.4092
2.20	.6662	.3710	2.156	1.796	1.176	.4255
2.25	.6536	.3593	2.268	1.819	1.184	.4413
2.30	.6412	.3482	2.388	1.842	1.192	.4566
2.35	.6290	.3375	2.517	1.864	1.200	.4715
2.40	.6170	.3273	2.654	1.885	1.208	.4860
2.45	.6051	.3175	2.800	1.906	1.215	.5000
2.50	.5935	.3082	2.954	1.926	1.223	.5136
2.55	.5822	.2992	3.119	1.946	1.230	.5267
2.60	.5711	.2906	3.295	1.965	1.237	.5394
2.65	.5601	.2824	3.482	1.983	1.244	.5517
2.70	.5493	.2745	3.681	2.001	1.250	.5636
2.75	.5388	.2669	3.892	2.019	1.257	.5752
2.80	.5285	.2596	4.116	2.036	1.263	.5864
2.85	.5184	.2526	4.354	2.052	1.270	.5972
2.90	.5085	.2459	4.607	2.068	1.276	.6077
2.95	.4988	.2394	4.875	2.084	1.282	.6179
3.00	.4894	.2332	5.160	2.099	1.288	.6277
3.50	.4053	.1819	9.110	2.228	1.338	.7110
4.00	.3382	.1454	15.94	2.326	1.378	.7726
4.50	.2848	.1186	27.39	2.402	1.409	.8189
5.00	.2421	.09841	45.95	2.460	1.433	.8543
6.00	.1797	.07065	120.1	2.543	1.468	.9037
7.00	.1377	.05302	285.3	2.598	1.491	.9355
8.00	.1085	.04117	625.2	2.635	1.507	.9570
9.00	.08745	.03286	1275	2.662	1.519	.9722
10.00	.07188	.02769	2438	2.681	1.527	.9832
∞	0	0	∞	2.769	1.565	1.0326

$\gamma = 1{\cdot}67$

M	$\frac{T}{T^*}$	$\frac{p}{p^*}$	$\frac{p_0}{p_0^*}$	$\frac{V}{V^*}$	$\frac{I}{I^*}$	$\frac{fL_{max}}{D}$
0	1.335	∞	∞	0	∞	∞
0.05	1.334	23.099	11.265	.05775	8.687	234.36
.10	1.331	11.535	5.661	.1154	4.392	55.83
.15	1.325	7.674	3.805	.1727	2.982	23.21
.20	1.317	5.739	2.887	.2296	2.293	12.11
.25	1.308	4.574	2.344	.2859	1.892	6.980
.30	1.296	3.795	1.989	.3415	1.635	4.337
.35	1.282	3.235	1.741	.3963	1.460	2.810
.40	1.267	2.814	1.560	.4502	1.336	1.868
.45	1.250	2.485	1.424	.5031	1.245	1.260
.50	1.232	2.220	1.320	.5549	1.178	.8549
.55	1.212	2.002	1.239	.6056	1.128	.5787
.60	1.191	1.819	1.176	.6548	1.0909	.3877
.65	1.169	1.664	1.126	.7029	1.0628	.2548
.70	1.146	1.530	1.0874	.7496	1.0418	.1625
.75	1.1233	1.413	1.0576	.7949	1.0265	.09870
.80	1.0993	1.311	1.0351	.8388	1.0155	.05576
.85	1.0748	1.220	1.0189	.8812	1.0080	.02780
.90	1.0501	1.139	1.0081	.9222	1.0033	.01106
.95	1.0251	1.0657	1.0019	.9618	1.0008	.00248
1.00	1.0000	1.0000	1.0000	1.0000	1.0000	0
1.05	.9749	.9404	1.0018	1.0368	1.0006	.00203
1.10	.9499	.8860	1.0070	1.0721	1.0024	.00740
1.15	.9251	.8364	1.0154	1.1061	1.0051	.01522
1.20	.9006	.7908	1.0266	1.1388	1.0084	.02481
1.25	.8763	.7489	1.0406	1.170	1.0124	.03564
1.30	.8524	.7102	1.0573	1.200	1.0167	.04733
1.35	.8289	.6744	1.0765	1.229	1.0213	.05957
1.40	.8059	.6412	1.0981	1.257	1.0262	.07212
1.45	.7833	.6104	1.1220	1.284	1.0313	.08481
1.50	.7612	.5817	1.148	1.309	1.0364	.09749
1.55	.7397	.5549	1.176	1.333	1.0416	.1101
1.60	.7187	.5298	1.207	1.356	1.0468	.1225
1.65	.6982	.5064	1.240	1.378	1.0520	.1346
1.70	.6783	.4845	1.275	1.400	1.0572	.1465

Table 4 (Continued)

$\gamma = 1{\cdot}67$

M	$\frac{T}{T^*}$	$\frac{p}{p^*}$	$\frac{p_0}{p_0^*}$	$\frac{V}{V^*}$	$\frac{I}{I^*}$	$\frac{fL_{max}}{D}$
1.75	.6590	.4639	1.312	1.421	1.0623	.1580
1.80	.6402	.4445	1.351	1.440	1.0673	.1692
1.85	.6219	.4263	1.392	1.459	1.0722	.1800
1.90	.6042	.4091	1.436	1.477	1.0770	.1905
1.95	.5871	.3929	1.482	1.494	1.0817	.2007
2.00	.5705	.3776	1.530	1.510	1.0863	.2105
2.05	.5544	.3632	1.580	1.526	1.0908	.2199
2.10	.5388	.3496	1.632	1.541	1.0952	.2290
2.15	.5238	.3367	1.687	1.556	1.0994	.2377
2.20	.5093	.3244	1.744	1.570	1.1035	.2461
2.25	.4952	.3128	1.803	1.583	1.107	.2542
2.30	.4816	.3017	1.865	1.596	1.111	.2620
2.35	.4684	.2912	1.929	1.608	1.115	.2694
2.40	.4557	.2813	1.995	1.620	1.119	.2766
2.45	.4434	.2718	2.064	1.631	1.122	.2835
2.50	.4315	.2628	2.135	1.642	1.126	.2901
2.55	.4200	.2542	2.209	1.653	1.129	.2965
2.60	.4089	.2460	2.285	1.663	1.132	.3026
2.65	.3982	.2381	2.364	1.672	1.135	.3085
2.70	.3878	.2306	2.445	1.682	1.138	.3141
2.75	.3778	.2235	2.529	1.691	1.141	.3196
2.80	.3681	.2167	2.616	1.699	1.144	.3248
2.85	.3587	.2102	2.705	1.707	1.146	.3299
2.90	.3497	.2039	2.797	1.715	1.149	.3348
2.95	.3410	.1979	2.892	1.723	1.152	.3395
3.00	.3325	.1922	2.990	1.730	1.154	.3440
3.50	.2616	.1461	4.134	1.790	1.174	.3810
4.00	.2099	.1145	5.608	1.833	1.189	.4071
4.50	.1715	.09203	7.456	1.864	1.200	.4261
5.00	.1424	.07547	9.721	1.887	1.208	.4402
6.00	.10222	.05329	15.68	1.918	1.220	.4594
7.00	.07666	.03955	23.85	1.938	1.227	.4714
8.00	.05949	.03049	34.58	1.951	1.232	.4793
9.00	.04745	.02420	48.24	1.960	1.235	.4849
10.00	.03870	.01996	65.18	1.967	1.238	.4889
∞	0	0	∞	1.996	1.249	.5064

Table 5

Rayleigh line—one-dimensional, frictionless, constant-area flow with stagnation temperature change for a perfect gas

$\gamma = 1{\cdot}4$

M	$\frac{T_0}{T_0^*}$	$\frac{T}{T^*}$	$\frac{p}{p^*}$	$\frac{p_0}{p_0^*}$	$\frac{V}{V^*}$	$\left(\frac{T_0}{T_0^*}\right)_{isoth}$
0	0	0	2.4000	1.2679	0	.87500
0.01	.000480	.000576	2.3997	1.2678	.000240	.87502
0.02	.00192	.00230	2.3987	1.2675	.000959	.87507
0.03	.00431	.00516	2.3970	1.2671	.00216	.87516
0.04	.00765	.00917	2.3946	1.2665	.00383	.87528
0.05	.01192	.01430	2.3916	1.2657	.00598	.87544
0.06	.01712	.02053	2.3880	1.2647	.00860	.87563
0.07	.02322	.02784	2.3837	1.2636	.01168	.87586
0.08	.03021	.03621	2.3787	1.2623	.01522	.87612
0.09	.03807	.04562	2.3731	1.2608	.01922	.87642
0.10	.04678	.05602	2.3669	1.2591	.02367	.87675
0.11	.05630	.06739	2.3600	1.2573	.02856	.87712
0.12	.06661	.07970	2.3526	1.2554	.03388	.87752
0.13	.07768	.09290	2.3445	1.2533	.03962	.87796
0.14	.08947	.10695	2.3359	1.2510	.04578	.87843
0.15	.10196	.12181	2.3267	1.2486	.05235	.87894
0.16	.11511	.13743	2.3170	1.2461	.05931	.87948
0.17	.12888	.15377	2.3067	1.2434	.06666	.88006
0.18	.14324	.17078	2.2959	1.2406	.07438	.88067
0.19	.15814	.18841	2.2845	1.2377	.08247	.88132
0.20	.17355	.20661	2.2727	1.2346	.09091	.88200
0.21	.18943	.22533	2.2604	1.2314	.09969	.88272
0.22	.20574	.24452	2.2477	1.2281	.10879	.88347
0.23	.22244	.26413	2.2345	1.2248	.11820	.88426
0.24	.23948	.28411	2.2209	1.2213	.12792	.88508
0.25	.25684	.30440	2.2069	1.2177	.13793	.88594
0.26	.27446	.32496	2.1925	1.2140	.14821	.88683
0.27	.29231	.34573	2.1777	1.2102	.15876	.88776
0.28	.31035	.36667	2.1626	1.2064	.16955	.88872
0.29	.32855	.38773	2.1472	1.2025	.18058	.88972

Table 5 (Continued)

$\gamma = 1{\cdot}4$

M	$\frac{T_0}{T_0^*}$	$\frac{T}{T^*}$	$\frac{p}{p^*}$	$\frac{p_0}{p_0^*}$	$\frac{V}{V^*}$	$\left(\frac{T_0}{T_0^*}\right)_{\text{isoth}}$
0.30	.34686	.40887	2.1314	1.1985	0.19183	.89075
0.31	.36525	.43004	2.1154	1.1945	0.20329	.89182
0.32	.38369	.45119	2.0991	1.1904	0.21494	.89292
0.33	.40214	.47228	2.0825	1.1863	0.22678	.89406
0.34	.42057	.49327	2.0657	1.1821	0.23879	.89523
0.35	.43894	.51413	2.0487	1.1779	0.25096	.89644
0.36	.45723	.53482	2.0314	1.1737	0.26327	.89768
0.37	.47541	.55530	2.0140	1.1695	0.27572	.89896
0.38	.49346	.57553	1.9964	1.1652	0.28828	.90027
0.39	.51134	.59549	1.9787	1.1609	0.30095	.90162
0.40	.52903	.61515	1.9608	1.1566	0.31372	.90300
0.41	.54651	.63448	1.9428	1.1523	0.32658	.90442
0.42	.56376	.65345	1.9247	1.1480	0.33951	.90587
0.43	.58075	.67205	1.9065	1.1437	0.35251	.90736
0.44	.59748	.69025	1.8882	1.1394	0.36556	.90888
0.45	.61393	.70803	1.8699	1.1351	0.37865	.91044
0.46	.63007	.72538	1.8515	1.1308	0.39178	.91203
0.47	.64589	.74228	1.8331	1.1266	0.40493	.91366
0.48	.66139	.75871	1.8147	1.1224	0.41810	.91532
0.49	.67655	.77466	1.7962	1.1182	0.43127	.91702
0.50	.69136	.79012	1.7778	1.1140	0.4445	.91875
0.51	.70581	.80509	1.7594	1.1099	0.45761	.92052
0.52	.71990	.81955	1.7410	1.1059	0.47075	.92232
0.53	.73361	.83351	1.7226	1.1019	0.48387	.92416
0.54	.74695	.84695	1.7043	1.0979	0.49696	.92603
0.55	.75991	.85987	1.6860	1.09397	0.51001	.92794
0.56	.77248	.87227	1.6678	1.09010	0.52302	.92988
0.57	.78467	.88415	1.6496	1.08630	0.53597	.93186
0.58	.79647	.89552	1.6316	1.08255	0.54887	.93387
0.59	.80789	.90637	1.6136	1.07887	0.56170	.93592
0.60	.81892	.91670	1.5957	1.07525	0.57447	.93800
0.61	.82956	.92653	1.5780	1.07170	0.58716	.94012
0.62	.83982	.93585	1.5603	1.06821	0.59978	.94227
0.63	.84970	.94466	1.5427	1.06480	0.61232	.94446
0.64	.85920	.95298	1.5253	1.06146	0.62477	.94668
0.65	.86833	.96081	1.5080	1.05820	.63713	.94894
0.66	.87709	.96816	1.4908	1.05502	.64941	.95123
0.67	.88548	.97503	1.4738	1.05192	.66159	.95356
0.68	.89350	.98144	1.4569	1.04890	.67367	.95592
0.69	.90117	.98739	1.4401	1.04596	.68564	.95832
0.70	.90850	.99289	1.4235	1.04310	.69751	.96075
0.71	.91548	.99796	1.4070	1.04033	.70927	.96322
0.72	.92212	1.00260	1.3907	1.03764	.72093	.96572
0.73	.92843	1.00682	1.3745	1.03504	.73248	.96826
0.74	.93442	1.01062	1.3585	1.03253	.74392	.97083
0.75	.94009	1.01403	1.3427	1.03010	.75525	.97344
0.76	.94546	1.01706	1.3270	1.02776	.76646	.97608
0.77	.95052	1.01971	1.3115	1.02552	.77755	.97876
0.78	.95528	1.02198	1.2961	1.02337	.78852	.98147
0.79	.95975	1.02390	1.2809	1.02131	.79938	.98422
0.80	.96394	1.02548	1.2658	1.01934	.81012	.98700
0.81	.96786	1.02672	1.2509	1.01746	.82075	.98982
0.82	.97152	1.02763	1.2362	1.01569	.83126	.99267
0.83	.97492	1.02823	1.2217	1.01399	.84164	.99556
0.84	.97807	1.02853	1.2073	1.01240	.85190	.99848
0.85	.98097	1.02854	1.1931	1.01091	.86204	1.00144
0.86	.98363	1.02826	1.1791	1.00951	.87206	1.00443
0.87	.98607	1.02771	1.1652	1.00819	.88196	1.00746
0.88	.98828	1.02690	1.1515	1.00698	.89175	1.01052
0.89	.99028	1.02583	1.1380	1.00587	.90142	1.01362
0.90	.99207	1.02451	1.1246	1.00485	.91097	1.01675
0.91	.99366	1.02297	1.1114	1.00393	.92039	1.01992
0.92	.99506	1.02120	1.09842	1.00310	.92970	1.02312
0.93	.99627	1.01921	1.08555	1.00237	93889	1.02636
0.94	.99729	1.01702	1.07285	1.00174	.94796	1.02963
0.95	.99814	1.01463	1.06030	1.00121	.95692	1.03294
0.96	.99883	1.01205	1.04792	1.00077	.96576	1.03628
0.97	.99935	1.00929	1.03570	1.00043	.97449	1.03966
0.98	.99972	1.00636	1.02364	1.00019	.98311	1.04307
0.99	.99993	1.00326	1.01174	1.00004	.99161	1.04652

Table 5 (Continued)

$\gamma = 1{\cdot}4$

M	$\frac{T_0}{T_0^*}$	$\frac{T}{T^*}$	$\frac{p}{p^*}$	$\frac{p_0}{p_0^*}$	$\frac{V}{V^*}$	$\left(\frac{T_0}{T_0^*}\right)_{isoth}$
1.00	1.00000	1.00000	1.00000	1.00000	1.00000	1.05000
1.01	.99993	.99659	.98841	1.00004	1.00828	1.05352
1.02	.99973	.99304	.97697	1.00019	1.01644	1.05707
1.03	.99940	.98936	.96569	1.00043	1.02450	1.06066
1.04	.99895	.98553	.95456	1.00077	1.03246	1.06428
1.05	.99838	.98161	.94358	1.00121	1.04030	1.06794
1.06	.99769	.97755	.93275	1.00175	1.04804	1.07163
1.07	.99690	.97339	.92206	1.00238	1.05567	1.07536
1.08	.99600	.96913	.91152	1.00311	1.06320	1.07912
1.09	.99501	.96477	.90112	1.00394	1.07062	1.08292
1.10	.99392	.96031	.89086	1.00486	1.07795	1.08675
1.11	.99274	.95577	.88075	1.00588	1.08518	1.09062
1.12	.99148	.95115	.87078	1.00699	1.09230	1.09452
1.13	.99013	.94646	.86094	1.00820	1.09933	1.09846
1.14	.98871	.94169	.85123	1.00951	1.10626	1.10243
1.15	.98721	.93685	.84166	1.01092	1.1131	1.10644
1.16	.98564	.93195	.83222	1.01243	1.1198	1.11048
1.17	.98400	.92700	.82292	1.01403	1.1264	1.11456
1.18	.98230	.92200	.81374	1.01572	1.1330	1.11867
1.19	.98054	.91695	.80468	1.01752	1.1395	1.12282
1.20	.97872	.91185	.79576	1.01941	1.1459	1.12700
1.21	.97685	.90671	.78695	1.02140	1.1522	1.13122
1.22	.97492	.90153	.77827	1.02348	1.1584	1.13547
1.23	.97294	.89632	.76971	1.02566	1.1645	1.13976
1.24	.97092	.89108	.76127	1.02794	1.1705	1.14408
1.25	.96886	.88581	.75294	1.03032	1.1764	1.14844
1.26	.96675	.88052	.74473	1.03280	1.1823	1.15283
1.27	.96461	.87521	.73663	1.03536	1.1881	1.15726
1.28	.96243	.86988	.72865	1.03803	1.1938	1.16172
1.29	.96022	.86453	.72078	1.04080	1.1994	1.16622
1.30	.95798	.85917	.71301	1.04365	1.2050	1.17075
1.31	.95571	.85380	.70535	1.04661	1.2105	1.17532
1.32	.95341	.84843	.69780	1.04967	1.2159	1.17992
1.33	.95108	.84305	.69035	1.05283	1.2212	1.18456
1.34	.94873	.83766	.68301	1.05608	1.2264	1.18923

$\gamma = 1{\cdot}4$

M	$\frac{T_0}{T_0^*}$	$\frac{T}{T^*}$	$\frac{p}{p^*}$	$\frac{p_0}{p_0^*}$	$\frac{V}{V^*}$	$\left(\frac{T_0}{T_0^*}\right)_{isoth}$
1.35	.94636	.83227	.67577	1.05943	1.2316	1.19394
1.36	.94397	.82698	.66863	1.06288	1.2367	1.19868
1.37	.94157	.82151	.66159	1.06642	1.2417	1.20346
1.38	.93915	.81613	.65464	1.07006	1.2467	1.20827
1.39	.93671	.81076	.64778	1.07380	1.2516	1.21312
1.40	.93425	.80540	.64102	1.07765	1.2564	1.21800
1.41	.93178	.80004	.63436	1.08159	1.2612	1.22292
1.42	.92931	.79469	.62779	1.08563	1.2659	1.22787
1.43	.92683	.78936	.62131	1.08977	1.2705	1.23286
1.44	.92434	.78405	.61491	1.09400	1.2751	1.23788
1.45	.92184	.77875	.60860	1.0983	1.2796	1.24294
1.46	.91933	.77346	.60237	1.1028	1.2840	1.24803
1.47	.91682	.76819	.59623	1.1073	1.2884	1.25316
1.48	.91431	.76294	.59018	1.1120	1.2927	1.25832
1.49	.91179	.75771	.58421	1.1167	1.2970	1.26352
1.50	.90928	.75250	.57831	1.1215	1.3012	1.26875
1.51	.90676	.74731	.57250	1.1264	1.3054	1.27402
1.52	.90424	.74215	.56677	1.1315	1.3095	1.27932
1.53	.90172	.73701	.56111	1.1367	1.3135	1.28466
1.54	.89920	.73189	.55553	1.1420	1.3175	1.29003
1.55	.89669	.72680	.55002	1.1473	1.3214	1.29544
1.56	.89418	.72173	.54458	1.1527	1.3253	1.30088
1.57	.89167	.71669	.53922	1.1582	1.3291	1.30636
1.58	.88917	.71168	.53393	1.1639	1.3329	1.31187
1.59	.88668	.70669	.52871	1.1697	1.3366	1.31742
1.60	.88419	.70173	.52356	1.1756	1.3403	1.32300
1.61	.88170	.69680	.51848	1.1816	1.3439	1.32862
1.62	.87922	.69190	.51346	1.1877	1.3475	1.33427
1.63	.87675	.68703	.50851	1.1939	1.3511	1.33996
1.64	.87429	.68219	.50363	1.2002	1.3546	1.34568
1.65	.87184	.67738	.49881	1.2066	1.3580	1.35144
1.66	.86940	.67259	.49405	1.2131	1.3614	1.35723
1.67	.86696	.66784	.48935	1.2197	1.3648	1.36306
1.68	.86453	.66312	.48471	1.2264	1.3681	1.36892
1.69	.86211	.65843	.48014	1.2332	1.3713	1.37482

Table 5 (Continued)

$\gamma = 1{\cdot}4$

M	$\frac{T_0}{T_0^*}$	$\frac{T}{T^*}$	$\frac{p}{p^*}$	$\frac{p_0}{p_0^*}$	$\frac{V}{V^*}$	$\left(\frac{T_0}{T_0^*}\right)_{isoth}$
1.70	.85970	.65377	.47563	1.2402	1.3745	1.38075
1.71	.85731	.64914	.47117	1.2473	1.3777	1.38672
1.72	.85493	.64455	.46677	1.2545	1.3809	1.39272
1.73	.85256	.63999	.46242	1.2618	1.3840	1.39876
1.74	.85020	.63546	.45813	1.2692	1.3871	1.40483
1.75	.84785	.63096	.45390	1.2767	1.3901	1.41094
1.76	.84551	.62649	.44972	1.2843	1.3931	1.41708
1.77	.84318	.62205	.44559	1.2920	1.3960	1.42326
1.78	.84087	.61765	.44152	1.2998	1.3989	1.42947
1.79	.83857	.61328	.43750	1.3078	1.4018	1.43572
1.80	.83628	.60894	.43353	1.3159	1.4046	1.44200
1.81	.83400	.60463	.42960	1.3241	1.4074	1.44832
1.82	.83174	.60036	.42573	1.3324	1.4102	1.45467
1.83	.82949	.59612	.42191	1.3408	1.4129	1.46106
1.84	.82726	.59191	.41813	1.3494	1.4156	1.46748
1.85	.82504	.58773	.41440	1.3581	1.4183	1.47394
1.86	.82283	.58359	.41072	1.3669	1.4209	1.48043
1.87	.82064	.57948	.40708	1.3758	1.4235	1.48696
1.88	.81846	.57540	.40349	1.3848	1.4261	1.49352
1.89	.81629	.57135	.39994	1.3940	1.4286	1.50012
1.90	.81414	.56734	.39643	1.4033	1.4311	1.50675
1.91	.81200	.56336	.39297	1.4127	1.4336	1.51342
1.92	.80987	.55941	.38955	1.4222	1.4360	1.52012
1.93	.80776	.55549	.38617	1.4319	1.4384	1.52686
1.94	.80567	.55160	.38283	1.4417	1.4408	1.53363
1.95	.80359	.54774	.37954	1.4516	1.4432	1.54044
1.96	.80152	.54391	.37628	1.4616	1.4455	1.54728
1.97	.79946	.54012	.37306	1.4718	1.4478	1.55416
1.98	.79742	.53636	.36988	1.4821	1.4501	1.56107
1.99	.79540	.53263	.36674	1.4925	1.4523	1.56802
2.00	.79339	.52893	.36364	1.5031	1.4545	1.57500
2.01	.79139	.52526	.36057	1.5138	1.4567	1.58202
2.02	.78941	.52161	.35754	1.5246	1.4589	1.58907
2.03	.78744	.51800	.35454	1.5356	1.4610	1.59616
2.04	.78549	.51442	.35158	1.5467	1.4631	1.60328

$\gamma = 1{\cdot}4$

M	$\frac{T_0}{T_0^*}$	$\frac{T}{T^*}$	$\frac{p}{p^*}$	$\frac{p_0}{p_0^*}$	$\frac{V}{V^*}$	$\left(\frac{T_0}{T_0^*}\right)_{isoth}$
2.05	.78355	.51087	.34866	1.5579	1.4652	1.61044
2.06	.78162	.50735	.34577	1.5693	1.4673	1.61763
2.07	.77971	.50386	.34291	1.5808	1.4694	1.62486
2.08	.77781	.50040	.34009	1.5924	1.4714	1.63212
2.09	.77593	.49697	.33730	1.6042	1.4734	1.63942
2.10	.77406	.49356	.33454	1.6161	1.4753	1.64675
2.11	.77221	.49018	.33181	1.6282	1.4773	1.65412
2.12	.77037	.48683	.32912	1.6404	1.4792	1.66152
2.13	.76854	.48351	.32646	1.6528	1.4811	1.66896
2.14	.76673	.48022	.32383	1.6653	1.4830	1.67643
2.15	.76493	.47696	.32122	1.6780	1.4849	1.68394
2.16	.76314	.47373	.31864	1.6908	1.4867	1.69148
2.17	.76137	.47052	.31610	1.7037	1.4885	1.69906
2.18	.75961	.46734	.31359	1.7168	1.4903	1.70667
2.19	.75787	.46419	.31110	1.7300	1.4921	1.71432
2.20	.75614	.46106	.30864	1.7434	1.4939	1.72200
2.21	.75442	.45796	.30621	1.7570	1.4956	1.72972
2.22	.75271	.45489	.30381	1.7707	1.4973	1.73747
2.23	.75102	.45184	.30143	1.7846	1.4990	1.74526
2.24	.74934	.44882	.29908	1.7986	1.5007	1.75308
2.25	.74767	.44582	.29675	1.8128	1.5024	1.76094
2.26	.74602	.44285	.29445	1.8271	1.5040	1.76883
2.27	.74438	.43990	.29218	1.8416	1.5056	1.77676
2.28	.74275	.43698	.28993	1.8562	1.5072	1.78472
2.29	.74114	.43409	.28771	1.8710	1.5088	1.79272
2.30	.73954	.43122	.28551	1.8860	1.5104	1.80075
2.31	.73795	.42837	.28333	1.9012	1.5119	1.80882
2.32	.73638	.42555	.28118	1.9165	1.5134	1.81692
2.33	.73482	.42276	.27905	1.9320	1.5150	1.82506
2.34	.73327	.41999	.27695	1.9476	1.5165	1.83323
2.35	.73173	.41724	.27487	1.9634	1.5180	1.84144
2.36	.73020	.41451	.27281	1.9794	1.5195	1.84968
2.37	.72868	.41181	.27077	1.9955	1.5209	1.85796
2.38	.72718	.40913	.26875	2.0118	1.5223	1.86627
2.39	.72569	.40647	.26675	2.0283	1.5237	1.87462

Table 5 (Continued)

$\gamma = 1{\cdot}4$

M	$\frac{T_0}{T_0^*}$	$\frac{T}{T^*}$	$\frac{p}{p^*}$	$\frac{p_0}{p_0^*}$	$\frac{V}{V^*}$	$\left(\frac{T_0}{T_0^*}\right)_{isoth}$
2.40	.72421	.40383	.26478	2.0450	1.5252	1.88300
2.41	.72274	.40122	.26283	2.0619	1.5266	1.89142
2.42	.72129	.39863	.26090	2.0789	1.5279	1.89987
2.43	.71985	.39606	.25899	2.0961	1.5293	1.90836
2.44	.71842	.39352	.25710	2.1135	1.5306	1.91688
2.45	.71700	.39100	.25523	2.1311	1.5320	1.92544
2.46	.71559	.38850	.25337	2.1489	1.5333	1.93403
2.47	.71419	.38602	.25153	2.1669	1.5346	1.94266
2.48	.71280	.38356	.24972	2.1850	1.5359	1.95132
2.49	.71142	.38112	.24793	2.2033	1.5372	1.96002
2.50	.71005	.37870	.24616	2.2218	1.5385	1.96875
2.51	.70870	.37630	.24440	2.2405	1.5398	1.97752
2.52	.70736	.37392	.24266	2.2594	1.5410	1.98632
2.53	.70603	.37157	.24094	2.2785	1.5422	1.99515
2.54	.70471	.36923	.23923	2.2978	1.5434	2.00403
2.55	.70340	.36691	.23754	2.3173	1.5446	2.01294
2.56	.70210	.36461	.23587	2.3370	1.5458	2.02188
2.57	.70081	.36233	.23422	2.3569	1.5470	2.03086
2.58	.69953	.36007	.23258	2.3770	1.5482	2.03987
2.59	.69825	.35783	.23096	2.3972	1.5494	2.04892
2.60	.69699	.35561	.22936	2.4177	1.5505	2.05800
2.61	.69574	.35341	.22777	2.4384	1.5516	2.06711
2.62	.69450	.35123	.22620	2.4593	1.5527	2.07627
2.63	.69327	.34906	.22464	2.4804	1.5538	2.08546
2.64	.69205	.34691	.22310	2.5017	1.5549	2.09468
2.65	.69084	.34478	.22158	2.5233	1.5560	2.10394
2.66	.68964	.34267	.22007	2.5451	1.5571	2.11323
2.67	.68845	.34057	.21857	2.5671	1.5582	2.12256
2.68	.68727	.33849	.21709	2.5892	1.5593	2.13192
2.69	.68610	.33643	.21562	2.6116	1.5603	2.14132
2.70	.68494	.33439	.21417	2.6342	1.5613	2.15075
2.71	.68378	.33236	.21273	2.6571	1.5623	2.16022
2.72	.68263	.33035	.21131	2.6802	1.5633	2.16972
2.73	.68150	.32836	.20990	2.7035	1.5644	2.17925
2.74	.68038	.32638	.20850	2.7270	1.5654	2.18883

$\gamma = 1{\cdot}4$

M	$\frac{T_0}{T_0^*}$	$\frac{T}{T^*}$	$\frac{p}{p^*}$	$\frac{p_0}{p_0^*}$	$\frac{V}{V^*}$	$\left(\frac{T_0}{T_0^*}\right)_{isoth}$
2.75	.67926	.32442	.20712	2.7508	1.5663	2.19844
2.76	.67815	.32248	.20575	2.7748	1.5673	2.20808
2.77	.67704	.32055	.20439	2.7990	1.5683	2.21776
2.78	.67595	.31864	.20305	2.8235	1.5692	2.22747
2.79	.67487	.31674	.20172	2.8482	1.5702	2.23722
2.80	.67380	.31486	.20040	2.8731	1.5711	2.24700
2.81	.67273	.31299	.19909	2.8982	1.5721	2.25682
2.82	.67167	.31114	.19780	2.9236	1.5730	2.26667
2.83	.67062	.30931	.19652	2.9493	1.5739	2.27655
2.84	.66958	.30749	.19525	2.9752	1.5748	2.28648
2.85	.66855	.30568	.19399	3.0013	1.5757	2.29644
2.86	.66752	.30389	.19274	3.0277	1.5766	2.30643
2.87	.66650	.30211	.19151	3.0544	1.5775	2.31646
2.88	.66549	.30035	.19029	3.0813	1.5784	2.32652
2.89	.66449	.29860	.18908	3.1084	1.5792	2.33662
2.90	.66350	.29687	.18788	3.1358	1.5801	2.34675
2.91	.66252	.29515	.18669	3.1635	1.5809	2.35692
2.92	.66154	.29344	.18551	3.1914	1.5818	2.36712
2.93	.66057	.29175	.18435	3.2196	1.5826	2.37735
2.94	.65961	.29007	.18320	3.2481	1.5834	2.38763
2.95	.65865	.28841	.18205	3.2768	1.5843	2.39794
2.96	.65770	.28676	.18091	3.3058	1.5851	2.40828
2.97	.65676	.28512	.17978	3.3351	1.5859	2.41865
2.98	.65583	.28349	.17867	3.3646	1.5867	2.42907
2.99	.65490	.28188	.17757	3.3944	1.5875	2.43952
3.00	.65398	.28028	.17647	3.4244	1.5882	2.45000
3.50	.61580	.21419	.13223	5.3280	1.6198	3.01875
4.00	.58909	.16831	.10256	8.2268	1.6410	3.67500
4.50	.56983	.13540	.08177	12.502	1.6559	4.41874
5.00	.55555	.11111	.06667	18.634	1.6667	5.25000
6.00	.53633	.07849	.04669	38.946	1.6809	7.17499
7.00	.52437	.05826	.03448	75.414	1.6896	9.45000
8.00	.51646	.04491	.02649	136.62	1.6954	12.07500
9.00	.51098	.03565	.02098	233.88	1.6993	15.05003
10.00	.50702	.02897	.01702	381.62	1.7021	18.37500
∞	.48980	0	0	∞	1.7143	∞

Table 5 (Continued)

$\gamma = 1{\cdot}0$

M	$\frac{T_0}{T_0^*} = \frac{T^*}{T}$	$\frac{p}{p^*}$	$\frac{p_0}{p_0^*}$	$\frac{V}{V^*}$
0	0	2.000	1.213	0
0.05	.00995	1.995	1.212	.00499
0.10	.03921	1.980	1.207	.01980
0.15	.08608	1.956	1.200	.04401
0.20	.14793	1.923	1.190	.07692
0.25	.2215	1.882	1.178	.1176
0.30	.3030	1.835	1.164	.1651
0.35	.3889	1.782	1.149	.2183
0.40	.4756	1.724	1.133	.2758
0.45	.5602	1.663	1.116	.3368
0.50	.6400	1.600	1.0997	.4000
0.55	.7132	1.536	1.0834	.4645
0.60	.7785	1.471	1.0679	.5294
0.65	.8352	1.406	1.0534	.5940
0.70	.8828	1.342	1.0402	.6577
0.75	.9216	1.280	1.0285	.7200
0.80	.9518	1.220	1.0186	.7805
0.85	.9740	1.161	1.0107	.8389
0.90	.9890	1.105	1.0048	.8950
0.95	.9974	1.0512	1.0012	.9488
1.00	1.0000	1.0000	1.0000	1.0000
1.05	.9976	.9512	1.0013	1.0488
1.10	.9910	.9049	1.0052	1.0951
1.15	.9807	.8611	1.0118	1.1389
1.20	.9675	.8197	1.0214	1.1802
1.25	.9518	.7805	1.0340	1.220
1.30	.9342	.7435	1.0498	1.257
1.35	.9151	.7086	1.0690	1.291
1.40	.8948	.6757	1.0919	1.324
1.45	.8737	.6447	1.1187	1.355
1.50	.8521	.6154	1.150	1.384
1.55	.8301	.5878	1.186	1.412
1.60	.8080	.5618	1.226	1.438
1.65	.7859	.5373	1.271	1.463
1.70	.7639	.5141	1.323	1.486

$\gamma = 1{\cdot}0$

M	$\frac{T_0}{T_0^*} = \frac{T}{T^*}$	$\frac{p}{p^*}$	$\frac{p_0}{p_0^*}$	$\frac{V}{V^*}$
1.75	0.7422	0.4923	1.381	1.508
1.80	0.7209	0.4717	1.446	1.528
1.85	0.7000	0.4522	1.519	1.547
1.90	0.6795	0.4338	1.600	i.566
1.95	0.6595	0.4164	1.691	1.584
2.00	0.6400	0.4000	1.793	1.601
2.05	0.6211	0.3844	1.907	1.616
2.10	0.6027	0.3697	2.034	1.630
2.15	0.5849	0.3557	2.176	1.644
2.20	0.5677	0.3425	2.336	1.657
2.25	0.5510	0.3299	2.515	1.670
2.30	0.5348	0.3179	2.716	1.682
2.35	0.5192	0.3066	2.942	1.693
2.40	0.5042	0.2959	3.197	1.704
2.45	0.4897	0.2857	3.484	1.714
2.50	0.4757	0.2759	3.808	1.724
2.55	0.4621	0.2666	4.175	1.733
2.60	0.4490	0.2577	4.591	1.742
2.65	0.4364	0.2493	5.064	1.751
2.70	0.4243	0.2413	5.602	1.759
2.75	0.4126	0.2336	6.215	1.766
2.80	0.4013	0.2262	6.916	1.774
2.85	0.3904	0.2192	7.719	1.781
2.90	0.3799	0.2125	8.640	1.787
2.95	0.3698	0.2061	9.699	1.794
3.00	0.3600	0.2000	10.92	1.800
3.50	0.2791	0.1509	41.85	1.849
4.00	0.2215	0.1176	212.71	1.882
4.50	0.1794	0.09412	1425	1.906
5.00	0.1479	0.07692	12519	1.923
6.00	0.10519	0.05405	$215(10)^4$	1.946
7.00	0.07840	0.04000	$106(10)^7$	1 960
8.00	0 06059	0 03077	$147(10)^{10}$	1.969
9.00	0.04818	0.02439	$574(10)^{13}$	1.976
10.00	0.03921	0.01980	$623(10)^{17}$	1.980
∞	0	0	∞	2.000

Table 5 (Continued)

$\gamma = 1{\cdot}1$

M	$\frac{T_0}{T_0^*}$	$\frac{T}{T^*}$	$\frac{p}{p^*}$	$\frac{p_0}{p_0^*}$	$\frac{V}{V^*}$	$\left(\frac{T_0}{T_0^*}\right)_{isoth}$
0	0	0	2.100	1.228	0	0.95652
0.05	.01044	0.01097	2.094	1.226	0.00524	0.95664
0.10	.04111	0.04315	2.077	1.221	0.02077	0.95700
0.15	.09009	0.09449	2.049	1.213	0.04611	0.95760
0.20	.15444	0.16184	2.011	1.203	0.08046	0.95843
0.25	.2305	0.2413	1.965	1.190	0.1228	0.95951
0.30	.3144	0.3286	1.911	1.174	0.1720	0.96083
0.35	.4020	0.4195	1.851	1.157	0.2267	0.96238
0.40	.4898	0.5102	1.786	1.140	0.2857	0.96417
0.45	.5746	0.5973	1.717	1.122	0.3478	0.96621
0.50	.6540	0.6782	1.647	1.1040	0.4118	0.96848
0.55	.7261	0.7510	1.576	1.0867	0.4766	0.97099
0.60	.7898	0.8147	1.504	1.0702	0.5416	0.97374
0.65	.8446	0.8684	1.434	1.0550	0.6057	0.97673
0.70	.8902	0.9123	1.365	1.0412	0.6686	0.97996
0.75	.9270	0.9467	1.297	1.0291	0.7297	0.98342
0.80	.9554	0.9720	1.232	1.0189	0.7887	0.98713
0.85	.9761	0.9892	1.170	1.0109	0.8453	0.99108
0.90	.9899	0.9989	1.111	1.0050	0.8995	0.99526
0.95	.9976	1.0023	1.0538	1.0013	0.9511	0.99968
1.00	1.0000	1.0000	1.0000	1.0000	1.0000	1.00435
1.05	.9979	0.9930	0.9490	1.0013	1.0463	1.00925
1.10	.9919	0.9821	0.9009	1.0051	1.0901	1.01439
1.15	.9827	0.9679	0.8555	1.0116	1.1314	1.01977
1.20	.9710	0.9511	0.8127	1.0209	1.1703	1.02539
1.25	.9572	0.9322	0.7724	1.0331	1.207	1.03125
1.30	.9418	0.9118	0.7345	1.0483	1.241	1.03735
1.35	.9251	0.8902	0.6989	1.0665	1.273	1.04368
1.40	.9074	0.8678	0.6654	1.0880	1.304	1.05026
1.45	.8892	0.8449	0.6339	1.1130	1.333	1.05708
1.50	.8706	0.8217	0.6043	1.141	1.360	1.06413
1.55	.8518	0.7984	0.5765	1.173	1.385	1.07142
1.60	.8329	0.7753	0.5503	1.210	1.409	1.07896
1.65	.8141	0.7524	0.5257	1.251	1.431	1.08673
1.70	.7955	0.7298	0.5025	1.297	1.452	1.09474

$\gamma = 1{\cdot}1$

M	$\frac{T_0}{T_0^*}$	$\frac{T}{T^*}$	$\frac{p}{p^*}$	$\frac{p_0}{p_0^*}$	$\frac{V}{V^*}$	$\left(\frac{T_0}{T_0^*}\right)_{isoth}$
1.75	.7771	.7076	.4807	1.347	1.472	1.10299
1.80	.7591	.6859	.4601	1.403	1.491	1.11148
1.85	.7415	.6648	.4407	1.465	1.508	1.12021
1.90	.7243	.6443	.4224	1.532	1.525	1.21917
1.95	.7076	.6243	.4052	1.607	1.541	1.13838
2.00	.6914	.6049	.3889	1.689	1.556	1.14783
2.05	.6756	.5862	.3735	1.780	1.570	1.15751
2.10	.6603	.5681	.3589	1.879	1.583	1.16743
2.15	.6456	.5506	.3451	1.987	1.595	1.17760
2.20	.6313	.5337	.3321	2.106	1.607	1.18800
2.25	.6175	.5174	.3197	2.237	1.618	1.19864
2.30	.6042	.5017	.3079	2.380	1.629	1.20952
2.35	.5914	.4866	.2968	2.537	1.639	1.22064
2.40	.5790	.4720	.2863	2.709	1.649	1.23200
2.45	.5671	.4580	.2763	2.897	1.658	1.24360
2.50	.5556	.4444	.2667	3.104	1.667	1.25543
2.55	.5445	.4314	.2576	3.332	1.675	1.26751
2.60	.5338	.4189	.2489	3.581	1.683	1.27983
2.65	.5235	.4068	.2406	3.855	1.690	1.29238
2.70	.5136	.3952	.2328	4.156	1.697	1.30517
2.75	.5041	.3840	.2253	4.487	1.704	1.31821
2.80	.4949	.3733	.2182	4.851	1.711	1.33148
2.85	.4860	.3629	.2114	5.251	1.717	1.34499
2.90	.4775	.3529	.2049	5.692	1.723	1.35874
2.95	.4693	.3433	.1986	6.176	1.729	1.37273
3.00	.4613	.3341	.1927	6.710	1.734	1.38696
3.50	.3960	.2578	.1451	16.26	1.777	1.54239
4.00	.3496	.2040	.1129	42.42	1.806	1.72174
4.50	.3160	.1648	.0902	115.70	1.827	1.92500
5.00	.2909	.1357	.0737	322.33	1.842	2.15217
6.00	.2568	.09631	.05172	2508	1.862	2.67826
7.00	.2356	.07169	.03825	18430	1.874	3.30000
8.00	.2215	.05536	.02941	$123(10)^3$	1 882	4.01739
9.00	.2116	.04400	.02331	$743(10)^3$	1.888	4.83044
10.00	.2045	.03579	.01892	$401(10)^4$	1.892	5.73913
∞	.1736	0	0	∞	1.909	∞

Table 5 (Continued)

$\gamma = 1{\cdot}2$

M	$\frac{T_0}{T_0^*}$	$\frac{T}{T^*}$	$\frac{p}{p^*}$	$\frac{p_0}{p_0^*}$	$\frac{V}{V^*}$	$\left(\frac{T_0}{T_0^*}\right)_{isoth}$
0	0	0	2.200	1.242	0	0.92308
0.05	.01094	.01203	2.193	1.239	.00548	0.92331
0.10	.04301	.04726	2.173	1.234	.02174	0.92400
0.15	.09408	.10325	2.141	1.226	.04820	0.92515
0.20	.16089	.17627	2.099	1.214	.08397	0.92677
0.25	.2395	.2618	2.047	1.199	.1279	0.92885
0.30	.3255	.3548	1.986	1.183	.1787	0.93138
0.35	.4147	.4507	1.918	1.165	.2350	0.93438
0.40	.5034	.5450	1.846	1.146	.2953	0.93785
0.45	.5884	.6343	1.770	1.127	.3584	0.94177
0.50	.6672	.7160	1.692	1.1078	.4231	0.94615
0.55	.7381	.7881	1.614	1.0895	.4884	0.95100
0.60	.8003	.8497	1.536	1.0722	.5531	0.95631
0.65	.8531	.9004	1.460	1.0563	.6168	0.96208
0.70	.8969	.9405	1.385	1.0420	.6788	0.96831
0.75	.9318	.9704	1.313	1.0296	.7388	0.97500
0.80	.9585	.9910	1.244	1.0191	.7964	0.98215
0.85	.9779	1.0032	1.178	1.0109	.8514	0.98977
0.90	.9907	1.0081	1.115	1.0049	.9037	0.99785
0.95	.9978	1.0067	1.0562	1.0012	.9532	1.00638
1.00	1.0000	1.0000	1.0000	1.0000	1.0000	1.01538
1.05	.9981	.9888	.9471	1.0013	1.0441	1.02485
1.10	.9927	.9741	.8972	1.0050	1.0856	1.03477
1.15	.9845	.9564	.8504	1.0114	1.1247	1.04515
1.20	.9740	.9365	.8065	1.0204	1.1613	1.05600
1.25	.9617	.9149	.7653	1.0322	1.196	1.06731
1.30	.9481	.8921	.7266	1.0467	1.228	1.07908
1.35	.9334	.8685	.6903	1.0640	1.258	1.09131
1.40	.9180	.8443	.6563	1.0843	1.286	1.10400
1.45	.9021	.8199	.6245	1.1077	1.313	1.11715
1.50	.8859	.7955	.5946	1.134	1.338	1.13077
1.55	.8695	.7712	.5666	1.164	1.361	1.14485
1.60	.8532	.7473	.5403	1.197	1.383	1.15938
1.65	.8370	.7237	.5156	1.234	1.404	1.17438
1.70	.8211	.7007	.4924	1.275	1.423	1.18985

$\gamma = 1{\cdot}2$

M	$\frac{T_0}{T_0^*}$	$\frac{T}{T^*}$	$\frac{p}{p^*}$	$\frac{p_0}{p_0^*}$	$\frac{V}{V^*}$	$\left(\frac{T_0}{T_0^*}\right)_{isoth}$
1.75	.8054	.6782	.4706	1.320	1.441	1.20577
1.80	.7900	.6563	.4501	1.369	1.458	1.22215
1.85	.7750	.6351	.4308	1.422	1.474	1.23900
1.90	.7604	.6146	.4126	1.480	1.490	1.25631
1.95	.7462	.5947	.3955	1.543	1.504	1.27408
2.00	.7325	.5755	.3793	1.612	1.517	1.29231
2.05	.7192	.5570	.3641	1.687	1.530	1.31100
2.10	.7063	.5391	.3497	1.767	1.542	1.33015
2.15	.6939	.5219	.3360	1.854	1.553	1.34977
2.20	.6819	.5054	.3231	1.948	1.564	1.36985
2.25	.6703	.4895	.3109	2.050	1.574	1.39039
2.30	.6591	.4742	.2994	2.159	1.584	1.41139
2.35	.6484	.4595	.2884	2.277	1.593	1.43285
2.40	.6381	.4453	.2780	2.405	1.602	1.45477
2.45	.6281	.4317	.2682	2.542	1.610	1.47715
2.50	.6185	.4187	.2588	2.690	1.618	1.50000
2.55	.6093	.4062	.2499	2.849	1.625	1.52331
2.60	.6004	.3941	.2414	3.021	1.632	1.54708
2.65	.5918	.3825	.2334	3.205	1.639	1.57131
2.70	.5836	.3713	.2257	3.403	1.645	1.59600
2.75	.5757	.3606	.2184	3.617	1.651	1.62115
2.80	.5681	.3503	.2114	3.847	1.657	1.64677
2.85	.5608	.3404	.2047	4.094	1.663	1.67285
2.90	.5537	.3309	.1983	4.359	1.668	1.69938
2.95	.5469	.3217	.1923	4.644	1.673	1.72638
3.00	.5404	.3128	.1864	4.951	1.678	1.75385
3.50	.4865	.2405	.1401	9.597	1.717	2.05385
4.00	.4486	.1898	.1089	18.99	1.743	2.40000
4.50	.4211	.1531	.08696	37.61	1.761	2.79230
5.00	.4006	.1259	.07097	73.64	1.774	3.23077
6.00	.3730	.08919	.04977	266.2	1.792	4.24615
7.00	.3557	.06632	.03679	875.9	1.803	5.44615
8.00	.3443	.05118	.02828	2621	1.810	6.83077
9.00	.3363	.04065	.02240	7181	1.815	8.40002
10.00	.3306	.03306	.01818	18182	1.818	10.15385
∞	0.3056	0	0	∞	1.833	∞

Table 5 (Continued)

$\gamma = 1{\cdot}3$

M	$\frac{T_0}{T_0^*}$	$\frac{T}{T^*}$	$\frac{p}{p^*}$	$\frac{p_0}{p_0^*}$	$\frac{V}{V^*}$	$\left(\frac{T_0}{T_0^*}\right)_{isoth}$
0	0	0	2.300	1.255	0	.89655
0.05	.01143	.01314	2.293	1.253	.00573	.89689
0.10	.04489	.05155	2.270	1.247	.02270	.89790
0.15	.09803	.11236	2.234	1.237	.05028	.89958
0.20	.16726	.19120	2.186	1.224	.08745	.90193
0.25	.2482	.2828	2.127	1.209	.1329	.90496
0.30	.3363	.3816	2.059	1.191	.1853	.90866
0.35	.4270	4822	1 984	1.172	.2430	.91303
0.40	.5165	.5800	1.904	1.152	.3046	.91807
0.45	.6015	.6713	1.821	1.131	.3687	.92378
0.50	.6796	.7533	1.736	1.1112	.4340	.93017
0.55	.7494	.8244	1.651	1.0919	.4994	.93723
0.60	.8099	.8837	1.567	1.0739	.5640	.94497
0.65	.8611	.9312	1.485	1.0574	.6272	.95337
0.70	.9029	.9673	1.405	1.0426	.6885	.96245
0.75	.9361	.9928	1.328	1.0299	.7473	.97220
0.80	.9614	1.0088	1.255	1.0193	.8035	.98262
0.85	.9795	1.0163	1.186	1.0109	.8569	.99372
0.90	.9914	1.0166	1.120	1.0049	.9075	1.00548
0.95	.9980	1.0108	1.0583	1.0012	.9552	1.01792
1.00	1.0000	1.0000	1.0000	1.0000	1.0000	1.03103
1.05	.9982	.9851	.9452	1.0012	1.0421	1.04482
1.10	.9933	.9669	.8939	1.0049	1.0816	1.05928
1.15	.9859	.9461	.8458	1.0111	1.1186	1.07441
1.20	.9765	.9235	.8008	1.0199	1.1532	1.09021
1.25	.9656	.8996	.7588	1.0312	1.186	1.10668
1.30	.9534	.8747	.7194	1.0451	1.216	1.12383
1.35	.9404	.8493	.6826	1.0617	1.244	1.14165
1.40	.9268	.8237	.6483	1.0809	1.270	1.16014
1.45	.9128	.7980	.6161	1.1028	1.295	1.17930
1.50	.8986	.7726	.5860	1.128	1.318	1.19914
1.55	.8843	.7475	.5578	1.155	1.340	1.21965
1.60	.8701	.7230	.5314	1.185	1.360	1.24083
1.65	.8560	.6990	.5067	1.219	1.379	1.26268
1.70	.8421	.6756	.4835	1.256	1.397	1.28521

$\gamma = 1{\cdot}3$

M	$\frac{T_0}{T_0^*}$	$\frac{T}{T^*}$	$\frac{p}{p^*}$	$\frac{p_0}{p_0^*}$	$\frac{V}{V^*}$	$\left(\frac{T_0}{T_0^*}\right)_{isoth}$
1.75	.8285	.6529	.4617	1.296	1.414	1.30841
1.80	.8153	.6309	.4413	1.340	1.430	1.33227
1.85	.8024	.6097	.4221	1.387	1.445	1.35682
1.90	.7898	.5892	.4040	1.438	1.459	1.38204
1.95	.7776	.5695	.3870	1.493	1.472	1.40792
2.00	.7659	.5505	.3710	1.552	1.484	1.43448
2.05	.7545	.5322	.3559	1.615	1.495	1.46172
2.10	.7435	.5146	.3416	1.683	1.506	1.48962
2.15	.7329	.4977	.3281	1.755	1.517	1.51820
2.20	.7227	.4815	.3154	1.832	1.527	1.54745
2.25	.7129	.4659	.3034	1.915	1.536	1.57737
2.30	.7034	.4510	.2920	2.003	1.545	1.60797
2.35	.6943	.4367	.2812	2.097	1.553	1.63923
2.40	.6855	.4229	.2710	2.197	1.561	1.67117
2.45	.6771	.4097	.2613	2.303	1.568	1.70378
2.50	.6690	.3971	.2521	2.416	1.575	1.73707
2.55	.6612	.3850	.2433	2.536	1.582	1.77103
2.60	.6537	.3733	.2350	2.664	1.588	1.80565
2.65	.6465	.3621	.2271	2.800	1.594	1.84096
2.70	.6396	.3513	.2195	2.944	1.600	1.87693
2.75	.6329	.3410	.2123	3.096	1.606	1.91358
2.80	.6265	.3311	.2055	3.258	1.611	1.95090
2.85	.6203	.3216	.1990	3.429	1.616	1.98889
2.90	.6144	.3124	.1928	3.611	1.621	2.02755
2.95	.6087	.3036	.1868	3.804	1.626	2.06689
3.00	.6032	.2952	.1811	4.007	1.630	2.10690
3.50	.5582	.2262	.1359	6.806	1.665	2.54397
4.00	.5265	.1781	.1055	11.57	1.688	3.04828
4.50	.5037	.1435	.08417	19.44	1.704	3.61982
5.00	.4867	.1178	.06866	32.06	1.716	4.25862
6.00	.4639	.08335	.04812	76.97	1.732	5.73793
7.00	.4496	.06192	.03555	191.3	1.742	7.48621
8.00	.4402	.04775	.02732	413.4	1.748	9.50345
9.00	.4336	.03792	.02164	833.4	1.753	11.78968
10.00	.4289	.03082	.01756	1582	1.756	14.34483
∞	.4083	0	0	∞	1.769	∞

Table 5 (Continued)

$\gamma = 1{\cdot}67$

M	$\frac{T_0}{T_0^*}$	$\frac{T}{T^*}$	$\frac{p}{p^*}$	$\frac{p_0}{p_0^*}$	$\frac{V}{V^*}$	$\left(\frac{T_0}{T_0^*}\right)_{isoth}$
0	0	0	2.670	1.299	0	.83292
0.05	.01325	.01767	2.659	1.297	.00665	.83362
0.10	.05183	.06896	2.626	1.289	.02626	.83571
0.15	.11243	.1490	2.573	1.276	.05790	.83920
0.20	.19020	.2506	2.503	1.259	.10011	.84408
0.25	.2794	.3653	2.418	1.239	.1511	.85036
0.30	.3742	.4849	2.321	1.216	.2089	.85803
0.35	.4693	.6018	2.216	1.192	.2715	.86710
0.40	.5606	.7103	2.107	1.168	.3371	.87756
0.45	.6448	.8062	1.995	1.144	.4040	.88942
0.50	.7201	.8870	1.884	1.1202	.4709	.90267
0.55	.7853	.9519	1.774	1.0981	.5366	.91732
0.60	.8402	1.0010	1.667	1.0778	.6003	.93337
0.65	.8853	1.0354	1.565	1.0597	.6614	.95081
0.70	.9213	1.0565	1.468	1.0438	.7195	.96964
0.75	.9491	1.0662	1.377	1.0303	.7744	.98987
0.80	.9697	1.0660	1.291	1.0193	.8260	1.01150
0.85	.9842	1.0578	1.210	1.0108	.8742	1.03452
0.90	.9935	1.0432	1.135	1.0048	.9192	1.05893
0.95	.9985	1.0235	1.0649	1.0012	.9611	1.08474
1.00	1.0000	1.0000	1.0000	1.0000	1.0000	1.11195
1.05	.9987	0.9736	.9398	1.0012	1.0361	1.14055
1.10	.9952	0.9454	.8839	1.0046	1.0695	1.17054
1.15	.9899	0.9158	.8321	1.0103	1.1005	1.20193
1.20	.9833	0.8855	.7842	1.0181	1.1292	1.23472
1.25	.9757	0.8550	.7397	1.0280	1.156	1.26890
1.30	.9674	0.8246	.6985	1.0400	1.181	1.30447
1.35	.9586	0.7946	.6603	1.0540	1.204	1.34145
1.40	.9495	0.7652	.6249	1.0700	1.225	1.37981
1.45	.9403	0.7365	.5919	1.0880	1.245	1.41957
1.50	.9310	0.7087	.5612	1.108	1.263	1.46073
1.55	.9217	0.6818	.5327	1.130	1.280	1.50328
1.60	.9125	0.6559	.5062	1.154	1.296	1.54723
1.65	.9035	0.6309	.4814	1.179	1.311	1.59257
1.70	.8947	0.6069	.4583	1.206	1.324	1.63931

$\gamma = 1{\cdot}67$

M	$\frac{T_0}{T_0^*}$	$\frac{T}{T^*}$	$\frac{p}{p^*}$	$\frac{p_0}{p_0^*}$	$\frac{V}{V^*}$	$\left(\frac{T_0}{T_0^*}\right)_{isoth}$
1.75	.8862	.5840	.4367	1.235	1.337	1.68744
1.80	.8779	.5620	.4165	1.266	1.349	1.73696
1.85	.8699	.5410	.3976	1.299	1.360	1.78789
1.90	.8621	.5209	.3799	1.334	1.371	1.84021
1.95	.8546	.5018	.3633	1.370	1.381	1.89392
2.00	.8474	.4835	.3477	1.408	1.391	1.94903
2.05	.8405	.4660	.3330	1.448	1.400	2.00553
2.10	.8338	.4493	.3192	1.490	1.408	2.06343
2.15	.8274	.4334	.3062	1.534	1.415	2.12273
2.20	.8213	.4183	.2940	1.580	1.423	2.18341
2.25	.8154	.4038	.2824	1.628	1.430	2.24550
2.30	.8097	.3899	.2715	1.678	1.436	2.30897
2.35	.8043	.3767	.2612	1.729	1.442	2.37385
2.40	.7991	.3641	.2514	1.783	1.448	2.44011
2.45	.7941	.3521	.2422	1.839	1.454	2.50778
2.50	.7893	.3406	.2334	1.897	1.459	2.57684
2.55	.7847	.3296	.2251	1.956	1.464	2.64729
2.60	.7803	.3191	.2173	2.018	1.469	2.71914
2.65	.7761	.3090	.2098	2.082	1.473	2.79239
2.70	.7721	.2994	.2027	2.148	1.477	2.86703
2.75	.7682	.2902	.1959	2.216	1.481	2.94307
2.80	.7644	.2814	.1895	2.287	1.485	3.02049
2.85	.7608	.2730	.1834	2.360	1.489	3.09932
2.90	.7574	.2649	.1775	2.435	1.493	3.17954
2.95	.7541	.2571	.1719	2.512	1.496	3.26115
3.00	.7509	.2497	.1666	2.587	1.499	3.34416
3.50	.7251	.1897	.1244	3.521	1.524	4.25100
4.00	.7072	.1484	.09632	4.716	1.541	5.29736
4.50	.6943	.1191	.07669	6.213	1.553	6.48321
5.00	.6848	.0975	.06246	8.044	1.561	7.80860
6.00	.6721	.06870	.04368	12.86	1.573	10.87789
7.00	.6642	.05092	.03224	19.44	1.580	14.50526
8.00	.6590	.03920	.02475	28.07	1.584	18.69067
9.00	.6553	.03110	.01959	39.05	1.587	23.43419
10.00	.6528	.02526	.01589	52.66	1.589	28.73566
∞	.6414	0	0	∞	1.599	∞

Oblique shocks: shock-wave angle versus flow-deflection angle

(Perfect gas, $\gamma = 1{\cdot}4$)

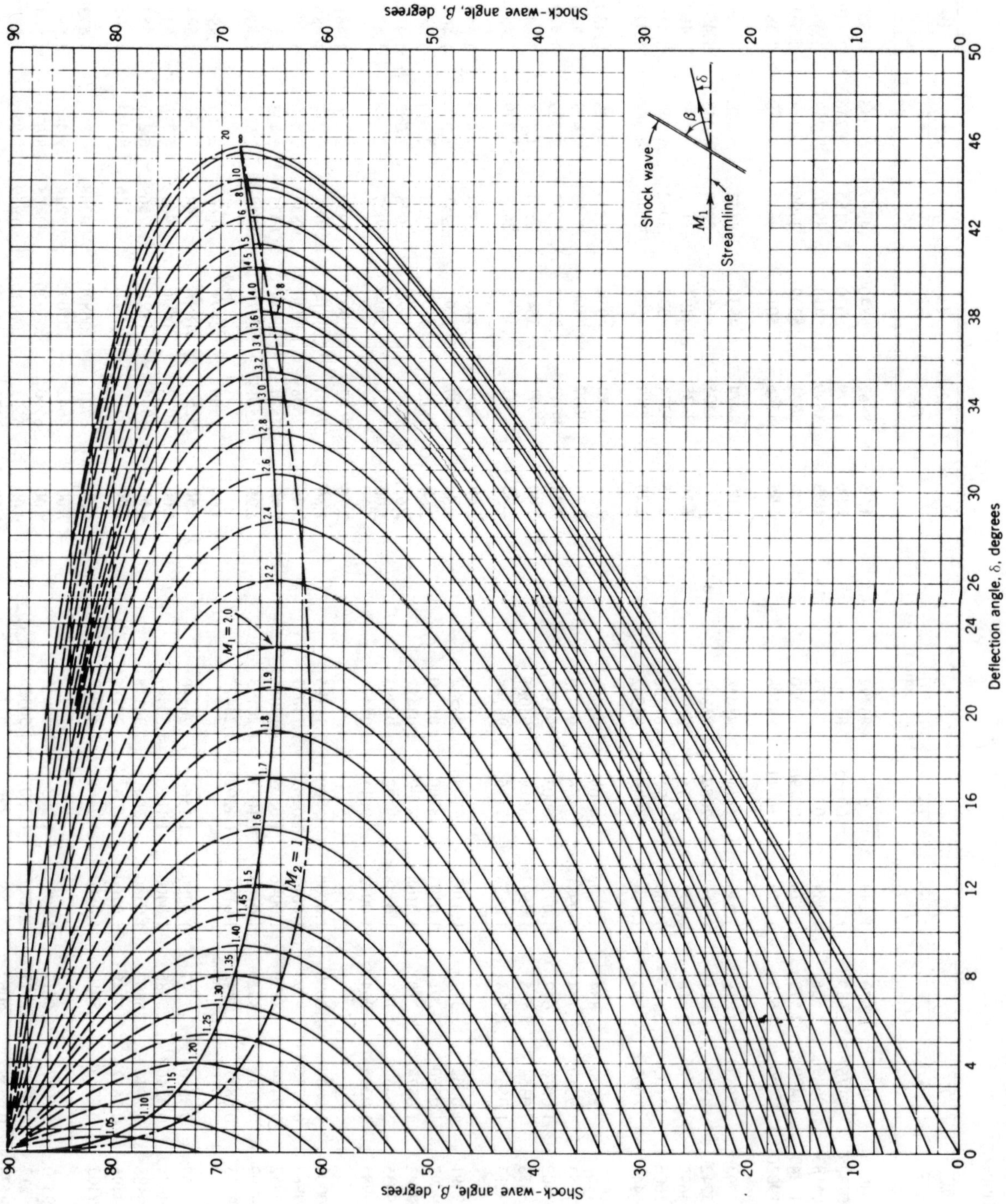

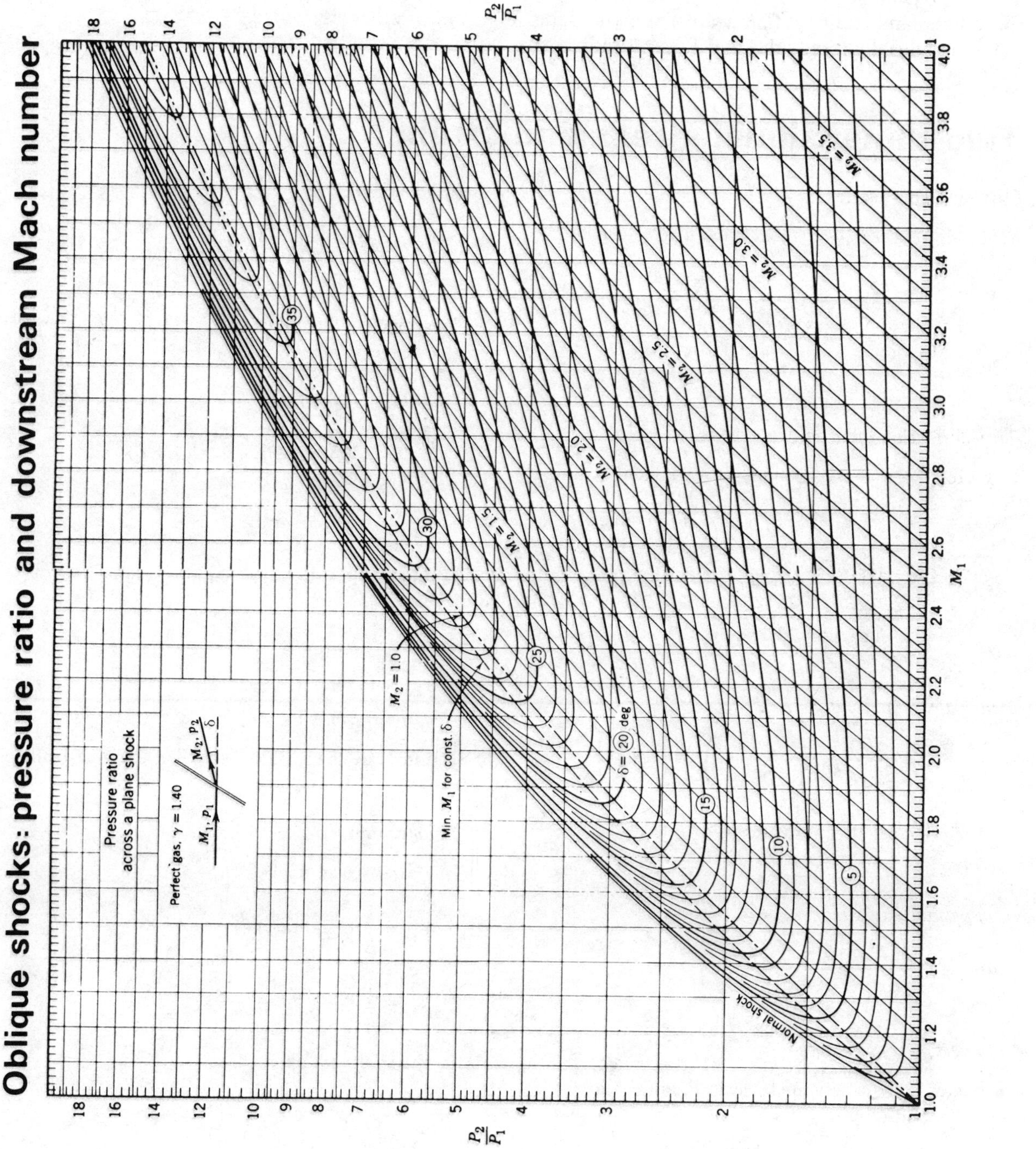
Oblique shocks: pressure ratio and downstream Mach number
Pressure ratio
across a plane shock
Perfect gas, γ = 1.40
M1, p1
M2, p2
δ
P2/P1
M1
Normal shock
Min. M1 for const. δ
M2 = 1.0
M2 = 1.5
M2 = 2.0
M2 = 2.5
M2 = 3.0
M2 = 3.5
δ = 20 deg
5
10
15
25
30
35

Solid mechanics and structures

In the following, u, v, w represent small displacements in the x, y, z directions (or as stated); σ, ϵ represent direct stress and strain (positive for tension), and τ, γ shear stress and strain; E, G, K, ν are Young modulus, shear modulus, bulk modulus and Poisson ratio; l, m, n are direction cosines and ψ a stress function; ρ is mass density. M, T are moment and torque; I is second moment or product moment of area; ω is a rotation unless otherwise stated.

Two-dimensional stress and strain

Cartesian coordinates

Relations between strains and small displacements

$$\epsilon_{xx} = \frac{\partial u}{\partial x} \qquad \epsilon_{yy} = \frac{\partial v}{\partial y} \qquad \gamma_{xy} = \frac{\partial v}{\partial x} + \frac{\partial u}{\partial y}$$

$$2\omega_z = \frac{\partial v}{\partial x} - \frac{\partial u}{\partial y}$$

Transformation of strain

For axes Ox' and Oy' inclined at θ to axes Ox and Oy:

$$\epsilon_{x'x'} = \frac{\epsilon_{xx} + \epsilon_{yy}}{2} + \frac{\epsilon_{xx} - \epsilon_{yy}}{2}\cos 2\theta + \frac{\gamma_{xy}}{2}\sin 2\theta$$

$$\epsilon_{y'y'} = \frac{\epsilon_{xx} + \epsilon_{yy}}{2} - \frac{\epsilon_{xx} - \epsilon_{yy}}{2}\cos 2\theta - \frac{\gamma_{xy}}{2}\sin 2\theta$$

$$\gamma_{x'y'} = (\epsilon_{yy} - \epsilon_{xx})\sin 2\theta + \gamma_{xy}\cos 2\theta$$

Principal strains

$$\begin{matrix}\epsilon_{max} \\ \epsilon_{min}\end{matrix} = \frac{\epsilon_{xx} + \epsilon_{yy}}{2} \pm \left\{\left(\frac{\epsilon_{xx} - \epsilon_{yy}}{2}\right)^2 + \left(\frac{\gamma_{xy}}{2}\right)^2\right\}^{1/2}$$

The principal directions are given by $\tan 2\theta_{xp} = \dfrac{\gamma_{xy}}{\epsilon_x - \epsilon_y}$

Compatibility of strains

$$\frac{\partial^2 \epsilon_{xx}}{\partial y^2} + \frac{\partial^2 \epsilon_{yy}}{\partial x^2} = \frac{\partial^2 \gamma_{xy}}{\partial x \partial y}$$

Transformation of stress

For axes Ox' and Oy' inclined at θ, anticlockwise, to axes Ox and Oy:

$$\sigma_{x'x'} = \frac{\sigma_{xx} + \sigma_{yy}}{2} + \frac{\sigma_{xx} - \sigma_{yy}}{2}\cos 2\theta + \tau_{xy}\sin 2\theta$$

$$\sigma_{y'y'} = \frac{\sigma_{xx} + \sigma_{yy}}{2} - \frac{\sigma_{xx} - \sigma_{yy}}{2}\cos 2\theta - \tau_{xy}\sin 2\theta$$

$$\tau_{x'y'} = \frac{\sigma_{yy} - \sigma_{xx}}{2}\sin 2\theta + \tau_{xy}\cos 2\theta$$

Principal stresses

$$\begin{matrix}\sigma_{max} \\ \sigma_{min}\end{matrix} = \frac{\sigma_{xx} + \sigma_{yy}}{2} \pm \left\{\left(\frac{\sigma_{xx} - \sigma_{yy}}{2}\right)^2 + \tau_{xy}{}^2\right\}^{1/2}$$

The principal directions are given by $\tan 2\theta_{xp} = \dfrac{2\tau_{xy}}{\sigma_{xx} - \sigma_{yy}}$

Equilibrium equations

$$\frac{\partial \sigma_{xx}}{\partial x} + \frac{\partial \tau_{xy}}{\partial y} + X = 0$$

$$\frac{\partial \tau_{yx}}{\partial x} + \frac{\partial \sigma_{yy}}{\partial y} + Y = 0$$

(X and Y are body forces per unit volume.)

Boundary conditions

$$\sigma_{xx}l + \tau_{xy}m = \overline{X}$$

$$\tau_{yx}l + \sigma_{yy}m = \overline{Y}$$

($\overline{X}$ and $\overline{Y}$ are the surface forces per unit area at the boundary.

Hooke's law

(i) *Plane stress*, $\sigma_{zz} = 0$

$$\epsilon_{xx} = \frac{1}{E}(\sigma_{xx} - \nu\sigma_{yy})$$

$$\epsilon_{yy} = \frac{1}{E}(\sigma_{yy} - \nu\sigma_{xx})$$

$$\epsilon_{zz} = -\frac{\nu}{E}(\sigma_{xx} + \sigma_{yy});$$

$$\sigma_{xx} = \frac{E}{(1-\nu^2)}(\epsilon_{xx} + \nu\epsilon_{yy})$$

$$\sigma_{yy} = \frac{E}{(1-\nu^2)}(\epsilon_{yy} + \nu\epsilon_{xx});$$

$$\gamma_{xy} = \frac{\tau}{G}$$

(ii) *Plane strain*, $\epsilon_{zz}=0$

$$\epsilon_{xx}=\frac{(1-v^2)}{E}\left(\sigma_{xx}-\frac{v}{1-v}\sigma_{yy}\right)$$

$$\epsilon_{yy}=\frac{1-v^2}{E}\left(\sigma_{yy}-\frac{v}{1-v}\sigma_{xx}\right);$$

$$\sigma_{xx}=\frac{E}{(1+v)(1-2v)}[(1-v)\epsilon_{xx}+v\epsilon_{yy}]$$

$$\sigma_{yy}=\frac{E}{(1+v)(1-2v)}[(1-v)\epsilon_{yy}+v\epsilon_{xx}]$$

$$\sigma_{zz}=v(\sigma_{xx}+\sigma_{yy})=\frac{E(\epsilon_{xx}+\epsilon_{yy})}{(1+v)(1-2v)};$$

$$\gamma_{xy}=\frac{\tau_{xy}}{G}$$

Stress function

If gravity is the only body force, the stresses are

$$\sigma_{xx}=\frac{\partial^2\psi}{\partial y^2}-\rho gy \qquad \sigma_{yy}=\frac{\partial^2\psi}{\partial x^2}-\rho gy$$

$$\tau_{xy}=-\frac{\partial^2\psi}{\partial x\partial y}$$

and the compatibility equation is

$$\frac{\partial^4\psi}{\partial x^4}+\frac{2\partial^4\psi}{\partial x^2\partial y^2}+\frac{\partial^4\psi}{\partial y^4}=0$$

or

$$\nabla^4\psi=0$$

Polar coordinates

Relations between strains and small displacements

$$\epsilon_{rr}=\frac{\partial u}{\partial r} \qquad \epsilon_{\theta\theta}=\frac{u}{r}+\frac{1}{r}\frac{\partial v}{\partial\theta} \qquad \gamma_{r\theta}=\frac{\partial v}{\partial r}-\frac{v}{r}+\frac{1}{r}\frac{\partial u}{\partial\theta}$$

(u and v are displacements in the radial and tangential directions).

Equilibrium equations

$$\frac{\partial\sigma_r}{\partial r}+\frac{1}{r}\frac{\partial\tau_{r\theta}}{\partial\theta}+\frac{\sigma_r-\sigma_\theta}{r}+F_r=0$$

$$\frac{1}{r}\frac{\partial\sigma_\theta}{\partial\theta}+\frac{\partial\tau_{r\theta}}{\partial r}+\frac{2\tau_{r\theta}}{r}+F_\theta=0$$

where F_r and F_θ are the body forces per unit volume.

Stress function

If the body forces are zero, the stresses are

$$\sigma_r=\frac{1}{r}\frac{\partial\psi}{\partial r}+\frac{1}{r^2}\frac{\partial^2\psi}{\partial\theta^2}$$

$$\sigma_\theta=\frac{\partial^2\psi}{\partial r^2}$$

$$\tau_{r\theta}=-\frac{\partial}{\partial r}\left(\frac{1}{r}\frac{\partial\psi}{\partial\theta}\right)$$

and the compatibility equation is

$$\left(\frac{\partial^2}{\partial r^2}+\frac{1}{r}\frac{\partial}{\partial r}+\frac{1}{r^2}\frac{\partial^2}{\partial\theta^2}\right)\left(\frac{\partial^2\psi}{\partial r^2}+\frac{1}{r}\frac{\partial\psi}{\partial r}+\frac{1}{r^2}\frac{\partial^2\psi}{\partial\theta^2}\right)=0$$

Thick cylinder under uniform pressure

$$\sigma_r=A+\frac{B}{r^2} \qquad \sigma_\theta=A-\frac{B}{r^2}$$

where A, B are constants.

Thin shell of revolution

For symmetrical loading with an outward normal component p per unit area, the hoop stress σ_θ and meridional stress σ_ϕ are related by

$$\sigma_\theta/r_\theta+\sigma_\phi/r_\phi=p/t$$

where r_θ and r_ϕ are the corresponding radii of curvature and t is the thickness.

Rotating discs and cylinders

If the angular velocity is ω, the body force is

$$F_r=\rho\omega^2 r$$

and the stresses are given by

$$r\sigma_r=\psi \qquad \sigma_\theta=\frac{\partial\psi}{\partial r}+\rho\omega^2r^2$$

Three-dimensional stress and strain

Cartesian coordinates

Relations between strains and small displacements

$$\epsilon_{xx}=\frac{\partial u}{\partial x} \qquad \epsilon_{yy}=\frac{\partial v}{\partial y} \qquad \epsilon_{zz}=\frac{\partial w}{\partial z}$$

$$\gamma_{xy}=\frac{\partial v}{\partial x}+\frac{\partial u}{\partial y} \qquad \gamma_{yz}=\frac{\partial w}{\partial y}+\frac{\partial v}{\partial z} \qquad \gamma_{zx}=\frac{\partial u}{\partial z}+\frac{\partial w}{\partial x}$$

$$2\omega_x=\frac{\partial w}{\partial y}-\frac{\partial v}{\partial z} \qquad 2\omega_y=\frac{\partial u}{\partial z}-\frac{\partial w}{\partial x} \qquad 2\omega_z=\frac{\partial v}{\partial x}-\frac{\partial u}{\partial y}$$

Transformation of strain

Original axes x, y, z; new axes x', y', z' with direction cosines as shown:

	x	y	z
x'	l_1	m_1	n_1
y'	l_2	m_2	n_2
z'	l_3	m_3	n_3

$$\epsilon_{x'x'} = \epsilon_{xx}l_1^2 + \epsilon_{yy}m_1^2 + \epsilon_{zz}n_1^2 + \gamma_{xy}l_1m_1 + \gamma_{yz}m_1n_1 + \gamma_{zx}n_1l_1$$

etc.

$$\gamma_{y'z'} = 2\epsilon_{xx}l_2l_3 + 2\epsilon_{yy}m_2m_3 + 2\epsilon_{zz}n_2n_3 + \gamma_{xy}(l_2m_3 + m_2l_3) + \gamma_{yz}(m_2n_3 + n_2m_3) + \gamma_{zx}(n_2l_3 + l_2n_3)$$

etc.

Compatibility of strains

$$\frac{\partial^2\epsilon_{xx}}{\partial y^2} + \frac{\partial^2\epsilon_{yy}}{\partial x^2} = \frac{\partial^2\gamma_{xy}}{\partial x\partial y}$$

and two similar equations.

$$2\frac{\partial^2\epsilon_{xx}}{\partial y\partial z} = \frac{\partial}{\partial x}\left(-\frac{\partial\gamma_{yz}}{\partial x} + \frac{\partial\gamma_{xz}}{\partial y} + \frac{\partial\gamma_{xy}}{\partial z}\right)$$

and two similar equations.

Principal stresses

The direction cosines of a principal plane having stress σ satisfy the equations

$$(\sigma_{xx} - \sigma)l + \tau_{xy}m + \tau_{xz}n = 0$$
$$\tau_{yx}l + (\sigma_{yy} - \sigma)m + \tau_{yz}n = 0$$
$$\tau_{zx}l + \tau_{zy}m + (\sigma_{zz} - \sigma)n = 0$$

By setting the determinant of the coefficients to zero a cubic in σ is obtained from which the principal stresses are obtained. The direction cosines can be found from the above and

$$l^2 + m^2 + n^2 = 1$$

Equilibrium equations

$$\frac{\partial\sigma_{xx}}{\partial x} + \frac{\partial\tau_{xy}}{\partial y} + \frac{\partial\tau_{xz}}{\partial z} + X = 0$$

$$\frac{\partial\tau_{yx}}{\partial x} + \frac{\partial\sigma_{yy}}{\partial y} + \frac{\partial\tau_{yz}}{\partial z} + Y = 0$$

$$\frac{\partial\tau_{zx}}{\partial x} + \frac{\partial\tau_{zy}}{\partial y} + \frac{\partial\sigma_{zz}}{\partial z} + Z = 0$$

(X, Y and Z are body forces per unit volume.)

Boundary conditions

$$\sigma_{xx}l + \tau_{xy}m + \tau_{xz}n = \bar{X}$$
$$\tau_{yx}l + \sigma_{yy}m + \tau_{yz}n = \bar{Y}$$
$$\tau_{zx}l + \tau_{zy}m + \sigma_{zz}n = \bar{Z}$$

($\bar{X}$, $\bar{Y}$ and $\bar{Z}$ are surface forces per unit area at the boundary.)

Hooke's law

$$\epsilon_{xx} = \frac{1}{E}\{\sigma_{xx} - \nu(\sigma_{yy} + \sigma_{zz})\}$$

etc.

$$\sigma_{xx} = \frac{\nu E}{(1+\nu)(1-2\nu)}\{\epsilon_{xx} + \epsilon_{yy} + \epsilon_{zz}\} + \frac{E}{1+\nu}\epsilon_{xx}$$

etc.

$$\gamma_{xy} = \frac{\tau_{xy}}{G} \quad \text{etc.}$$

Relations between elastic constants

$$G = \frac{E}{2(1+\nu)} \qquad K = \frac{E}{3(1-2\nu)}$$

The *Lamé constants* denoted μ and λ signify respectively the shear modulus G and $K - \frac{2}{3}G$.

Cylindrical coordinates

Relations between strains and small displacements

$$\epsilon_{rr} = \frac{\partial u}{\partial r} \qquad \epsilon_{\theta\theta} = \frac{u}{r} + \frac{1}{r}\frac{\partial v}{\partial\theta} \qquad \epsilon_{zz} = \frac{\partial w}{\partial z}$$

$$\gamma_{r\theta} = \frac{\partial v}{\partial r} - \frac{v}{r} + \frac{1}{r}\frac{\partial u}{\partial\theta} \qquad \gamma_{rz} = \frac{\partial w}{\partial r} + \frac{\partial u}{\partial z}$$

$$\gamma_{\theta z} = \frac{1}{r}\frac{\partial w}{\partial\theta} + \frac{\partial v}{\partial z}$$

(u, v, w are displacements in the radial, tangential and axial directions.)

Spherical coordinates

Relations between strains and small displacements

For displacements u_r, u_θ, u_ϕ:

$$\epsilon_{rr} = \frac{\partial u_r}{\partial r}$$

$$\epsilon_{\theta\theta} = \frac{1}{r}\frac{\partial u_\theta}{\partial \theta} + \frac{u_r}{r}$$

$$\epsilon_{\phi\phi} = \frac{1}{r\sin\theta}\frac{\partial u_\phi}{\partial \phi} + \frac{u_\theta}{r}\cot\theta + \frac{u_r}{r}$$

$$\epsilon_{\theta\phi} = \frac{1}{r}\frac{\partial u_\phi}{\partial \theta} - \frac{u_\phi}{r}\cot\theta + \frac{1}{r\sin\theta}\frac{\partial u_\theta}{\partial \phi}$$

$$\epsilon_{\phi r} = \frac{1}{r\sin\theta}\frac{\partial u_r}{\partial \phi} + \frac{\partial u_\phi}{\partial r} - \frac{u_\phi}{r}$$

$$\epsilon_{r\theta} = \frac{\partial u_\theta}{\partial r} - \frac{u_\theta}{r} + \frac{1}{r}\frac{\partial u_r}{\partial \theta}$$

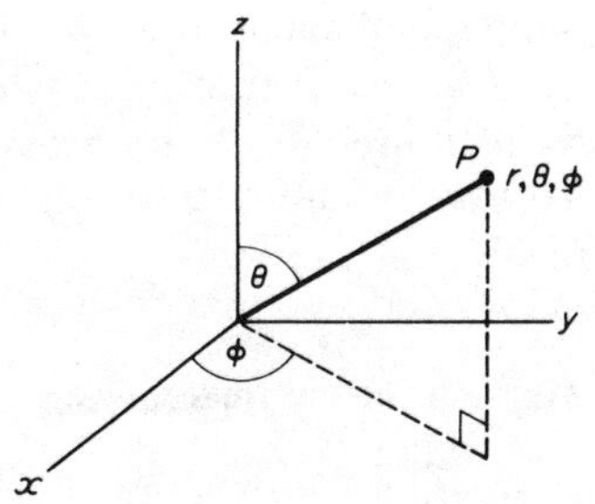

Bending of laterally loaded plates

For a plate in the x–y plane, on the assumption that $\sigma_{zz} = 0$,

$$\sigma_{xx} = -\frac{Ez}{1-\nu^2}\left(\frac{\partial^2 w}{\partial x^2} + \nu\frac{\partial^2 w}{\partial y^2}\right)$$

$$\sigma_{yy} = -\frac{Ez}{1-\nu^2}\left(\frac{\partial^2 w}{\partial y^2} + \nu\frac{\partial^2 w}{\partial x^2}\right)$$

$$\tau_{xy} = -\frac{Ez}{1+\nu}\frac{\partial^2 w}{\partial x\partial y}$$

$$M_x = -D\left(\frac{\partial^2 w}{\partial x^2} + \nu\frac{\partial^2 w}{\partial y^2}\right)$$

$$M_y = -D\left(\frac{\partial^2 w}{\partial y^2} + \nu\frac{\partial^2 w}{\partial x^2}\right)$$

$$M_{xy} = -D(1-\nu)\frac{\partial^2 w}{\partial x\partial y}$$

where, for plate thickness t,

$$D = \frac{Et^3}{12(1-\nu^2)}$$

The differential equation for deflection is

$$\frac{\partial^4 w}{\partial x^4} + 2\frac{\partial^4 w}{\partial x^2\partial y^2} + \frac{\partial^4 w}{\partial y^4} = \frac{p}{D}$$

where p is the load per unit area in the z direction.

Circular plates

$$M_r = -D\left[\frac{\partial^2 w}{\partial r^2} + \nu\left(\frac{1}{r}\frac{\partial w}{\partial r} + \frac{1}{r^2}\frac{\partial^2 w}{\partial \theta^2}\right)\right]$$

$$M_\theta = -D\left[\frac{1}{r}\frac{\partial w}{\partial r} + \frac{1}{r^2}\frac{\partial^2 w}{\partial \theta^2} + \nu\frac{\partial^2 w}{\partial r^2}\right]$$

$$M_{r\theta} = -(1-\nu)D\left(\frac{1}{r}\frac{\partial^2 w}{\partial r\partial\theta} - \frac{1}{r^2}\frac{\mathrm{d}w}{\partial\theta}\right)$$

The differential equation for deflection is

$$\left(\frac{\partial^2}{\partial r^2} + \frac{1}{r}\frac{\partial}{\partial r} + \frac{1}{r^2}\frac{\partial^2}{\partial\theta^2}\right)^2 w = \frac{p}{D}$$

which for radial symmetry becomes

$$\frac{1}{r}\frac{\mathrm{d}}{\mathrm{d}r}\left\{r\frac{\mathrm{d}}{\mathrm{d}r}\left[\frac{1}{r}\frac{\mathrm{d}}{\mathrm{d}r}\left(r\frac{\mathrm{d}w}{\mathrm{d}r}\right)\right]\right\} = \frac{p}{D}$$

Yield and failure criteria

Yield criteria

In the following σ_1, σ_2 and σ_3 are principal stresses and σ_Y the yield stress in uniaxial tension.

Von Mises

$$(\sigma_1-\sigma_2)^2 + (\sigma_2-\sigma_3)^2 + (\sigma_3-\sigma_1)^2 = 2\sigma_Y^2$$

For plane stress with $\sigma_3 = 0$,

$$\sigma_1^2 - \sigma_1\sigma_2 + \sigma_2^2 = \sigma_Y^2$$

and in pure shear yielding occurs at the shear stress $\tau = \sigma_Y/\sqrt{3}$.

Tresca

If $\sigma_1 \geqslant \sigma_2 \geqslant \sigma_3$, then yielding occurs when

$$\sigma_1 - \sigma_3 = \sigma_Y$$

and the shear stress is then $\sigma_Y/2$.

Ultimate tensile strength

On the assumption that necking occurs at constant volume, the point of instability (i.e. maximum load) is that point on the curve of true stress against nominal strain ϵ_n touched by the tangent which intersects the axis at $\epsilon_n = -1$; this is *Considère's construction.* If true stress varies as the nth power of true strain ϵ_t this point occurs at $\epsilon_t = n$. True strain is given by $\ln(l/l_0)$ when the length l has the unstrained value l_0.

Griffith theory for cracks

A crack of half-length c in a material of Young's modulus E will lead to tensile failure at a stress

$$\sigma = \sqrt{\frac{2\gamma E}{\pi c}}$$

where γ is the free surface energy per unit area, or

$$\sigma = \sqrt{\frac{2E(\gamma + p)}{\pi c}}$$

where p is an additional energy per unit area due to plastic deformation at the crack.

Fatigue

In the following N_f is the number of cycles to failure and a, b are constants. The relations are empirical and of varying accuracy.

Basquin's law: If stress is always less than the yield value and has a peak-to-peak variation $\Delta\sigma$ about a zero mean, then

$$\Delta\sigma N_f^a = \text{constant}.$$

If the mean stress is σ_m, the value of $\Delta\sigma$ for given N_f is reduced by the factor $(1 - |\sigma_m|/\sigma_f)$ where σ_f is the tensile strength of the material (Goodman's rule).

Coffin–Manson law: If the peak stress exceeds the yield value and the plastic strain has a peak-to-peak range $\Delta\epsilon_p$, then

$$\Delta\epsilon_p N_f^b = \text{constant}.$$

Miner's rule: If a stress range $\Delta\sigma_i$ which would cause failure in N_i cycles is applied for n_i cycles, then failure occurs when

$$\frac{n_1}{N_1} + \frac{n_2}{N_2} + \cdots = 1$$

Elastic behaviour of structural members

In this section the origin of right-handed axes x, y, z is at the centroid of a cross-section of area A and the member has length L, aligned with the x-axis; unless otherwise stated, y and z are principal axes. I_{xx} is denoted I_x, etc. and I without subscript refers to a principal axis about which the moment M acts. A bending moment at any cross-section is taken to be positive if a positive couple acts on the positive x-face. A positive couple is one which acts in a right-handed sense about its axis. For sections without circular symmetry I_x denotes an effective polar moment such that GI_x is the torsional rigidity.

Bending stress for straight beams

For any orientation of the y, z axes the stress due to elastic bending is

$$\sigma_{xx} = \frac{(M_y I_z + M_z I_{yz})z - (M_z I_y + M_y I_{yz})y}{I_y I_z - I_{yz}^2}.$$

For bending about a principal axis z this becomes
$$\sigma_{xx} = My/I$$

Winkler theory for curved beams

The direct stress due to a moment M about the z-axis is

$$\sigma_{xx} = -\frac{M}{AR_0}\left[1 + \frac{R_0^2 y}{h^2(R_0 + y)}\right]$$

where

$$h^2 = \frac{R_0}{A}\int \frac{y^2}{R_0 + y}\,dA$$

and R_0 is the original radius of curvature.

Deflection of beams

The curvature due to a displacement $v(x)$ is

$$\frac{1}{R} = \frac{d^2v/dx^2}{\{1 + (dv/dx)^2\}^{3/2}} \approx \frac{d^2v}{dx^2}$$

for small v.

The curvatures due to bending, for any orientation of x and y, are

$$\frac{1}{R_z} = \frac{M_z I_y + M_y I_{yz}}{E(I_y I_z - I_{yz}^2)}; \qquad \frac{1}{R_y} = \frac{M_y I_z + M_z I_{yz}}{E(I_y I_z - I_{yz}^2)}$$

For bending about a principal axis the curvature is

$$\frac{1}{R} = \frac{M}{EI}$$

The end slope θ_e end deflection δ_e and central deflection δ_c are given below for the loadings shown. In each case the length is L.

	θ_e	δ_e	δ_c
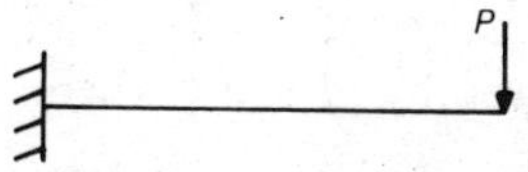	$\frac{PL^2}{2EI}$	$\frac{PL^3}{3EI}$	$\frac{5PL^3}{48EI}$
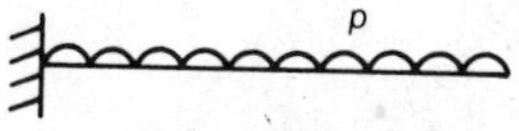	$\frac{pL^3}{6EI}$	$\frac{pL^4}{8EI}$	$\frac{17pL^4}{384EI}$
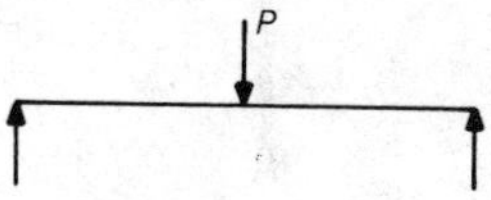	$\frac{PL^2}{16EI}$	0	$\frac{PL^3}{48EI}$
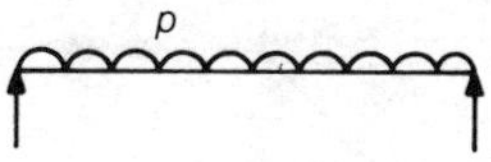	$\frac{pL^3}{24EI}$	0	$\frac{5pL^4}{384EI}$
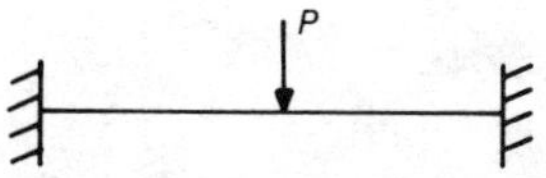	0	0	$\frac{PL^3}{192EI}$
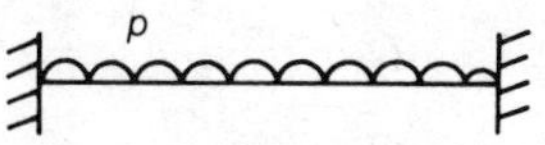	0	0	$\frac{pL^4}{384EI}$
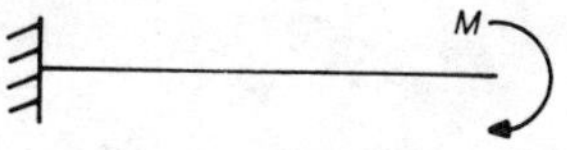	$\frac{ML}{EI}$	$\frac{ML^2}{2EI}$	$\frac{ML^2}{8EI}$

Shear stress in bending

On any cross-section the shear force is given by

$$F_y = -\frac{dM_z}{dx}; \qquad F_z = \frac{dM_y}{dx}$$

In a beam of open thin-walled section, for any orientation of the x, y axes a shear force F_y produces at any point P where the thickness is t a shear stress given by

$$\tau = \frac{F_y A_P(I_y\bar{y} - I_{yz}\bar{z})}{t(I_y I_z - I_{yz}^2)}$$

where A_P is the area of the section beyond the point P and $\bar{y}$, $\bar{z}$ define the centroid of A_P.

When y is a principal axis this becomes

$$\tau = \frac{F_y A_P \bar{y}}{t I_z}$$

and this expression applies also to a section of any shape if t signifies the breadth at any y and if τ is assumed to be constant over the breadth.

Torsion

In the following, θ is the angle of twist per unit length.

General

The displacements in a cross-section rotated by θx are

$$u = \theta\phi(y, z) \qquad v = -\theta xz \qquad w = \theta xy$$

where ϕ is a function which satisfies the equation

$$\frac{\partial^2\phi}{\partial x^2} + \frac{\partial^2\phi}{\partial y^2} = 0.$$

The shear stresses are given by

$$\tau_{xy} = \frac{\partial\psi}{\partial z} \qquad \tau_{xz} = -\frac{\partial\psi}{\partial y}$$

where ψ is a stress function which satisfies the equation

$$\frac{\partial^2\psi}{\partial y^2} + \frac{\partial^2\psi}{\partial z^2} = -2G\theta.$$

The torque is given by

$$T = 2\iint \psi \, dy\, dz$$

Circular sections

The shear stress at radius r is given by

$$\frac{\tau}{r} = \frac{T}{I_x} = G\theta$$

and the torsional rigidity is

$$\frac{T}{\theta} = GI_x \quad (\text{or } GJ)$$

Closed thin-walled sections

For any closed thin-walled section in torsion the shear flow is constant and given by

$$\tau t = \frac{T}{2A}$$

where A is the area *enclosed* and t the wall thickness. The torsional rigidity is

$$\frac{T}{\theta} = \frac{4GA^2}{\int ds/t}$$

where ds is an element of perimeter. The quantity $4A^2/\int ds/t$ is the effective polar moment of area, and except for circular tubes differs from the actual polar moment I_x (or J).

Open thin sections

The torsional rigidity of a thin rectangular section of thickness t and breadth b is

$$\frac{T}{\theta} = \frac{1}{3}bt^3G$$

The same expression can be applied to any thin open section if b is taken to be the edge-to-edge perimeter.

Flexibility matrix: one-dimensional member

For a cantilever the deflections which result from the loading $W = P_x \cdots M_z$ shown are given by $[F]\{W\}$ where

$$
[F] = \begin{array}{c} \begin{matrix} P_x & P_y & P_z & M_x & M_y & M_z \end{matrix} \\ \begin{bmatrix} \dfrac{L}{EA} & 0 & 0 & 0 & 0 & 0 \\ 0 & \dfrac{L^3}{3EI_z} & 0 & 0 & 0 & \dfrac{L^2}{2EI_z} \\ 0 & 0 & \dfrac{L^3}{3EI_y} & 0 & \dfrac{-L^2}{2EI_y} & 0 \\ 0 & 0 & 0 & \dfrac{L}{GI_x} & 0 & 0 \\ 0 & 0 & \dfrac{-L^2}{2EI_y} & 0 & \dfrac{L}{EI_y} & 0 \\ 0 & \dfrac{L^2}{2EI_z} & 0 & 0 & 0 & \dfrac{L}{EI_z} \end{bmatrix} \end{array}
$$

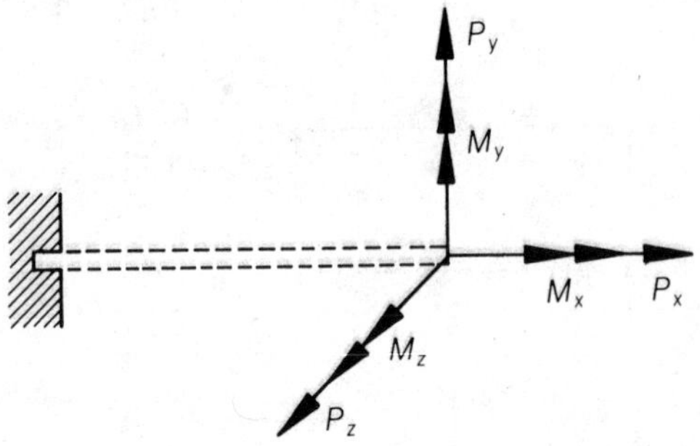

Stiffness matrix: one-dimensional member

The loading to produce deflection $\delta = u_1 \cdots \theta_{z2}$ as shown is given by $[K]\{\delta\}$ where

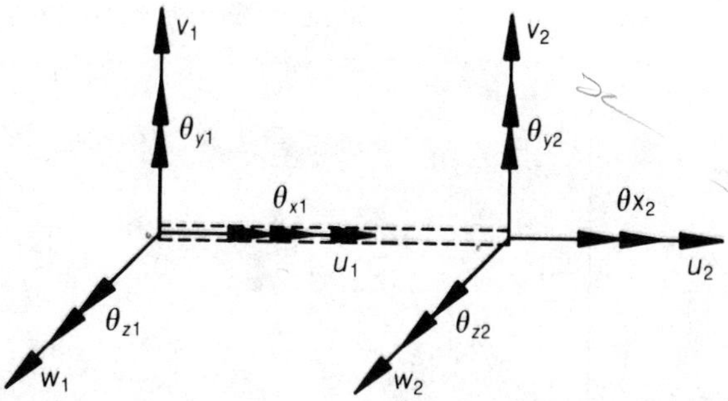

$$[K] = \begin{array}{c} \\ \end{array}\left[\begin{array}{cccccc|cccccc}
u_1 & v_1 & w_1 & \theta_{x1} & \theta_{y1} & \theta_{z1} & u_2 & v_2 & w_2 & \theta_{x2} & \theta_{y2} & \theta_{z2} \\
\frac{AE}{L} & 0 & 0 & 0 & 0 & 0 & -\frac{AE}{L} & 0 & 0 & 0 & 0 & 0 \\
0 & \frac{12EI_z}{L^3} & 0 & 0 & 0 & \frac{6EI_z}{L^2} & 0 & -\frac{12EI_z}{L^3} & 0 & 0 & 0 & \frac{6EI_z}{L^2} \\
0 & 0 & \frac{12EI_y}{L^3} & 0 & -\frac{6EI_y}{L^2} & 0 & 0 & 0 & -\frac{12EI_y}{L^3} & 0 & -\frac{6EI_y}{L^2} & 0 \\
0 & 0 & 0 & \frac{GI_x}{L} & 0 & 0 & 0 & 0 & 0 & -\frac{GI_x}{L} & 0 & 0 \\
0 & 0 & -\frac{6EI_y}{L^2} & 0 & \frac{4EI_y}{L} & 0 & 0 & 0 & \frac{6EI_y}{L^2} & 0 & \frac{2EI_y}{L} & 0 \\
0 & \frac{6EI_z}{L^2} & 0 & 0 & 0 & \frac{4EI_z}{L} & 0 & -\frac{6EI_z}{L^2} & 0 & 0 & 0 & \frac{2EI_z}{L} \\
\hline
-\frac{AE}{L} & 0 & 0 & 0 & 0 & 0 & \frac{AE}{L} & 0 & 0 & 0 & 0 & 0 \\
0 & -\frac{12EI_z}{L^3} & 0 & 0 & 0 & -\frac{6EI_z}{L^2} & 0 & \frac{12EI_z}{L^3} & 0 & 0 & 0 & -\frac{6EI_z}{L^2} \\
0 & 0 & -\frac{12EI_y}{L^3} & 0 & \frac{6EI_y}{L^2} & 0 & 0 & 0 & \frac{12EI_y}{L^3} & 0 & \frac{6EI_y}{L^2} & 0 \\
0 & 0 & 0 & -\frac{GI_x}{L} & 0 & 0 & 0 & 0 & 0 & \frac{GI_x}{L} & 0 & 0 \\
0 & 0 & -\frac{6EI_y}{L^2} & 0 & \frac{2EI_y}{L} & 0 & 0 & 0 & \frac{6EI_y}{L^2} & 0 & \frac{4EI_y}{L} & 0 \\
0 & \frac{6EI_z}{L^2} & 0 & 0 & 0 & \frac{2EI_z}{L} & 0 & -\frac{6EI_z}{L^2} & 0 & 0 & 0 & \frac{4EI_z}{L}
\end{array}\right]$$

For deflection in the xy plane only, this becomes

$$[K]=\begin{bmatrix} u_1 & v_1 & \theta_{z1} & u_2 & v_2 & \theta_{z2} \\ \frac{AE}{L} & 0 & 0 & -\frac{AE}{L} & 0 & 0 \\ 0 & \frac{12EI_z}{L^3} & \frac{6EI_z}{L^2} & 0 & -\frac{12EI_z}{L^3} & \frac{6EI_z}{L^2} \\ 0 & \frac{6EI_z}{L^2} & \frac{4EI_z}{L} & 0 & -\frac{6EI_z}{L^2} & \frac{2EI_z}{L} \\ -\frac{AE}{L} & 0 & 0 & \frac{AE}{L} & 0 & 0 \\ 0 & -\frac{12EI_z}{L^3} & -\frac{6EI_z}{L^2} & 0 & \frac{12EI_z}{L^3} & -\frac{6EI_z}{L^2} \\ 0 & \frac{6EI_z}{L^2} & \frac{2EI_z}{L} & 0 & -\frac{6EI_z}{L^2} & \frac{4EI_z}{L} \end{bmatrix}$$

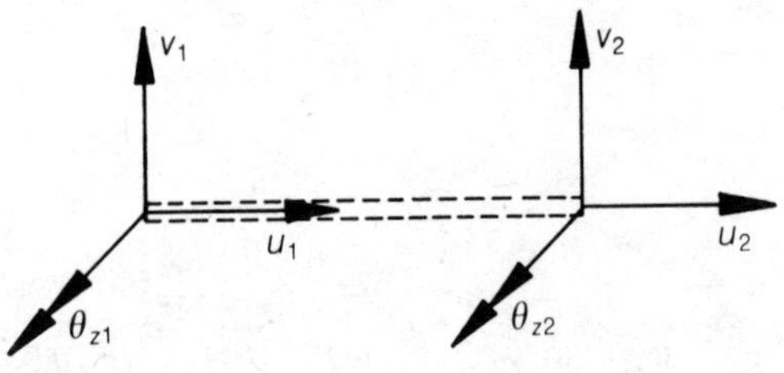

Slope-deflection relations

The slope-deflection equations for a uniform straight member AB of length L are

$$M_{AB}=\frac{2EI}{L}\left(2\theta_A+\theta_B-\frac{3\delta}{L}\right)+M_{FA}$$

$$M_{BA}=\frac{2EI}{L}\left(2\theta_B+\theta_A-\frac{3\delta}{L}\right)+M_{FB}$$

or

$$\begin{Bmatrix} M_{AB} \\ M_{BA} \end{Bmatrix}=\frac{EI}{L}\begin{bmatrix} 4 & 2 & -6 \\ 2 & 4 & -6 \end{bmatrix}\begin{Bmatrix} \theta_A \\ \theta_B \\ \delta/L \end{Bmatrix}+\begin{Bmatrix} M_{FA} \\ M_{FB} \end{Bmatrix}$$

where δ is the deflection of B relative to A and M_{FA}, M_{FB} are the fixed-end moments on AB due to transverse loading, if any (see below). Moments and deflections are taken to be positive if clockwise. The equations give the following end moments for the cases shown with $M_{FA}=M_{FB}=0$.

		θ_A	θ_B	δ	M_{AB}	M_{BA}
End A fixed	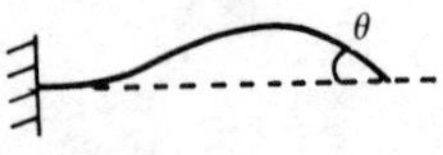	0	θ	0	$\frac{2EI\theta}{L}$	$\frac{4EI\theta}{L}$
End A pinned	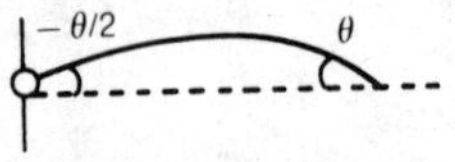	$-\frac{\theta}{2}$	θ	0	0	$\frac{3EI\theta}{L}$
Symmetry	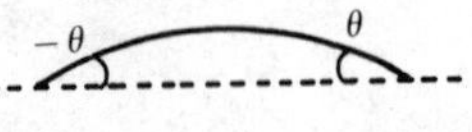	$-\theta$	θ	0	$-\frac{2EI\theta}{L}$	$\frac{2EI\theta}{L}$
Skew symmetry		θ	θ	0	$\frac{6EI\theta}{L}$	$\frac{6EI\theta}{L}$
Sidesway	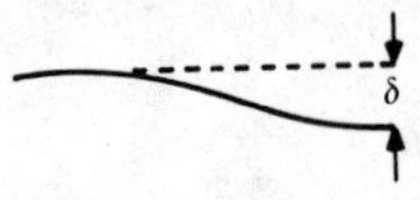	0	0	δ	$-\frac{6EI\delta}{L^2}$	$-\frac{6EI\delta}{L^2}$

Fixed-end moments

For a clamped member ($\theta_A = \theta_B = 0$, $\delta = 0$) the end moments due to transverse loading are as shown.

Uniformly distributed load p:

$$M_{AB} = M_{BA} = \frac{pL^2}{12}$$

Concentrated load

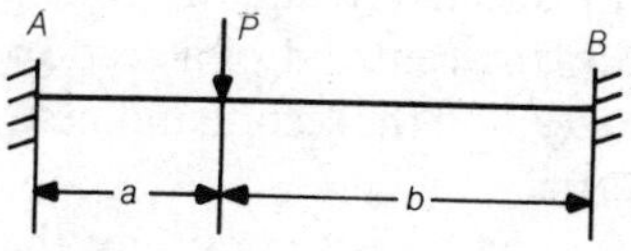

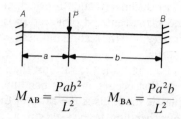

$$M_{AB} = \frac{Pab^2}{L^2} \qquad M_{BA} = \frac{Pa^2b}{L^2}$$

General loading

If the loading would give, when applied to the same span with free ends, a bending-moment diagram of area X with centroid distant a from A and b from B, then it produces fixed-end moments

$$M_{AB} = \frac{2X(2b - a)}{L^2} \qquad M_{BA} = \frac{2X(2a - b)}{L^2}$$

Stability of struts

Euler critical loads

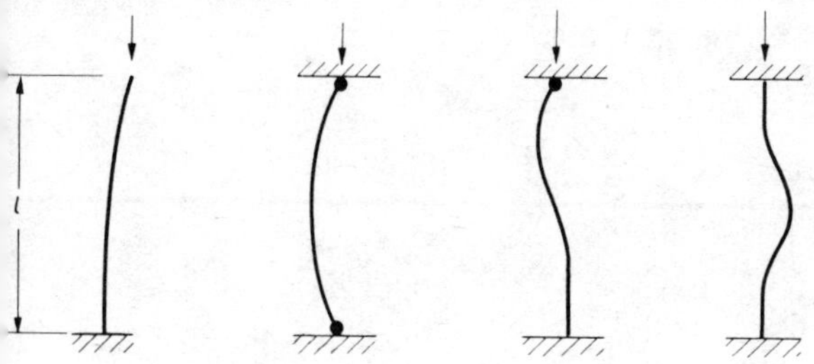

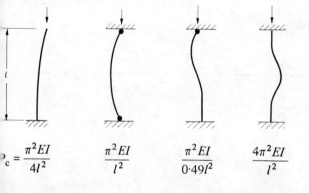

$$P_c = \frac{\pi^2 EI}{4l^2} \qquad \frac{\pi^2 EI}{l^2} \qquad \frac{\pi^2 EI}{0{\cdot}49l^2} \qquad \frac{4\pi^2 EI}{l^2}$$

Energy methods

Rayleigh

$$P_c = \frac{\int_0^l EI\left(\frac{d^2y}{dx^2}\right)^2 dx}{\int_0^l \left(\frac{dy}{dx}\right)^2 dx}$$

Timoshenko

$$P_c = \frac{\int_0^l \left(\frac{dy}{dx}\right)^2 dx}{\int_0^l \frac{y^2}{EI}\, dx}$$

Dimensions and properties of British Standard sections

In the following tables the symbols are not always consistent with those defined earlier. The centroidal axes of a cross-section which are parallel to the flange and the web are denoted by x and y respectively, and the principal pair, if different, by u and v; the centroid of a section is referred to as 'centre of gravity' and the second moment of area as 'moment of inertia'.

The elastic modulus of a section is the ratio of the moment at which yielding first occurs, in pure bending, to the yield stress; the plastic modulus is the ratio of the fully plastic moment, in pure bending, to the yield stress. The buckling parameter u and torsion index x govern the load for the onset of lateral-torsional buckling in bending (see British Standard 5950 Part I: 1985). The warping constant H governs the effect of warping on torsion: the torque required for a twist θ per unit length z is

$$T = GJ\theta - EH\frac{d^2\theta}{dz^2}$$

where GJ is the torsional rigidity and J is here referred to as the torsional constant (p. 113). For beams, columns, joists and channels the warping constant is given approximately by

$$H = \frac{(D-T)^2 B^3 T}{24}$$

and for Tees by

$$H = \frac{I_y}{4}(A - \tfrac{1}{2}T)^2,$$

in which A, B, D and T are dimensions defined in the tables.

JOISTS – DIMENSIONS AND PROPERTIES
To BS 4: Part 1: 1980

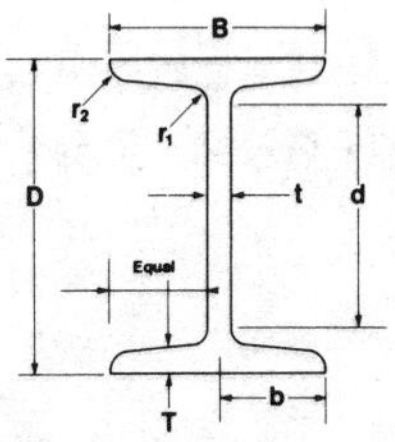

Inside slope 8°

Designation		Depth of Section	Width of Section	Thickness		Root radius	Toe radius	Web	Ratios for Local Buckling		Moment of inertia		Radius of gyration	
Serial size	Mass per metre	D	B	Web t	Flange T	r_1	r_2	Depth between fillets d	Flange b/T	Web d/t	Axis x-x	Axis y-y	Axis x-x	Axis y-y
mm	kg	mm	mm	mm	mm	mm	mm	mm			cm^4	cm^4	cm	cm
254x203	81·85	254·0	203·2	10·2	19·9	19·6	9·7	166·6	5·11	16·3	12016	2278	10·7	4·67
254x114	37·20	254·0	114·3	7·6	12·8	12·4	6·1	199·1	4·46	26·2	5092	270·1	10·4	2·39
203x152	52·09	203·2	152·4	8·9	16·5	15·5	7·6	133·4	4·62	15·0	4789	813·3	8·48	3·51
152x127	37·20	152·4	127·0	10·4	13·2	13·5	6·6	94·5	4·81	9·08	1818	378·8	6·20	2·82
127x114	29·76	127·0	114·3	10·2	11·5	12·4	4·8	71·9	4·97	7·78	979·0	241·9	5·12	2·55
127x114	26·79	127·0	114·3	7·4	11·4	9·9	5·0	79·2	5·01	10·7	944·8	235·4	5·26	2·63
114x114	26·79	114·3	114·3	9·5	10·7	14·2	3·2	61·1	5·34	6·41	735·4	223·1	4·62	2·54
102x102	23·07	101·6	101·6	9·5	10·3	11·1	3·2	54·0	4·93	5·81	486·1	154·4	4·06	2·29
89x89	19·35	88·9	88·9	9·5	9·9	11·1	3·2	45·2	4·49	4·65	306·7	101·1	3·51	2·01
76x76	12·65	76·2	76·2	5·1	8·4	9·4	4·6	38·1	4·54	7·45	158·6	52·03	3·12	1·80

JOISTS – DIMENSIONS AND PROPERTIES
BS 4: Part 1: 1980

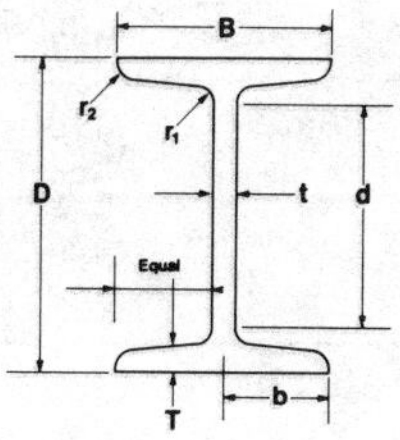

Elastic modulus		Plastic modulus		Buckling Parameter	Torsional Index	Warping Constant	Torsional Constant	Area of Section	Designation	
Axis x-x	Axis y-y	Axis x-x	Axis y-y	u	x	H	J		Mass per metre	Serial size
cm^3	cm^3	cm^3	cm^3			dm^6	cm^4	cm^2	kg	mm
946·1	224·3	1076	370·4	0·890	11·0	0·312	153	104·4	81·85	**254x203**
401·0	47·19	460·0	79·30	0·885	18·6	0·0393	25·5	47·4	37·20	**254x114**
471·4	106·7	539·8	175·6	0·891	10·7	0·0709	64·9	66·4	52·09	**203x152**
238·7	59·65	278·6	99·85	0·866	9·29	0·0183	34·2	47·5	37·20	**152x127**
154·2	42·32	180·9	70·85	0·853	8·74	0·00807	20·9	37·3	29·76	**127x114**
148·8	41·19	171·9	68·07	0·869	9·30	0·00787	16·9	34·1	26·79	**127x114**
128·6	39·00	151·2	65·63	0·841	7·90	0·00599	19·0	34·4	26·79	**114x114**
95·72	30·32	113·4	50·70	0·836	7·39	0·00321	14·4	29·4	23·07	**102x102**
69·04	22·78	82·77	38·03	0·830	6·54	0·00158	11·6	24·9	19·35	**89x89**
41·62	13·60	48·84	22·51	0·852	7·16	0·000597	4·67	16·3	12·65	**76x76**

UNIVERSAL BEAMS – DIMENSIONS AND PROPERTIES
To BS 4: Part 1: 1980

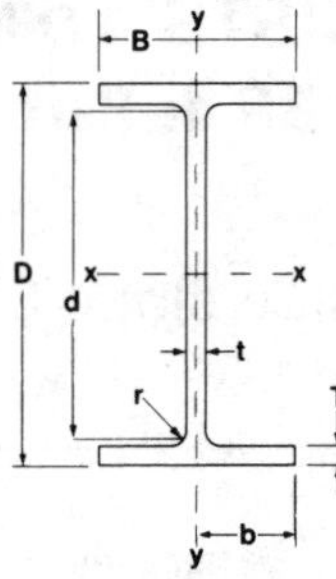

Designation		Depth of Section	Width of Section	Thickness		Root Radius	Web	Ratios For Local Buckling		Moment of inertia		Radius of gyration	
Serial size	Mass per metre	D	B	Web t	Flange T	r	Depth between fillets d	Flange b/T	Web d/t	Axis x-x	Axis y-y	Axis x-x	Axis y-y
mm	kg	mm	mm	mm	mm	mm	mm			cm^4	cm^4	cm	cm
914 x 419	388	920·5	420·5	21·5	36·6	24·1	799·0	5·74	37·2	718742	45407	38·1	9·58
	343	911·4	418·5	19·4	32·0	24·1	799·0	6·54	41·2	625282	39150	37·8	9·46
914 x 305	289	926·6	307·8	19·6	32·0	19·1	824·4	4·81	42·1	504594	15610	37·0	6·51
	253	918·5	305·5	17·3	27·9	19·1	824·4	5·47	47·7	436610	13318	36·8	6·42
	224	910·3	304·1	15·9	23·9	19·1	824·4	6·36	51·9	375924	11223	36·3	6·27
	201	903·0	303·4	15·2	20·2	19·1	824·4	7·51	54·2	325529	9427	35·6	6·06
838 x 292	226	850·9	293·8	16·1	26·8	17·8	761·7	5·48	47·3	339747	11353	34·3	6·27
	194	840·7	292·4	14·7	21·7	17·8	761·7	6·74	51·8	279450	9069	33·6	6·06
	176	834·9	291·6	14·0	18·8	17·8	761·7	7·76	54·4	246029	7792	33·1	5·90
762 x 267	197	769·6	268·0	15·6	25·4	16·5	685·8	5·28	44·0	239894	8174	30·9	5·71
	173	762·0	266·7	14·3	21·6	16·5	685·8	6·17	48·0	205177	6846	30·5	5·57
	147	753·9	265·3	12·9	17·5	16·5	685·8	7·58	53·2	168966	5468	30·0	5·39
686 x 254	170	692·9	255·8	14·5	23·7	15·2	615·0	5·40	42·4	170147	6621	28·0	5·53
	152	687·6	254·5	13·2	21·0	15·2	615·0	6·06	46·6	150319	5782	27·8	5·46
	140	683·5	253·7	12·4	19·0	15·2	615·0	6·68	49·6	136276	5179	27·6	5·38
	125	677·9	253·0	11·7	16·2	15·2	615·0	7·81	52·6	118003	4379	27·2	5·24
610 x 305	238	633·0	311·5	18·6	31·4	16·5	537·2	4·96	28·9	207571	15838	26·1	7·22
	179	617·5	307·0	14·1	23·6	16·5	537·2	6·50	38·1	151631	11412	25·8	7·08
	149	609·6	304·8	11·9	19·7	16·5	537·2	7·74	45·1	124660	9300	25·6	6·99
610 x 229	140	617·0	230·1	13·1	22·1	12·7	547·2	5·21	41·8	111844	4512	25·0	5·03
	125	611·9	229·0	11·9	19·6	12·7	547·2	5·84	46·0	98579	3933	24·9	4·96
	113	607·3	228·2	11·2	17·3	12·7	547·2	6·60	48·9	87431	3439	24·6	4·88
	101	602·2	227·6	10·6	14·8	12·7	547·2	7·69	51·6	75720	2912	24·2	4·75
533 x 210	122	544·6	211·9	12·8	21·3	12·7	476·5	4·97	37·2	76207	3393	22·1	4·67
	109	539·5	210·7	11·6	18·8	12·7	476·5	5·60	41·1	66739	2937	21·9	4·60
	101	536·7	210·1	10·9	17·4	12·7	476·5	6·04	43·7	61659	2694	21·8	4·56
	92	533·1	209·3	10·2	15·6	12·7	476·5	6·71	46·7	55353	2392	21·7	4·51
	82	528·3	208·7	9·6	13·2	12·7	476·5	7·91	49·6	47491	2005	21·3	4·38
457 x 191	98	467·4	192·8	11·4	19·6	10·2	407·9	4·92	35·8	45717	2343	19·1	4·33
	89	463·6	192·0	10·6	17·7	10·2	407·9	5·42	38·5	41021	2086	19·0	4·28
	82	460·2	191·3	9·9	16·0	10·2	407·9	5·98	41·2	37103	1871	18·8	4·23
	74	457·2	190·5	9·1	14·5	10·2	407·9	6·57	44·8	33388	1671	18·7	4·19
	67	453·6	189·9	8·5	12·7	10·2	407·9	7·48	48·0	29401	1452	18·5	4·12

UNIVERSAL BEAMS – DIMENSIONS AND PROPERTIES
To BS 4: Part 1: 1980

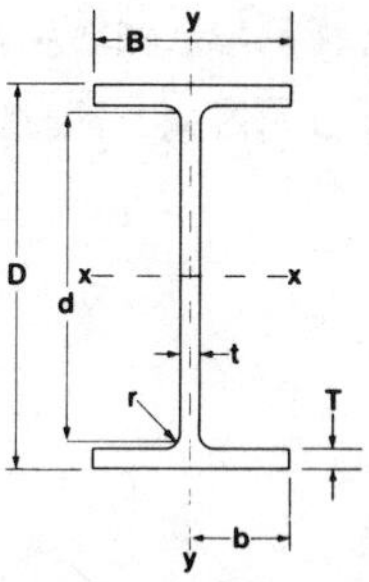

Elastic modulus		Plastic modulus		Buckling Parameter **u**	Torsional Index **x**	Warping Constant **H**	Torsional Constant **J**	Area of Section	Designation	
Axis x-x	Axis y-y	Axis x-x	Axis y-y						Mass per metre	Serial size
cm^3	cm^3	cm^3	cm^3			dm^6	cm^4	cm^2	kg	mm
15616	2160	17657	3339	0·884	26·7	88·7	1730	494·5	388	**914 x 419**
13722	1871	15474	2890	0·883	30·1	75·7	1190	437·5	343	
10891	1014	12583	1603	0·867	31·9	31·2	929	368·8	289	**914 x 305**
9507	871·9	10947	1372	0·866	36·2	26·4	627	322·8	253	
8259	738·1	9522	1162	0·861	41·3	22·0	421	285·3	224	
7210	621·4	8362	982·5	0·853	46·8	18·4	293	256·4	201	
7986	772·9	9157	1211	0·870	35·0	19·3	514	288·7	226	**838 x 292**
6648	620·4	7648	974·4	0·862	41·6	15·2	307	247·2	194	
5894	534·4	6809	841·5	0·856	46·5	13·0	222	224·1	176	
6234	610·0	7167	958·7	0·869	33·2	11·3	405	250·8	197	**762 x 267**
5385	513·4	6197	807·3	0·864	33·1	9·38	267	220·5	173	
4483	412·3	5174	649·0	0·857	45·1	7·41	161	188·1	147	
4911	517·7	5624	810·3	0·872	31·8	7·41	307	216·6	170	**686 x 254**
4372	454·5	4997	710·0	0·871	35·5	6·42	219	193·8	152	
3988	408·2	4560	637·8	0·868	38·7	5·72	169	178·6	140	
3481	346·1	3996	542·0	0·862	43·9	4·79	116	159·6	125	
6559	1017	7456	1574	0·886	21·1	14·3	788	303·8	238	**610 x 305**
4911	743·3	5521	1144	0·886	27·5	10·1	341	227·9	179	
4090	610·3	4572	936·8	0·886	32·5	8·09	200	190·1	149	
3626	392·1	4146	612·5	0·875	30·5	3·99	217	178·4	140	**610 x 229**
3222	343·5	3677	535·7	0·873	34·0	3·45	155	159·6	125	
2879	301·4	3288	470·2	0·870	37·9	2·99	112	144·5	113	
2515	255·9	2882	400·0	0·863	43·0	2·51	77·2	129·2	101	
2799	320·2	3203	500·6	0·876	27·6	2·32	180	155·8	122	**533 x 210**
2474	278·8	2824	435·1	0·875	30·9	1·99	126	138·6	109	
2298	256·6	2620	400·0	0·874	33·1	1·82	102	129·3	101	
2076	228·6	2366	356·2	0·872	36·4	1·60	76·2	117·8	92	
1798	192·2	2056	300·1	0·865	41·6	1·33	51·3	104·4	82	
1956	243·0	2232	378·3	0·880	25·8	1·17	121	125·3	98	**457 x 191**
1770	217·4	2014	337·9	0·879	28·3	1·04	90·5	113·9	89	
1612	195·6	1833	304·0	0·877	30·9	0·923	69·2	104·5	82	
1461	175·5	1657	272·2	0·876	33·9	0·819	52·0	95·0	74	
1296	152·9	1471	237·3	0·873	37·9	0·706	37·1	85·4	67	

UNIVERSAL BEAMS – DIMENSIONS AND PROPERTIES
To BS 4: Part 1: 1980

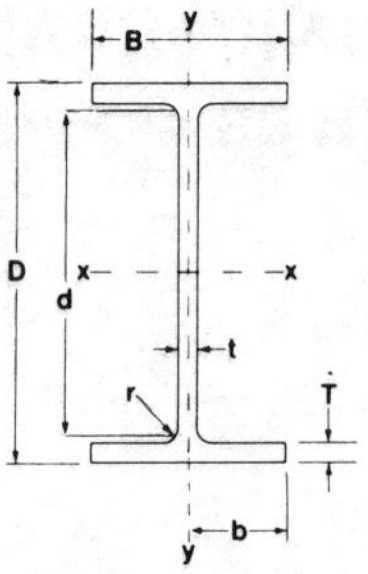

Designation		Depth of Section	Width of Section	Thickness		Root Radius	Web	Ratios For Local Buckling		Moment of inertia		Radius of gyration	
Serial size	Mass per metre	D	B	Web t	Flange T	r	Depth between fillets d	Flange b/T	Web d/t	Axis x-x	Axis y-y	Axis x-x	Axis y-y
mm	kg	mm	mm	mm	mm	mm	mm			cm^4	cm^4	cm	cm
457 x 152	82	465·1	153·5	10·7	18·9	10·2	406·9	4·06	38·0	36215	1143	18·6	3·31
	74	461·3	152·7	9·9	17·0	10·2	406·9	4·49	41·1	32435	1012	18·5	3·26
	67	457·2	151·9	9·1	15·0	10·2	406·9	5·06	44·7	28577	878	18·3	3·21
	60	454·7	152·9	8·0	13·3	10·2	407·7	5·75	51·0	25464	794	18·3	3·23
	52	449·8	152·4	7·6	10·9	10·2	407·7	6·99	53·6	21345	645	17·9	3·11
406 x 178	74	412·8	179·7	9·7	16·0	10·2	360·5	5·62	37·2	27329	1545	17·0	4·03
	67	409·4	178·8	8·8	14·3	10·2	360·5	6·25	41·0	24329	1365	16·9	4·00
	60	406·4	177·8	7·8	12·8	10·2	360·5	6·95	46·2	21508	1199	16·8	3·97
	54	402·6	177·6	7·6	10·9	10·2	360·5	8·15	47·4	18626	1017	16·5	3·85
406 x 140	46	402·3	142·4	6·9	11·2	10·2	359·6	6·36	52·1	15647	539	16·3	3·02
	39	397·3	141·8	6·3	8·6	10·2	359·6	8·24	57·1	12452	411	15·9	2·89
356 x 171	67	364·0	173·2	9·1	15·7	10·2	312·2	5·52	34·3	19522	1362	15·1	3·99
	57	358·6	172·1	8·0	13·0	10·2	312·2	6·62	39·0	16077	1109	14·9	3·92
	51	355·6	171·5	7·3	11·5	10·2	312·2	7·46	42·8	14156	968	14·8	3·87
	45	352·0	171·0	6·9	9·7	10·2	312·2	8·81	45·3	12091	812	14·6	3·78
356 x 127	39	352·8	126·0	6·5	10·7	10·2	311·1	5·89	47·9	10087	357	14·3	2·69
	33	348·5	125·4	5·9	8·5	10·2	311·1	7·38	52·7	8200	280	14·0	2·59
305 x 165	54	310·9	166·8	7·7	13·7	8·9	265·6	6·09	34·5	11710	1061	13·1	3·94
	46	307·1	165·7	6·7	11·8	8·9	265·6	7·02	39·7	9948	897	13·0	3·90
	40	303·8	165·1	6·1	10·2	8·9	265·6	8·09	43·6	8523	763	12·9	3·85
305 x 127	48	310·4	125·2	8·9	14·0	8·9	264·6	4·47	29·7	9504	460	12·5	2·75
	42	306·6	124·3	8·0	12·1	8·9	264·6	5·14	33·1	8143	388	12·4	2·70
	37	303·8	123·5	7·2	10·7	8·9	264·6	5·77	36·7	7162	337	12·3	2·67
305 x 102	33	312·7	102·4	6·6	10·8	7·6	275·8	4·74	41·8	6487	193	12·5	2·15
	28	308·9	101·9	6·1	8·9	7·6	275·8	5·72	45·2	5421	157	12·2	2·08
	25	304·8	101·6	5·8	6·8	7·6	275·8	7·47	47·6	4387	120	11·8	1·96
254 x 146	43	259·6	147·3	7·3	12·7	7·6	218·9	5·80	30·0	6558	677	10·9	3·51
	37	256·0	146·4	6·4	10·9	7·6	218·9	6·72	34·2	5556	571	10·8	3·47
	31	251·5	146·1	6·1	8·6	7·6	218·9	8·49	35·9	4439	449	10·5	3·35
254 x 102	28	260·4	102·1	6·4	10·0	7·6	225·0	5·10	35·2	4008	178	10·5	2·22
	25	257·0	101·9	6·1	8·4	7·6	225·0	6·07	36·9	3408	148	10·3	2·14
	22	254·0	101·6	5·8	6·8	7·6	225·0	7·47	38·8	2867	120	10·0	2·05
203 x 133	30	206·8	133·8	6·3	9·6	7·6	172·3	6·97	27·3	2887	384	8·72	3·18
	25	203·2	133·4	5·8	7·8	7·6	172·3	8·55	29·7	2356	310	8·54	3·10
203 x 102	23	203·2	101·6	5·2	9·3	7·6	169·4	5·46	32·6	2090	163	8·49	2·37
178 x 102	19	177·8	101·6	4·7	7·9	7·6	146·8	6·43	31·2	1360	138	7·49	2·39
152 x 89	16	152·4	88·9	4·6	7·7	7·6	121·8	5·77	26·5	838	90·4	6·40	2·10
127 x 76	13	127·0	76·2	4·2	7·6	7·6	96·6	5·01	23·0	477	56·2	5·33	1·83

UNIVERSAL BEAMS – DIMENSIONS AND PROPERTIES
To BS 4: Part 1: 1980

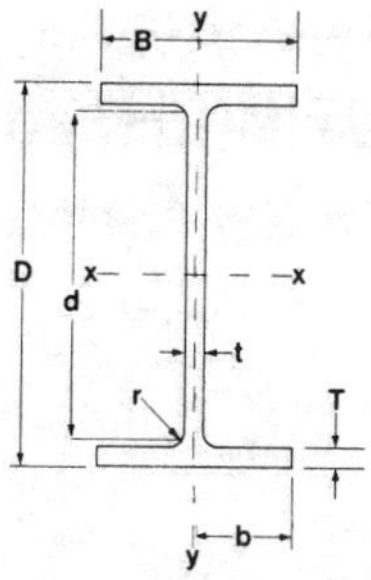

Elastic modulus		Plastic modulus		Buckling Parameter u	Torsional Index x	Warping Constant H	Torsional Constant J	Area of Section	Designation	
Axis x-x	Axis y-y	Axis x-x	Axis y-y						Mass per metre	Serial size
cm^3	cm^3	cm^3	cm^3			dm^6	cm^4	cm^2	kg	mm
1557	149·0	1800	235·4	0·872	27·3	0·569	89·3	104·5	82	**457 x 152**
1406	132·5	1622	209·1	0·870	30·0	0·499	66·6	95·0	74	
1250	115·5	1441	182·2	0·867	33·6	0·429	47·5	85·4	67	
1120	103·9	1284	162·9	0·869	37·5	0·387	33·6	75·9	60	
949	84·6	1094	133·2	0·859	43·9	0·311	21·3	66·5	52	
1324	172·0	1504	266·9	0·881	27·6	0·608	63·0	95·0	74	**406 x 178**
1188	152·7	1346	236·5	0·880	30·5	0·533	46·0	85·5	67	
1058	134·8	1194	208·3	0·880	33·9	0·464	32·9	76·0	60	
925·3	114·5	1048	177·5	0·872	38·5	0·390	22·7	68·4	54	
777·8	75·7	888·4	118·3	0·870	38·8	0·206	19·2	59·0	46	**406 x 140**
626·9	58·0	720·8	91·08	0·859	47·4	0·155	10·6	49·4	39	
1073	157·3	1212	243·0	0·887	24·4	0·413	55·5	85·4	67	**356 x 171**
896·5	128·9	1009	198·8	0·884	28·9	0·331	33·1	72·2	57	
796·2	112·9	894·9	174·1	0·882	32·2	0·286	23·6	64·6	51	
686·9	95·0	773·7	146·7	0·875	36·9	0·238	15·7	57·0	45	
571·8	56·6	653·6	88·68	0·872	35·3	0·104	14·9	49·4	39	**356 x 127**
470·6	44·7	534·8	70·24	0·864	42·2	0·081	8·68	41·8	33	
753·3	127·3	844·8	195·3	0·890	23·7	0·234	34·5	68·4	54	**305 x 165**
647·9	108·3	722·7	165·8	0·890	27·2	0·196	22·3	58·9	46	
561·2	92·4	624·5	141·5	0·888	31·1	0·164	14·7	51·5	40	
612·4	73·5	706·1	115·7	0·874	23·3	0·101	31·4	60·8	48	**305 x 127**
531·2	62·5	610·5	98·24	0·872	26·5	0·0842	21·0	53·2	42	
471·5	54·6	540·5	85·66	0·871	29·6	0·0724	14·9	47·5	37	
415·0	37·8	479·9	59·85	0·866	31·7	0·0441	12·1	41·8	33	**305 x 102**
351·0	30·8	407·2	48·92	0·858	37·0	0·0353	7·63	36·3	28	
287·9	23·6	337·8	37·98	0·844	43·8	0·0266	4·65	31·4	25	
505·3	92·0	568·2	141·2	0·889	21·1	0·103	24·1	55·1	43	**254 x 146**
434·0	78·1	485·3	119·6	0·889	24·3	0·0858	15·5	47·5	37	
353·1	61·5	395·6	94·52	0·879	29·4	0·0662	8·73	40·0	31	
307·9	34·9	353·4	54·84	0·873	27·5	0·0279	9·64	36·2	28	**254 x 102**
265·2	29·0	305·6	45·82	0·864	31·4	0·0228	6·45	32·2	25	
225·7	23·6	261·9	37·55	0·854	35·9	0·0183	4·31	28·4	22	
279·3	57·4	313·3	88·05	0·882	21·5	0·0373	10·2	38·0	30	**203 x 133**
231·9	46·4	259·8	71·39	0·876	25·4	0·0295	6·12	32·3	25	
206·0	32·1	232·0	49·50	0·890	22·6	0·0153	6·87	29·0	23	**203 x 102**
153·0	27·2	171·0	41·90	0·889	22·6	0·00998	4·37	24·2	19	**178 x 102**
110·0	20·3	124·0	31·40	0·889	19·5	0·00473	3·61	20·5	16	**152 x 89**
75·1	14·7	85·0	22·70	0·893	16·2	0·002	2·92	16·8	13	**127 x 76**

UNIVERSAL COLUMNS – DIMENSIONS AND PROPERTIES

To BS 4: Part 1: 1980

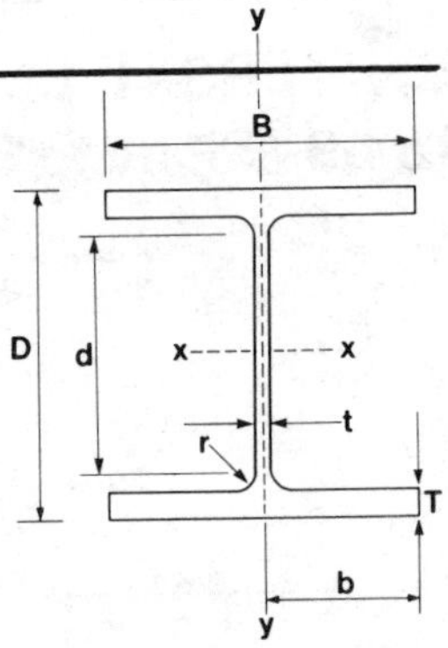

Designation		Depth of Section	Width of Section	Thickness		Root Radius	Web	Ratios For Local Buckling		Moment of inertia		Radius of gyration	
Serial size	Mass per metre	D	B	Web t	Flange T	r	Depth between fillets d	Flange b/T	Web d/t	Axis x-x	Axis y-y	Axis x-x	Axis y-y
mm	kg	mm	mm	mm	mm	mm	mm			cm^4	cm^4	cm	cm
356 x 406	634	474·7	424·1	47·6	77·0	15·2	290·1	2·75	6·10	275140	98211	18·5	11·0
	551	455·7	418·5	42·0	67·5	15·2	290·1	3·10	6·91	227023	82665	18·0	10·9
	467	436·6	412·4	35·9	58·0	15·2	290·1	3·56	8·08	183118	67905	17·5	10·7
	393	419·1	407·0	30·6	49·2	15·2	290·1	4·14	9·48	146765	55410	17·1	10·5
	340	406·4	403·0	26·5	42·9	15·2	290·1	4·70	11·0	122474	46816	16·8	10·4
	287	393·7	399·0	22·6	36·5	15·2	290·1	5·47	12·8	99994	38714	16·5	10·3
	235	381·0	395·0	18·5	30·2	15·2	290·1	6·54	15·7	79110	31008	16·2	10·2
356 x 368	202	374·7	374·4	16·8	27·0	15·2	290·1	6·93	17·3	66307	23632	16·0	9·57
	177	368·3	372·1	14·5	23·8	15·2	290·1	7·82	20·0	57153	20470	15·9	9·52
	153	362·0	370·2	12·6	20·7	15·2	290·1	8·94	23·0	48525	17469	15·8	9·46
	129	355·6	368·3	10·7	17·5	15·2	290·1	10·5	27·1	40246	14555	15·6	9·39
305 x 305	283	365·3	321·8	26·9	44·1	15·2	246·6	3·65	9·17	78777	24545	14·8	8·25
	240	352·6	317·9	23·0	37·7	15·2	246·6	4·22	10·7	64177	20239	14·5	8·14
	198	339·9	314·1	19·2	31·4	15·2	246·6	5·00	12·8	50832	16230	14·2	8·02
	158	327·2	310·6	15·7	25·0	15·2	246·6	6·21	15·7	38740	12524	13·9	7·89
	137	320·5	308·7	13·8	21·7	15·2	246·6	7·11	17·9	32838	10672	13·7	7·82
	118	314·5	306·8	11·9	18·7	15·2	246·6	8·20	20·7	27601	9006	13·6	7·75
	97	307·8	304·8	9·9	15·4	15·2	246·6	9·90	24·9	22202	7268	13·4	7·68
254 x 254	167	289·1	264·5	19·2	31·7	12·7	200·2	4·17	10·4	29914	9796	11·9	6·79
	132	276·4	261·0	15·6	25·3	12·7	200·2	5·16	12·8	22575	7519	11·6	6·68
	107	266·7	258·3	13·0	20·5	12·7	200·2	6·30	15·4	17510	5901	11·3	6·57
	89	260·4	255·9	10·5	17·3	12·7	200·2	7·40	19·1	14307	4849	11·2	6·52
	73	254·0	254·0	8·6	14·2	12·7	200·2	8·94	23·3	11360	3873	11·1	6·46
203 x 203	86	222·3	208·8	13·0	20·5	10·2	160·8	5·09	12·4	9462	3119	9·27	5·32
	71	215·9	206·2	10·3	17·3	10·2	160·8	5·96	15·6	7647	2536	9·16	5·28
	60	209·6	205·2	9·3	14·2	10·2	160·8	7·23	17·3	6088	2041	8·96	5·19
	52	206·2	203·9	8·0	12·5	10·2	160·8	8·16	20·1	5263	1770	8·90	5·16
	46	203·2	203·2	7·3	11·0	10·2	160·8	9·24	22·0	4564	1539	8·81	5·11
152 x 152	37	161·8	154·4	8·1	11·5	7·6	123·4	6·71	15·2	2218	709	6·84	3·87
	30	157·5	152·9	6·6	9·4	7·6	123·4	8·13	18·7	1742	558	6·75	3·82
	23	152·4	152·4	6·1	6·8	7·6	123·4	11·2	20·2	1263	403	6·51	3·68

UNIVERSAL COLUMNS – DIMENSIONS AND PROPERTIES

To BS4: Part I: 1980

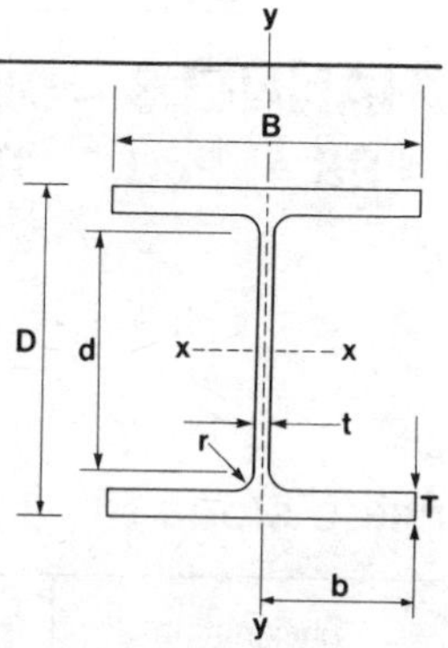

Elastic modulus		Plastic modulus		Buckling Parameter u	Torsional Index x	Warping Constant H	Torsional Constant J	Area of Section	Designation	
Axis x-x	Axis y-y	Axis x-x	Axis y-y						Mass per metre	Serial size
cm^3	cm^3	cm^3	cm^3			dm^6	cm^4	cm^2	kg	mm
11592	4632	14247	7114	0·843	5·46	38·8	13700	808·1	634	**356 x 406**
9964	3951	12078	6058	0·841	6·05	31·1	9240	701·8	551	
8388	3293	10009	5038	0·839	6·86	24·3	5820	595·5	467	
7004	2723	8229	4157	0·837	7·86	19·0	3550	500·9	393	
6027	2324	6994	3541	0·836	8·85	15·5	2340	432·7	340	
5080	1940	5818	2952	0·835	10·2	12·3	1440	366·0	287	
4153	1570	4689	2384	0·834	12·1	9·54	812	299·8	235	
3540	1262	3977	1917	0·844	13·3	7·14	560	257·9	202	**356 x 368**
3104	1100	3457	1668	0·844	15·0	6·07	383	225·7	177	
2681	943·8	2964	1430	0·844	17·0	5·09	251	195·2	153	
2264	790·4	2482	1196	0·843	19·9	4·16	153	164·9	129	
4314	1525	5101	2337	0·855	7·65	6·33	2030	360·4	283	**305 x 305**
3641	1273	4245	1947	0·854	8·73	5·01	1270	305·6	240	
2991	1034	3436	1576	0·854	10·2	3·86	734	252·3	198	
2368	806·3	2680	1228	0·852	12·5	2·86	379	201·2	158	
2049	691·4	2298	1052	0·851	14·1	2·38	250	174·6	137	
1755	587·0	1953	891·7	0·851	16·2	1·97	160	149·8	118	
1442	476·9	1589	723·5	0·850	19·3	1·55	91·1	123·3	97	
2070	740·6	2417	1132	0·852	8·49	1·62	625	212·4	167	**254 x 254**
1634	576·2	1875	878·6	0·850	10·3	1·18	322	168·9	132	
1313	456·9	1485	695·5	0·848	12·4	0·894	173	136·6	107	
1099	378·9	1228	575·4	0·849	14·4	0·716	104	114·0	89	
894·5	305·0	988·6	462·4	0·849	17·3	0·557	57·3	92·9	73	
851·5	298·7	978·8	455·9	0·850	10·2	0·317	138	110·1	86	**203 x 203**
708·4	246·0	802·4	374·2	0·852	11·9	0·250	81·5	91·1	71	
581·1	199·0	652·0	302·8	0·847	14·1	0·195	46·6	75·8	60	
510·4	173·6	568·1	263·7	0·848	15·8	0·166	32·0	66·4	52	
449·2	151·5	497·4	230·0	0·846	17·7	0·142	22·2	58·8	46	
274·2	91·78	310·1	140·1	0·848	13·3	0·040	19·5	47·4	37	**152 x 152**
221·2	73·06	247·1	111·2	0·848	16·0	0·0306	10·5	38·2	30	
165·7	52·95	184·3	80·87	0·837	20·4	0·0214	4·87	29·8	23	

CHANNELS – DIMENSIONS AND PROPERTIES

To BS 4: Part 1: 1980

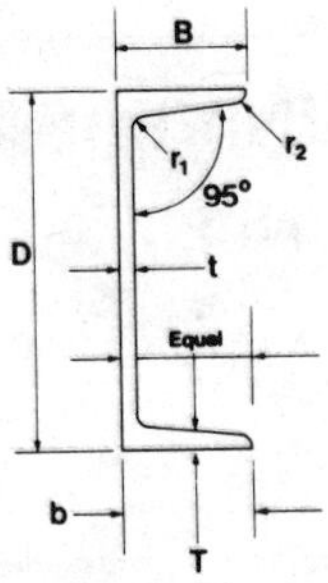

Inside slope 5°

Designation: Serial size	Designation: Mass per metre	Depth of Section D	Width of Section B	Thickness: Web t	Thickness: Flange T	Distance of Cy	Root radius r_1	Toe radius r_2	Web: Depth between fillets d	Ratios for Local Buckling: Flange b/T	Ratios for Local Buckling: Web d/t	Moment of inertia: Axis x-x	Moment of inertia: Axis y-y
mm	kg	mm	mm	mm	mm	cm	mm	mm	mm			cm^4	cm^4
432 x 102	65·54	431·8	101·6	12·2	16·8	2·32	15·2	4·8	362·5	6·05	29·7	21399	628·6
381 x 102	55·10	381·0	101·6	10·4	16·3	2·52	15·2	4·8	312·6	6·23	30·0	14894	579·7
305 x 102	46·18	304·8	101·6	10·2	14·8	2·66	15·2	4·8	239·3	6·86	23·5	8214	499·5
305 x 89	41·69	304·8	88·9	10·2	13·7	2·18	13·7	3·2	245·4	6·49	24·1	7061	325·4
254 x 89	35·74	254·0	88·9	9·1	13·6	2·42	13·7	3·2	194·7	6·54	21·4	4448	302·4
254 x 76	28·29	254·0	76·2	8·1	10·9	1·86	12·2	3·2	203·9	6·99	25·2	3367	162·6
229 x 89	32·76	228·6	88·9	8·6	13·3	2·53	13·7	3·2	169·9	6·68	19·7	3387	285·0
229 x 76	26·06	228·6	76·2	7·6	11·2	2·00	12·2	3·2	177·8	6·80	23·4	2610	158·7
203 x 89	29·78	203·2	88·9	8·1	12·9	2·65	13·7	3·2	145·2	6·89	17·9	2491	264·4
203 x 76	23·82	203·2	76·2	7·1	11·2	2·13	12·2	3·2	152·4	6·80	21·5	1950	151·3
178 x 89	26·81	177·8	88·9	7·6	12·3	2·76	13·7	3·2	121·0	7·23	15·9	1753	241·0
178 x 76	20·84	177·8	76·2	6·6	10·3	2·20	12·2	3·2	128·8	7·40	19·5	1337	134·0
152 x 89	23·84	152·4	88·9	7·1	11·6	2·86	13·7	3·2	96·9	7·66	13·7	1166	215·1
152 x 76	17·88	152·4	76·2	6·4	9·0	2·21	12·2	2·4	105·9	8·47	16·5	851·5	113·8
127 x 64	14·90	127·0	63·5	6·4	9·2	1·94	10·7	2·4	84·0	6·90	13·1	482·5	67·23

[229 + 254] 6336.6

or 254 + 254. 25784

254
89

CHANNELS – DIMENSIONS AND PROPERTIES
To BS 4: Part 1: 1980

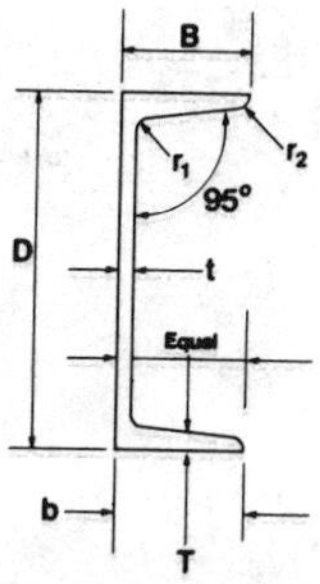

Radius of gyration		Elastic modulus		Plastic modulus		Buckling Parameter	Torsional Index	Warping Constant	Torsional Constant	Area of Section	Designation	
Axis x-x	Axis y-y	Axis x-x	Axis y-y	Axis x-x	Axis y-y	u	x	H	J		Mass per metre	Serial size
cm	cm	cm^3	cm^3	cm^3	cm^3			dm^6	cm^4	cm^2	kg	mm
16·0	2·74	991·1	80·14	1207	153·1	0·876	24·6	0·217	61·0	83·49	65·54	**432 x 102**
14·6	2·87	781·8	75·86	932·7	144·4	0·895	22·7	0·153	46·0	70·19	55·10	**381 x 102**
11·8	2·91	539·0	66·59	638·3	128·1	0·900	18·9	0·0842	35·4	58·83	46·18	**305 x 102**
11·5	2·48	463·3	48·49	557·1	92·60	0·887	20·4	0·0551	27·6	53·11	41·69	**305 x 89**
9·88	2·58	350·2	46·70	414·4	89·56	0·906	17·1	0·0347	22·9	45·52	35·74	**254 x 89**
9·67	2·12	265·1	28·21	317·4	54·14	0·886	21·2	0·0194	12·3	36·03	28·29	**254 x 76**
9·01	2·61	296·4	44·82	348·4	86·38	0·912	15·5	0·0263	20·4	41·73	32·76	**229 x 89**
8·87	2·19	228·3	28·22	270·3	54·24	0·900	18·8	0·0151	11·4	33·20	26·06	**229 x 76**
8·10	2·64	245·2	42·34	286·6	81·62	0·915	14·1	0·0192	17·8	37·94	29·78	**203 x 89**
8·02	2·23	192·0	27·59	225·2	53·32	0·911	16·7	0·0112	10·4	30·34	23·82	**203 x 76**
7·16	2·66	197·2	39·29	229·6	75·44	0·915	12·7	0·0134	15·1	34·15	26·81	**178 x 89**
7·10	2·25	150·4	24·72	175·4	48·07	0·911	15·5	0·00764	8·13	26·54	20·84	**178 x 76**
6·20	2·66	153·0	35·70	177·7	68·12	0·909	11·3	0·00881	12·4	30·36	23·84	**152 x 89**
6·11	2·24	111·8	21·05	130·0	41·26	0·902	14·5	0·00486	5·94	22·77	17·88	**152 x 76**
5·04	1·88	75·99	15·25	89·4	29·31	0·910	11·7	0·00187	4·92	18·98	14·90	**127 x 64**

EQUAL ANGLES – DIMENSIONS AND PROPERTIES
To BS 4848: Part 4: 1972

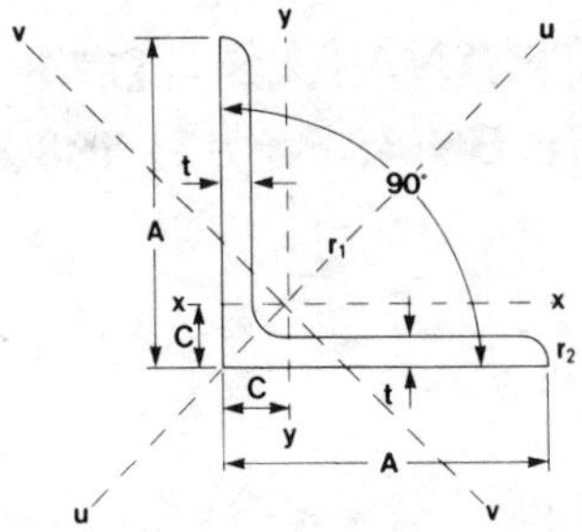

Designation			Radius				Moment of inertia		
Size A	Thickness t	Mass per metre	Root r_1	Toe r_2	Area of Section	Distance of centre of gravity c_x and c_y	Axis x-x, y-y	Axis u-u Max.	Axis v-v Min.
mm	mm	kg	mm	mm	cm^2	cm	cm^4	cm^4	cm^4
250 x 250	35	128·0	20·0	10·0	163·0	7·49	9252	14600	3860
	32	118·0	20·0	10·0	150·0	7·38	8596	13600	3560
	28	104·0	20·0	10·0	133·0	7·23	7686	12200	3170
	25	93·6	20·0	10·0	119·0	7·12	6975	11100	2860
200 x 200	24	71·1	18·0	4·8	90·6	5·84	3330	5280	1380
	20	59·9	18·0	4·8	76·3	5·68	2850	4530	1170
	18	54·2	18·0	4·8	69·1	5·60	2600	4130	1070
	16	48·5	18·0	4·8	61·8	5·52	2340	3720	960
150 x 150	18	40·1	16·0	4·8	51·0	4·37	1050	1670	435
	15	33·8	16·0	4·8	43·0	4·25	898	1430	370
	12	27·3	16·0	4·8	34·8	4·12	737	1170	303
	10	23·0	16·0	4·8	29·3	4·03	624	991	258
120 x 120	15	26·6	13·0	4·8	33·9	3·51	445	705	185
	12	21·6	13·0	4·8	27·5	3·40	368	584	152
	10	18·2	13·0	4·8	23·2	3·31	313	497	129
	8	14·7	13·0	4·8	18·7	3·23	255	405	105
100 x 100†	15	21·9	12·0	4·8	27·9	3·02	249	393	104
	12	17·8	12·0	4·8	22·7	2·90	207	328	85·7
	8	12·2	12·0	4·8	15·5	2·74	145	230	59·9
90 x 90	12	15·9	11·0	4·8	20·3	2·66	148	234	61·7
	10	13·4	11·0	4·8	17·1	2·58	127	201	52·6
	8	10·9	11·0	4·8	13·9	2·50	104	166	43·1
	6	8·30	11·0	4·8	10·6	2·41	80·3	127	33·3

† 100 x 100 x 10, an ISO size, is also frequently rolled

100 x 100	10	15·0	12·0	4·8	19·2	2·82	177	280	72·9

Other non-standard sections, particularly other thicknesses of the standard range, may also be available.

EQUAL ANGLES – DIMENSIONS AND PROPERTIES
To BS 4848: Part 4: 1972

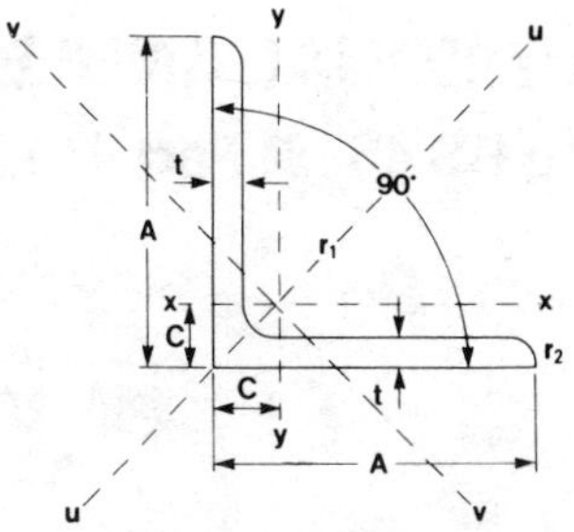

Radius of gyration			Elastic modulus	Designation	
Axis x-x, y-y	Axis u-u Max.	Axis v-v Min.	Axis x-x, y-y	Thickness **t**	Size **A**
cm	cm	cm	cm^3	mm	mm
7·53	9·47	4·86	529	35	**250 x 250**
7·57	9·53	4·87	498	32	
7·61	9·59	4·89	433	28	
7·65	9·64	4·90	390	25	
6·06	7·64	3·90	235	24	**200 x 200**
6·11	7·70	3·92	199	20	
6·13	7·73	3·93	181	18	
6·16	7·76	3·94	162	16	
4·54	5·71	2·92	98·7	18	**150 x 150**
4·57	5·76	2·93	83·5	15	
4·60	5·80	2·95	67·7	12	
4·62	5·82	2·97	56·9	10	
3·62	4·56	2·33	52·4	15	**120 x 120**
3·65	4·60	2·35	42·7	12	
3·67	4·63	2·36	36·0	10	
3·69	4·65	2·37	29·1	8	
2·98	3·75	1·93	35·6	15	**†100 x 100**
3·02	3·80	1·94	29·1	12	
3·06	3·85	1·96	19·9	8	
2·70	3·40	1·75	23·3	12	**90 x 90**
2·72	3·42	1·76	19·8	10	
2·74	3·45	1·76	16·1	8	
2·76	3·47	1·78	12·2	6	

3·04	3·83	1·95	24·6	10	**100 x 100**

UNEQUAL ANGLES – DIMENSIONS AND PROPERTIES

To BS 4848: Part 4: 1972 except where marked †

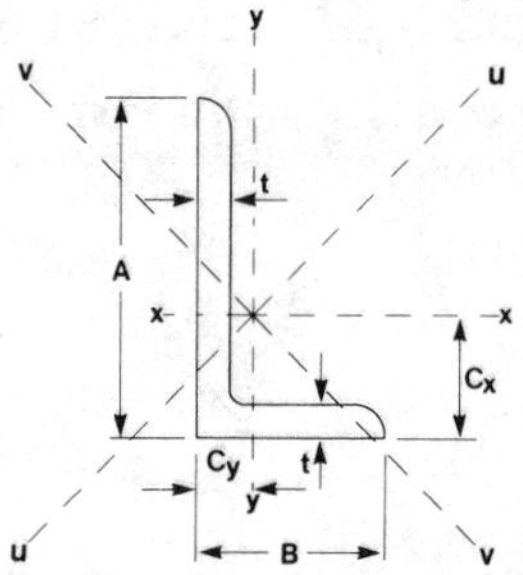

Designation		Mass per metre	Radius		Area of Section	Distance of centre of gravity		Moment of inertia			
Size A x B	Thickness t		Root r_1	Toe r_2		c_x	c_y	Axis x-x	Axis y-y	Axis u-u Max.	Axis v-v Min.
mm	mm	kg	mm	mm	cm^2	cm	cm	cm^4	cm^4	cm^4	cm^4
200 x 150	18	47·1	15·0	4·8	60·0	6·33	3·85	2376	1146	2900	618
	15	39·6	15·0	4·8	50·5	6·21	3·73	2022	979	2480	526
	12	32·0	15·0	4·8	40·8	6·08	3·61	1652	803	2030	430
200 x 100	15	33·7	15·0	4·8	43·0	7·16	2·22	1758	299	1860	194
	12	27·3	15·0	4·8	34·8	7·03	2·10	1440	247	1530	159
	10	23·0	15·0	4·8	29·2	6·93	2·01	1220	210	1290	135
150 x 90	15	26·6	12·0	4·8	33·9	5·21	2·23	761	205	841	126
	12	21·6	12·0	4·8	27·5	5·08	2·12	627	171	694	104
	10	18·2	12·0	4·8	23·2	5·00	2·04	533	146	591	88·3
150 x 75	15	24·8	11·0	4·8	31·6	5·53	1·81	713	120	754	78·8
	12	20·2	11·0	4·8	25·7	5·41	1·69	589	99·9	624	64·9
	10	17·0	11·0	4·8	21·6	5·32	1·61	501	85·8	532	55·3
137 x 102 †	9·5	17·3	11·0	4·8	21·9	4·22	2·49	417	194	506	107
	7·9	14·5	11·0	4·8	18·4	4·17	2·44	353	169	428	90·4
	6·4	11·7	11·0	4·8	14·8	4·09	2·36	287	137	352	74·7
125 x 75	12	17·8	11·0	4·8	22·7	4·31	1·84	354	95·5	391	58·5
	10	15·0	11·0	4·8	19·1	4·23	1·76	302	82·1	334	49·9
	8	12·2	11·0	4·8	15·5	4·14	1·68	247	67·6	274	40·9
	6·5	9·98	11·0	4·8	12·7	4·06	1·61	204	56·1	228	34·3
100 x 75	12	15·4	10·0	4·8	19·7	3·27	2·03	189	90·2	230	49·5
	10	13·0	10·0	4·8	16·6	3·19	1·95	162	77·6	197	42·2
	8	10·6	10·0	4·8	13·5	3·10	1·87	133	64·1	162	34·6
100 x 65	10	12·3	10·0	4·8	15·6	3·36	1·63	154	51·0	175	30·1
	8	9·94	10·0	4·8	12·7	3·27	1·55	127	42·2	144	24·8
	7	8·77	10·0	4·8	11·2	3·23	1·51	113	37·6	128	22·0

Note: Non-standard sections, particularly other thicknesses of the standard range may also be available.

UNEQUAL ANGLES – DIMENSIONS AND PROPERTIES
To BS 4848: Part 4: 1972 except where marked †

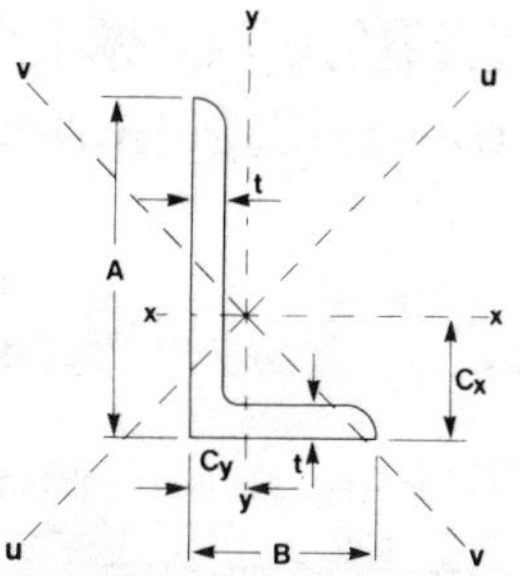

Radius of gyration				Elastic modulus		Angle x-x Axis to u-u Axis	Designation	
Axis x-x	Axis y-y	Axis u-u Max.	Axis v-v Min.	Axis x-x	Axis y-y		Thickness t	Size A x B
cm	cm	cm	cm	cm³	cm³	Tan ∝	mm	mm
6·29	4·37	6·96	3·21	174·0	103·0	0·548	18	**200 x 150**
6·33	4·40	7·00	3·23	147·0	86·9	0·551	15	
6·36	4·44	7·04	3·25	119·0	70·5	0·552	12	
6·40	2·64	6·58	2·12	137·0	38·4	0·260	15	**200 x 100**
6·43	2·67	6·63	2·14	111·0	31·3	0·262	12	
6·46	2·68	6·65	2·15	93·2	26·3	0·263	10	
4·74	2·46	4·98	1·93	77·7	30·4	0·354	15	**150 x 90**
4·77	2·49	5·02	1·94	63·3	24·8	0·358	12	
4·80	2·51	5·05	1·95	53·3	21·0	0·360	10	
4·75	1·94	4·88	1·58	75·3	21·0	0·254	15	**150 x 75**
4·79	1·97	4·93	1·59	61·4	17·2	0·259	12	
4·81	1·99	4·96	1·60	51·8	14·6	0·261	10	
4·34	3·00	4·80	2·20	43·8	25·8	0·543	9·5	**† 137 x 102**
4·34	3·00	4·82	2·22	37·3	21·8	0·544	7·9	
4·37	3·02	4·84	2·23	29·9	17·6	0·544	6·4	
3·95	2·05	4·15	1·61	43·2	16·9	0·354	12	**125 x 75**
3·97	2·07	4·18	1·61	36·5	14·3	0·357	10	
4·00	2·09	4·21	1·63	29·6	11·6	0·360	8	
4·01	2·10	4·23	1·64	24·2	9·5	0·361	6·5	
3·10	2·14	3·42	1·59	28·0	16·5	0·540	12	**100 x 75**
3·12	2·16	3·45	1·59	23·8	14·0	0·544	10	
3·14	2·18	3·47	1·60	19·3	11·4	0·547	8	
3·14	1·81	3·35	1·39	23·2	10·5	0·410	10	**100 x 65**
3·16	1·83	3·37	1·40	18·9	8·54	0·413	8	
3·17	1·83	3·39	1·40	16·6	7·53	0·415	7	

STRUCTURAL TEES SPLIT FROM UNIVERSAL BEAMS
DIMENSIONS AND PROPERTIES

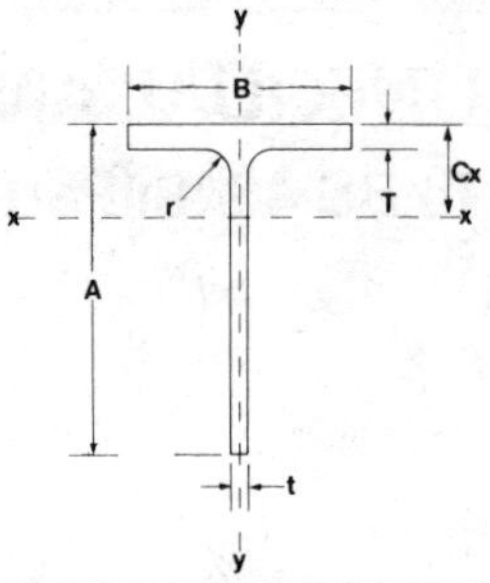

Properties have been calculated assuming that there is no loss of material due to splitting.

Designation		Split From		Width of Section	Depth of Section	Thickness		Root radius	Dimension	Moment of inertia		Radius of gyration	
Serial size	Mass per metre	Serial size	Mass per metre	**B**	**A**	Web **t**	Flange **T**	**r**	**Cx**	Axis x-x	Axis y-y	Axis x-x	Axis y-y
mm	kg	mm	kg	mm	mm	mm	mm	mm	cm	cm^4	cm^4	cm	cm
305 x 305	90	**610 x 305**	179	307·0	308·7	14·1	23·6	16·5	6·65	8950	5710	8·86	7·08
	75		149	304·8	304·8	11·9	19·7	16·5	6·43	7370	4650	8·80	6·99
254 x 343	63	**686 x 254**	125	253·0	339·0	11·7	16·2	15·2	8·87	9000	2190	10·6	5·24
229 x 305	70	**610 x 229**	140	230·1	308·5	13·1	22·1	12·7	7·61	7750	2260	9·32	5·03
	63		125	229·0	305·9	11·9	19·6	12·7	7·55	6910	1970	9·31	4·96
	57		113	228·2	303·7	11·2	17·3	12·7	7·60	6290	1720	9·34	4·88
	51		101	227·6	301·1	10·6	14·8	12·7	7·80	5710	1460	9·40	4·75
210 x 267	61	**533 x 210**	122	211·9	272·3	12·8	21·3	12·7	6·67	5180	1700	8·16	4·67
	55		109	210·7	269·7	11·6	18·8	12·7	6·60	4590	1470	8·14	4·60
	51		101	210·1	268·4	10·9	17·4	12·7	6·57	4280	1350	8·14	4·56
	46		92	209·3	266·6	10·2	15·6	12·7	6·56	3910	1200	8·14	4·51
	41		82	208·7	264·2	9·6	13·2	12·7	6·73	3520	1000	8·21	4·38
191 x 229	49	**457 x 191**	98	192·8	233·7	11·4	19·6	10·2	5·55	2980	1170	6·90	4·33
	45		89	192·0	231·8	10·6	17·7	10·2	5·49	2700	1040	6·89	4·28
	41		82	191·3	230·1	9·9	16·0	10·2	5·48	2480	936	6·89	4·23
	37		74	190·5	228·6	9·1	14·5	10·2	5·42	2250	836	6·88	4·19
	34		67	189·9	226·8	8·5	12·7	10·2	5·47	2040	726	6·90	4·12
152 x 229	41	**457 x 152**	82	153·5	232·5	10·7	18·9	10·2	6·03	2610	572	7·07	3·31
	37		74	152·7	230·6	9·9	17·0	10·2	5·98	2360	506	7·06	3·26
	34		67	151·9	228·6	9·1	15·0	10·2	5·99	2130	439	7·06	3·21
	30		60	152·9	227·3	8·0	13·3	10·2	5·82	1870	397	7·02	3·23
	26		52	152·4	224·9	7·6	10·9	10·2	6·03	1670	322	7·08	3·11
178 x 203	37	**406 x 178**	74	179·7	206·4	9·7	16·0	10·2	4·80	1760	772	6·08	4·03
	34		67	178·8	204·7	8·8	14·3	10·2	4·73	1570	682	6·07	4·00
	30		60	177·8	203·2	7·8	12·8	10·2	4·62	1380	599	6·03	3·97
	27		54	177·6	201·3	7·6	10·9	10·2	4·81	1280	508	6·12	3·85
140 x 203	23	**406 x 140**	46	142·4	201·2	6·9	11·2	10·2	5·05	1130	269	6·19	3·02
	20		39	141·8	198·6	6·3	8·6	10·2	5·27	968	206	6·26	2·89

STRUCTURAL TEES SPLIT FROM UNIVERSAL BEAMS
DIMENSIONS AND PROPERTIES

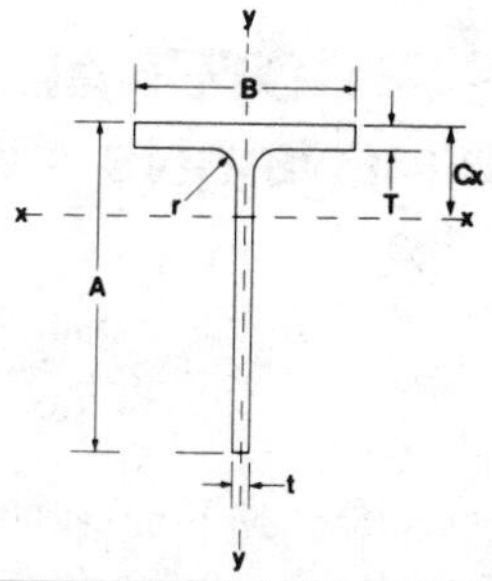

Properties have been calculated assuming that there is no loss of material due to splitting.

Elastic Modulus Axis x-x Flange	Elastic Modulus Axis x-x Toe	Elastic Modulus Axis y-y	Plastic Modulus Axis x-x	Plastic Modulus Axis y-y	Buckling Parameter u	Torsional Index x	Warping Constant H	Torsional Constant J	Area of Section	Designation Mass per metre	Designation Serial size
cm³	cm³	cm³	cm³	cm³			dm⁶	cm⁴	cm²	kg	mm
1350	369	372	652	572	0·482	13·8	1·26	170	114	90	**305 x 305**
1140	306	305	538	468	0·483	16·3	1·01	99·9	95·1	75	
1010	359	173	645	271	0·652	22·0	0·599	58·0	79·8	63	**254 x 343**
1020	333	196	592	306	0·613	15·3	0·499	108	89·2	70	**229 x 305**
916	300	172	533	268	0·618	17·0	0·431	77·1	79·8	63	
828	277	151	492	235	0·628	19·0	0·374	56·0	72·2	57	
732	256	128	458	200	0·646	21·5	0·314	38·5	64·6	51	
777	252	160	448	250	0·601	13·8	0·290	89·5	77·9	61	**210 x 267**
696	225	139	400	218	0·605	15·5	0·249	62·8	69·3	55	
652	211	128	375	200	0·608	16·6	0·227	50·8	64·6	51	
595	194	114	345	178	0·614	18·2	0·200	38·0	58·9	46	
522	179	96·1	319	150	0·633	20·8	0·166	25·6	52·2	41	
536	167	122	297	189	0·574	12·9	0·147	60·3	62·6	49	**191 x 229**
492	153	109	271	169	0·578	14·2	0·130	45·1	57·0	45	
453	142	97·8	251	152	0·584	15·5	0·115	34·5	52·3	41	
415	129	87·7	228	136	0·586	17·0	0·102	25·9	47·5	37	
373	118	76·5	210	119	0·597	19·0	0·0882	18·5	42·7	34	
433	151	74·5	270	118	0·640	13·7	0·0711	44·5	52·2	41	**152 x 229**
395	138	66·3	247	105	0·645	15·0	0·0624	33·2	47·5	37	
355	126	57·8	226	91·1	0·653	16·8	0·0536	23·7	42·7	34	
321	111	52·0	197	81·5	0·647	18·8	0·0484	16·8	38·0	30	
276	101	42·3	183	66·6	0·670	22·0	0·0388	10·6	33·2	26	
366	111	86·0	197	133	0·559	13·8	0·076	31·4	47·5	37	**178 x 203**
333	100·0	76·3	177	118	0·561	15·3	0·0666	23·0	42·7	34	
299	88·1	67·4	155	104	0·559	17·0	0·058	16·4	38·0	30	
267	83·7	57·2	148	88·7	0·586	19·3	0·0487	11·3	34·2	27	
224	75·0	37·8	134	59·2	0·636	19·4	0·0258	9·56	29·5	23	**140 x 203**
184	66·4	29·0	120	45·5	0·665	23·7	0·0194	5·30	24·7	20	

STRUCTURAL TEES SPLIT FROM UNIVERSAL BEAMS
DIMENSIONS AND PROPERTIES

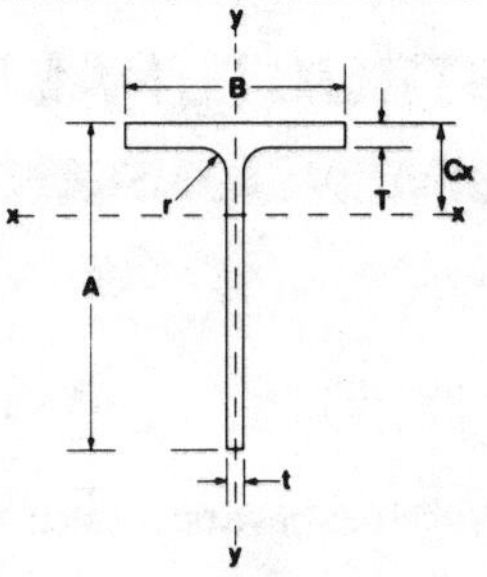

Properties have been calculated assuming that there is no loss of material due to splitting.

Designation		Split From		Width of Section	Depth of Section	Thickness		Root radius	Dimension	Moment of inertia		Radius of gyration	
Serial size	Mass per metre	Serial size	Mass per metre	B	A	Web t	Flange T	r	Cx	Axis x-x	Axis y-y	Axis x-x	Axis y-y
mm	kg	mm	kg	mm	mm	mm	mm	mm	cm	cm^4	cm^4	cm	cm
171 x 178	34	**356 x 171**	67	173·2	182·0	9·1	15·7	10·2	4·01	1160	681	5·21	3·99
	29		57	172·1	179·3	8·0	13·0	10·2	3·95	979	554	5·21	3·92
	26		51	171·5	177·8	7·3	11·5	10·2	3·92	878	484	5·21	3·87
	23		45	171·0	176·0	6·9	9·7	10·2	4·02	792	406	5·27	3·78
127 x 178	20	**356 x 127**	39	126·0	176·4	6·5	10·7	10·2	4·41	720	178	5·40	2·69
	17		33	125·4	174·2	5·9	8·5	10·2	4·52	618	140	5·44	2·59
165 x 152	27	**305 x 165**	54	166·8	155·4	7·7	13·7	8·9	3·19	636	531	4·31	3·94
	23		46	165·7	153·5	6·7	11·8	8·9	3·09	541	449	4·29	3·90
	20		40	165·1	151·9	6·1	10·2	8·9	3·07	476	381	4·30	3·85
127 x 152	24	**305 x 127**	48	125·2	155·2	8·9	14·0	8·9	3·91	654	230	4·64	2·75
	21		42	124·3	153·3	8·0	12·1	8·9	3·85	568	194	4·62	2·70
	19		37	123·5	151·9	7·2	10·7	8·9	3·80	504	169	4·61	2·67
102 x 152	17	**305 x 102**	33	102·4	156·3	6·6	10·8	7·6	4·14	487	96·7	4·83	2·15
	14		28	101·9	154·4	6·1	8·9	7·6	4·23	427	78·4	4·85	2·08
	13		25	101·6	152·4	5·8	6·8	7·6	4·48	376	60·0	4·89	1·96
146 x 127	22	**254 x 146**	43	147·3	129·8	7·3	12·7	7·6	2·67	349	339	3·56	3·51
	19		37	146·4	128·0	6·4	10·9	7·6	2·58	297	286	3·54	3·47
	16		31	146·1	125·7	6·1	8·6	7·6	2·68	263	224	3·63	3·35
102 x 127	14	**254 x 102**	28	102·1	130·2	6·4	10·0	7·6	3·25	279	89·1	3·93	2·22
	13		25	101·9	128·5	6·1	8·4	7·6	3·35	253	73·9	3·96	2·14
	11		22	101·6	127·0	5·8	6·8	7·6	3·49	227	60·0	4·00	2·05
133 x 102	15	**203 x 133**	30	133·8	103·4	6·3	9·6	7·6	2·10	153	192	2·83	3·18
	13		25	133·4	101·6	5·8	7·8	7·6	2·12	134	155	2·88	3·10

STRUCTURAL TEES SPLIT FROM UNIVERSAL BEAMS
DIMENSIONS AND PROPERTIES

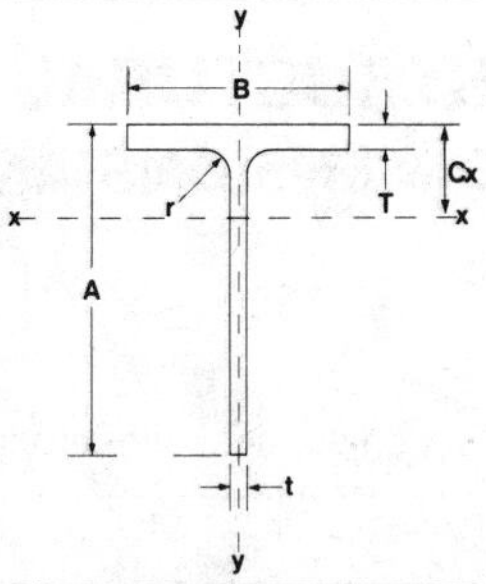

Properties have been calculated assuming that there is no loss of material due to splitting.

Elastic Modulus			Plastic Modulus		Buckling Parameter u	Torsional Index x	Warping Constant H	Torsional Constant J	Area of Section	Designation	
Axis x-x		Axis y-y	Axis x-x	Axis y-y						Mass per metre	Serial size
Flange	Toe										
cm³	cm³	cm³	cm³	cm³			dm⁶	cm⁴	cm²	kg	mm
289	81·6	78·6	145	121	0·500	12·2	0·0516	27·7	42·7	34	**171 x 178**
248	70·0	64·4	124	99·4	0·511	14·5	0·0414	16·5	36·1	29	
224	63·4	56·5	112	87·0	0·519	16·1	0·0358	11·8	32·3	26	
197	58·3	47·5	103	73·3	0·542	18·5	0·0297	7·81	28·5	23	
163	54·4	28·3	96·9	44·3	0·631	17·7	0·0130	7·41	24·7	20	**127 x 178**
137	47·9	22·4	86·0	35·1	0·652	21·2	0·0101	4·33	20·9	17	
199	51·5	63·6	91·6	97·7	0·383	11·8	0·0293	17·2	34·2	27	**165 x 152**
175	44·1	54·1	77·8	82·9	0·385	13·6	0·0244	11·1	29·5	23	
155	39·3	46·2	68·9	70·7	0·403	15·6	0·0206	7·35	25·8	20	
167	56·3	36·7	101	57·9	0·599	11·7	0·0126	15·6	30·4	24	**127 x 152**
147	49·5	31·2	88·2	49·1	0·605	13·3	0·0105	10·4	26·6	21	
133	44·3	27·3	78·9	42·8	0·608	14·8	0·00905	7·44	23·7	19	
117	42·4	18·9	75·9	29·9	0·657	15·9	0·00551	6·05	20·9	17	**102 x 152**
101	38·0	15·4	68·6	24·5	0·675	18·6	0·00441	3·80	18·2	14	
83·9	34·9	11·8	63·5	19·0	0·706	22·0	0·00333	2·31	15·7	13	
131	33·9	46·0	60·6	70·6	0·248	10·6	0·0129	12·0	27·6	22	**146 x 127**
115	29·0	39·0	51·5	59·8	0·262	12·1	0·0107	7·75	23·7	19	
98·1	26·6	30·7	46·8	47·3	0·385	14·7	0·00827	4·35	20·0	16	
85·7	28·6	17·5	50·9	27·4	0·609	13·8	0·00349	4·80	18·1	14	**102 x 127**
75·3	26·6	14·5	47·6	22·9	0·633	15·7	0·00286	3·21	16·1	13	
65·2	24·7	11·8	44·6	18·8	0·660	18·0	0·00229	2·14	14·2	11	
72·8	18·5	28·7	33·1	44·0	0	0	0·00466	5·09	19·0	15	**133 x 102**
63·0	16·6	23·2	29·4	35·7	0	0	0·00369	3·05	16·2	13	

Where values for u and x have been given as 0 there is no possibility of lateral torsional buckling.

STRUCTURAL TEES SPLIT FROM UNIVERSAL COLUMNS
DIMENSIONS AND PROPERTIES

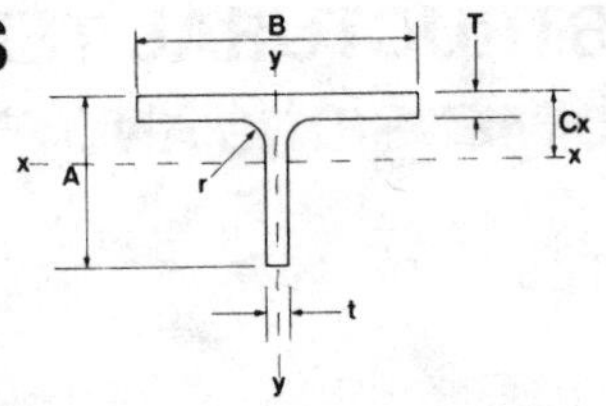

Properties have been calculated assuming that there is no loss of material due to splitting.

Designation		Split From		Width of Section	Depth of Section	Thickness		Root radius	Dimension	Moment of inertia		Radius of gyration	
Serial size	Mass per metre	Serial size	Mass per metre	**B**	**A**	Web **t**	Flange **T**	**r**	**Cx**	Axis x-x	Axis y-y	Axis x-x	Axis y-y
mm	kg	mm	kg	mm	mm	mm	mm	mm	cm	cm^4	cm^4	cm	cm
305 x 152	79	**305 x 305**	158	310·6	163·6	15·7	25·0	15·2	3·04	1530	6260	3·90	7·89
	69		137	308·7	160·3	13·8	21·7	15·2	2·86	1290	5340	3·85	7·82
	59		118	306·8	157·2	11·9	18·7	15·2	2·69	1070	4500	3·79	7·75
	49		97	304·8	153·9	9·9	15·4	15·2	2·50	858	3630	3·73	7·68
254 x 127	66	**254 x 254**	132	261·0	138·2	15·6	25·3	12·7	2·72	887	3760	3·24	6·67
	54		107	258·3	133·4	13·0	20·5	12·7	2·47	683	2950	3·16	6·57
	45		89	255·9	130·2	10·5	17·3	12·7	2·24	535	2420	3·06	6·52
	37		73	254·0	127·0	8·6	14·2	12·7	2·06	419	1940	3·00	6·46
203 x 102	43	**203 x 203**	86	208·8	111·1	13·0	20·5	10·2	2·22	380	1560	2·63	5·32
	36		71	206·2	108·0	10·3	17·3	10·2	1·98	288	1270	2·52	5·28
	30		60	205·2	104·8	9·3	14·2	10·2	1·88	241	1020	2·52	5·19
	26		52	203·9	103·1	8·0	12·5	10·2	1·76	203	885	2·47	5·16
	23		46	203·2	101·6	7·3	11·0	10·2	1·71	180	769	2·47	5·11
152 x 76	19	**152 x 152**	37	154·4	80·9	8·1	11·5	7·6	1·55	94·7	354	2·00	3·87
	15		30	152·9	78·7	6·6	9·4	7·6	1·41	72·8	279	1·95	3·82
	12		23	152·4	76·2	6·1	6·8	7·6	1·43	61·1	202	2·03	3·68

STRUCTURAL TEES SPLIT FROM JOISTS

The following sizes of structural tees split from joists may be supplied.

127 x 203, split from **254 x 203**
127 x 114, split from **254 x 114**
102 x 152, split from **203 x 152**

STRUCTURAL TEES SPLIT FROM UNIVERSAL COLUMNS
DIMENSIONS AND PROPERTIES

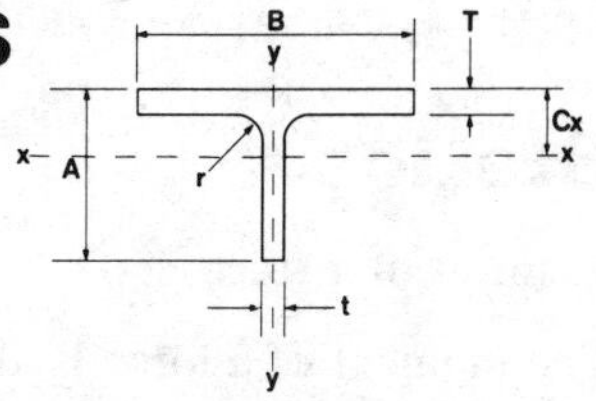

Properties have been calculated assuming that there is no loss of material due to splitting.

Elastic Modulus			Plastic Modulus		Warping Constant H	Torsional Constant J	Area of Section	Designation	
Axis x-x		Axis y-y	Axis x-x	Axis y-y				Mass per metre	Serial size
Flange	Toe								
cm³	cm³	cm³	cm³	cm³	dm⁶	cm⁴	cm²	kg	mm
503	115	403	225	614	0·357	189	101	79	**305 x 152**
451	98·1	346	188	526	0·298	125	87·3	69	
400	82·4	294	156	446	0·246	80·0	74·9	59	
343	66·6	238	123	362	0·194	45·4	61·6	49	
326	79·9	288	162	439	0·148	161	84·5	66	**254 x 127**
277	62·8	228	123	348	0·112	86·4	68·3	54	
239	49·6	189	96·0	288	0·0895	51·8	57·0	45	
204	39·3	152	74·3	231	0·0696	28·6	46·4	37	
171	42·7	149	86·0	228	0·0397	68·7	55·0	43	**203 x 102**
145	32·7	123	65·2	187	0·0312	40·6	45·5	36	
128	28·1	99·5	53·8	151	0·0244	23·2	37·9	30	
115	23·8	86·8	45·1	132	0·0208	15·9	33·2	26	
105	21·2	75·7	39·6	115	0·0178	11·1	29·4	23	
61·2	14·5	45·9	27·6	70·0	0·005	9·69	23·7	19	**152 x 76**
51·5	11·3	36·5	21·1	55·6	0·00383	5·22	19·1	15	
42·7	9·87	26·5	17·7	40·4	0·00267	2·42	14·9	12	

Values for u and x have not been given as there is no possibility of lateral torsional buckling.

Mechanics

Statics

Moment of a force

The moment of a force **F** about any axis is (**r** × **F**.**e**) where **r** is the vector from any point on the axis to any point on the line of **F**, and **e** is the unit vector along the axis.

Laws of Coulomb friction

(1) The friction force developed is independent of the magnitude of the area of contact.
(2) The limiting static friction force is proportional to the normal force.
(3) At low velocity of sliding the kinetic friction force is independent of the velocity and proportional to the normal force.

Belt friction

The limiting ratio of tensions at the ends of an arc of flat belt subtending angle θ is

$$T_1/T_2 = e^{\mu\theta}$$

where μ is the coefficient of friction. For a belt in a V-groove of semiangle α,

$$T_1/T_2 = e^{\mu\theta\operatorname{cosec}\alpha}$$

The funicular curve

The funicular curve for a load p per unit length is given by

$$\frac{d^2y}{dx^2} = \frac{p}{H}$$

where H is the polar distance. This is also the equation for the shape of a flexible cable carrying a vertical load p per unit of x; when H is the horizontal reaction.

Kinematics

In the following, v and a are velocity and acceleration, s is arc length and ω is angular velocity. Unit vectors are **i**, **j**, **k** for Cartesian, and **e** with the relevant subscript for other coordinates.

Rectangular coordinates

$$\mathbf{v} = v_x\mathbf{i} + v_y\mathbf{j} + v_z\mathbf{k}$$
$$\mathbf{a} = a_x\mathbf{i} + a_y\mathbf{j} + a_z\mathbf{k}$$

Normal and tangential components

$$\mathbf{v} = \frac{ds}{dt}\mathbf{e}_t$$

$$\mathbf{a} = \frac{d^2s}{dt^2}\mathbf{e}_t - \frac{(ds/dt)^2}{R}\mathbf{e}_n$$

Cylindrical coordinates

$$\mathbf{v} = \dot{r}\mathbf{e}_r + r\dot{\theta}\mathbf{e}_\theta + \dot{z}\mathbf{e}_z$$
$$\mathbf{a} = (\ddot{r} - r\dot{\theta}^2)\mathbf{e}_r + (r\ddot{\theta} + 2\dot{r}\dot{\theta})\mathbf{e}_\theta + \ddot{z}\mathbf{e}_z$$

Spherical polar coordinates

$$\mathbf{v} = \dot{r}\mathbf{e}_r + r\dot{\theta}\mathbf{e}_\theta + r\dot{\phi}\sin\theta\,\mathbf{e}_\phi$$

$$\begin{aligned}\mathbf{a} = {} & [\ddot{r} - r\dot{\theta}^2 - r\dot{\phi}^2\sin^2\theta]\mathbf{e}_r \\ & + [r\ddot{\theta} + 2\dot{r}\dot{\theta} - r\dot{\phi}^2\sin\theta\cos\theta]\mathbf{e}_\theta \\ & + [(r\ddot{\phi} + 2\dot{r}\dot{\phi})\sin\theta + 2r\dot{\phi}\dot{\theta}\cos\theta]\mathbf{e}_\phi\end{aligned}$$

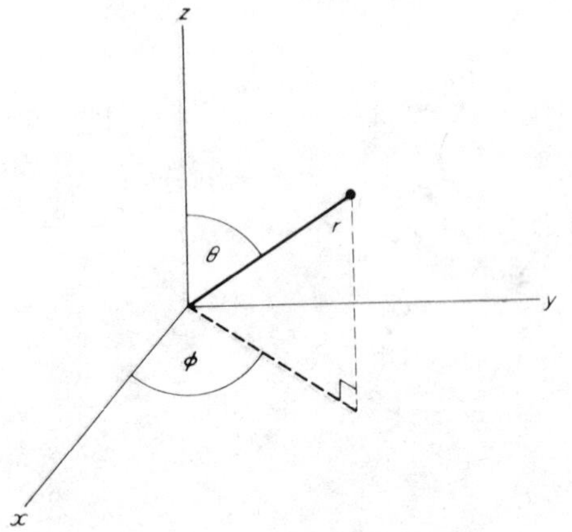

Motion referred to a moving coordinate system

$\mathbf{r} = \mathbf{R} + \boldsymbol{\rho}$

$\dot{\mathbf{r}} = \dot{\mathbf{R}} + \dot{\boldsymbol{\rho}} + \boldsymbol{\omega} \times \boldsymbol{\rho}$

$\ddot{\mathbf{r}} = \ddot{\mathbf{R}} + \boldsymbol{\omega} \times (\boldsymbol{\omega} \times \boldsymbol{\rho}) + \dot{\boldsymbol{\omega}} \times \boldsymbol{\rho} + \ddot{\boldsymbol{\rho}} + 2\boldsymbol{\omega} \times \dot{\boldsymbol{\rho}}$

$\dot{\boldsymbol{\rho}}$ is the velocity of P measured relative to $O'xyz$, which has angular velocity $\boldsymbol{\omega}$ relative to $OXYZ$.

In matrix form (see p. 19),

$\{\dot{r}\} = \{\dot{R}\} + \{\dot{\rho}\} + [\omega]\{\rho\}$

$\{\ddot{r}\} = \{\ddot{R}\} + [\omega]^2\{\rho\} + [\dot{\omega}]\{\rho\} + \{\ddot{\rho}\} + 2[\omega]\{\dot{\rho}\}$

The final term in the second equation is known as the *Coriolis acceleration.*

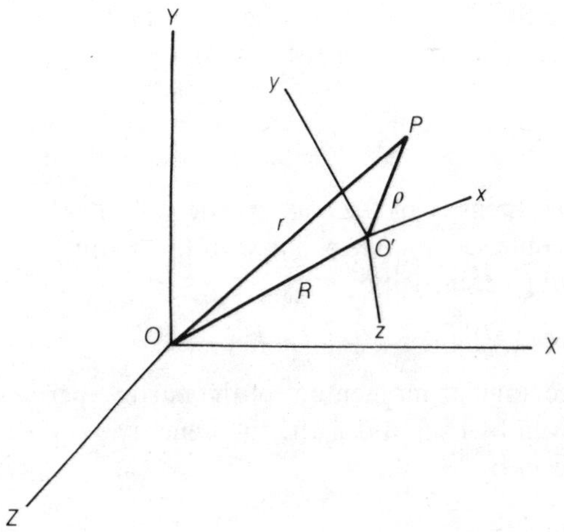

Dynamics

In the following, m is mass, F force, H angular momentum and M moment except where specified; V, T, and E are respectively potential, kinetic and total energy.

Newton's Laws

(1) Every body stays in a state of rest or uniform motion in a straight line unless it is acted on by a force which may change that state.

(2) The rate of change of momentum with respect to time is equal to the force producing it. The change takes place in the direction of the force.

(3) To every action there is an equal and opposite reaction.

Particle dynamics

Impulse and momentum

$$\int_{t_1}^{t_2} \mathbf{F}\, dt = m\mathbf{v}_2 - m\mathbf{v}_1.$$

Moment of momentum

$$\mathbf{r} \times \mathbf{F} = \frac{d\mathbf{H}_0}{dt}$$

where $\mathbf{H}_0 = \mathbf{r} \times m\mathbf{v}$.

Conservation of momentum

If in a system of particles only mutual interactions are involved, the momentum of the system is constant.

Kinetic energy

$$\int_{r_1}^{r_2} \mathbf{F}.d\mathbf{r} = \tfrac{1}{2}mv_2^2 - \tfrac{1}{2}mv_1^2 = T_2 - T_1$$

Potential energy

If $\mathbf{F} = \text{grad}\, \phi$, $\mathbf{F}$ is conservative. Then

$$\int_{r_1}^{r_2} \mathbf{F}.d\mathbf{r} = \phi_2 - \phi_1$$

and the change in potential energy is

$$-\int_{r_1}^{r_2} \mathbf{F}.d\mathbf{r} = V_2 - V_1.$$

For an inverse square law

$$\mathbf{F} = -\frac{\mu m}{r^2}\mathbf{e}_r$$

where μ is a constant of the field of force acting on m, and

$$V = \frac{\mu m}{r}$$

if $V = 0$ at $r \to \infty$.

Conservation of energy

For a conservative system

$$V + T = \text{constant}$$

Central force motion

Kepler's Laws

(1) Each planet has an elliptical orbit with the sun at a focus.

(2) The radius vector drawn from the sun to the planet sweeps out equal areas in equal times.

(3) The squares of the periods of the planets are proportional to the cubes of the semi-major axes of the elliptical orbits.

The attractive force on a particle of mass m and distance r from a fixed source of attraction may be written

$$F = \frac{\mu m}{r^2}$$

where μ is the intensity of the source. The orbit of the particle is a conic section (see p. 20), with the source as focus, having eccentricity

$$\epsilon = (1 + 2h^2 E/\mu^2)^{1/2}$$

where h is the angular momentum of the particle per unit mass, given by $r^2\omega$, and E the total energy per unit mass, given by

$$E = \tfrac{1}{2}v^2 - \mu/r.$$

In Cartesian coordinates a particle having velocity components v_{x0}, v_{y0} at the point $x_0, 0$ has $h = x_0 v_{y0}$ and its orbit has the equation

$$r = \frac{h^2}{\mu} + x\left(1 - \frac{h}{\mu}v_{y0}\right) + y\frac{h}{\mu}v_{x0}.$$

An elliptical orbit ($E < 0$, $\epsilon < 1$) of area A has period

$$\tau = 2A/h$$

Rigid-body dynamics

In the following, x, y, z are Cartesian axes fixed on a rotating body with inertia matrix $[I]$ defined on p. 25; for matrix notation see p. 19.

Moment of momentum

$$H_x = I_{xx}\omega_x - I_{xy}\omega_y - I_{xz}\omega_z$$
$$H_y = -I_{yx}\omega_x + I_{yy}\omega_y - I_{yz}\omega_z$$
$$H_z = -I_{zx}\omega_x - I_{zy}\omega_y + I_{zz}\omega_z$$

or

$$\{H\} = [I]\{\omega\}$$

General equations of motion

If the origin is either fixed or at the centre of mass, then

$$M_x = \dot{H}_x - \omega_z H_y + \omega_y H_z$$
$$M_y = \dot{H}_y - \omega_x H_z + \omega_z H_x$$
$$M_z = \dot{H}_z - \omega_y H_x + \omega_x H_y$$

or

$$\{M\} = \{\dot{H}\} + [\omega]\{H\}$$
$$= [I]\{\dot{\omega}\} + [\omega][I]\{\omega\}$$

Euler's equations

If x, y, z are principal axes,

$$M_x = I_{xx}\dot{\omega}_x + (I_{zz} - I_{yy})\omega_y\omega_z$$
$$M_y = I_{yy}\dot{\omega}_y + (I_{xx} - I_{zz})\omega_z\omega_x$$
$$M_z = I_{zz}\dot{\omega}_z + (I_{yy} - I_{xx})\omega_x\omega_y$$

Kinetic energy

$$T = \tfrac{1}{2}mv_C^2 + \tfrac{1}{2}\boldsymbol{\omega}.\mathbf{H}_C$$
$$= \tfrac{1}{2}m\{v_C\}^T\{v_C\} + \tfrac{1}{2}\{\omega\}^T[I]\{\omega\}$$

Gyroscopic motion

If ω is the angular velocity of the housing and Ω is that of the rotor relative to the housing,

$$\{H\} = [I]\{\omega\} + [I]\{\Omega\}.$$

For principal axes

$$\{M\} = \{\dot{H}\} + [\omega]\{H\}$$
$$= [I]\{\dot{\omega}\} + [\omega][I]\{\omega\} + [I]\{\dot{\Omega}\} + [\omega][I]\{\Omega\}$$

Lagrange's equations

$$\frac{d}{dt}\left(\frac{\partial T}{\partial \dot{q}_j}\right) - \frac{\partial T}{\partial q_j} = Q_j$$

where q_j is a generalized coordinate and Q_j is a generalized force.

For a conservative system

$$\frac{d}{dt}\left(\frac{\partial L}{\partial \dot{q}_j}\right) - \frac{\partial L}{\partial q_j} = 0$$

where $L = (T - V)$.

Euler's differential equation

$$\frac{d}{dx}\left(\frac{\partial f}{\partial y'}\right) - \frac{\partial f}{\partial y} = 0$$

where

$$y' = \frac{dy}{dx}$$

Hamilton's principle

$$\delta \int_{t_1}^{t_2} (T - V)\, dt = 0$$

Vibrations

In the following, k is spring stiffness, c a viscous damping constant, ω_0 an undamped natural or resonant frequency and m, M masses.

Free vibration with viscous damping

For a mass m the undamped natural frequency is

$$\omega_0 = \sqrt{(k/m)},$$

the critical damping constant is

$$c_c = 2\sqrt{(km)},$$

the damping ratio is $\zeta = c/c_c$, and the logarithmic decrement is

$$\ln d = 2\pi n\zeta/\sqrt{(1-\zeta^2)} \approx 2\pi n\zeta$$

for small ζ, where d is the ratio of the amplitude at any instant to that n periods later.

Steady-state vibration with viscous damping

The governing equation is

$$\ddot{x} + 2\zeta\omega_0\dot{x} + \omega_0^2 x = \frac{P\sin\omega t}{m}$$

for an applied force $P\sin\omega t$.
The ratio of peak amplitude X to the steady displacement $X_0 = P/k$ is

$$\frac{X}{X_0} = \frac{1}{[\{1-(\omega/\omega_0)^2\}^2 + \{2\zeta\omega/\omega_0\}^2]^{1/2}}$$

and the phase angle ϕ is given by

$$\tan\phi = \frac{2\zeta\omega/\omega_0}{1-(\omega/\omega_0)^2}$$

These relations yield the curves given below. Except for the linear scales they are identical to those on p. 149. For $\zeta < 1/\sqrt{2}$ the amplitude has a peak at $\omega/\omega_0 = \sqrt{1-2\zeta^2}$

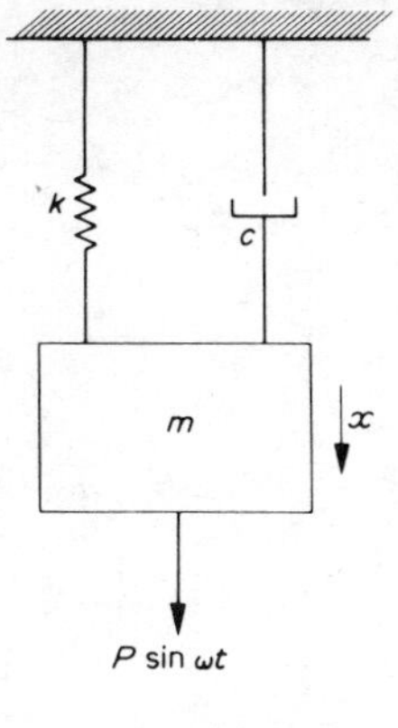

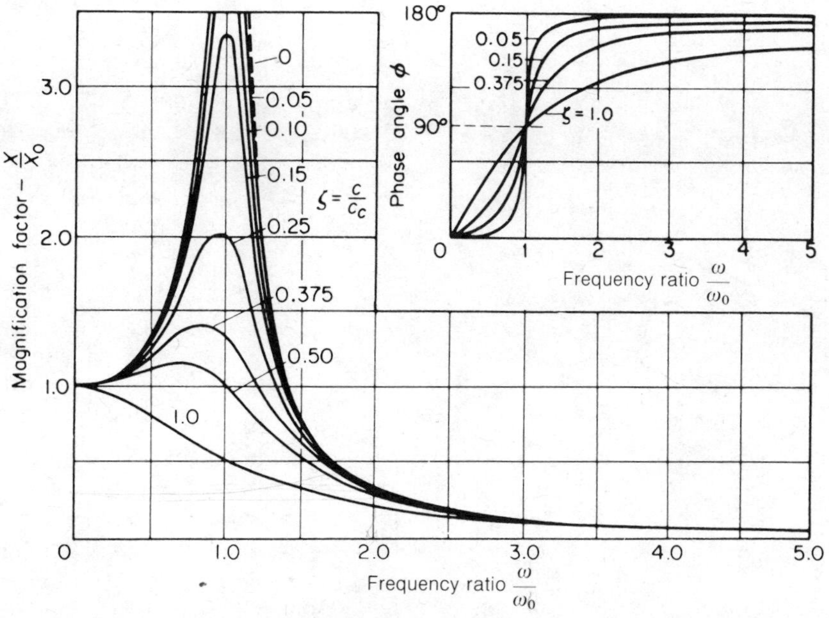

Rotating unbalance

The governing equation is

$$\ddot{x}+2\zeta\omega_o\dot{x}+\omega_0^2 x=\frac{m\omega^2 e}{M}\sin\omega t.$$

The peak amplitude X and the phase ϕ are given by

$$\frac{MX}{me}=\frac{(\omega/\omega_0)^2}{[\{1-(\omega/\omega_0)^2\}^2+\{2\zeta\omega/\omega_0\}^2]^{1/2}}$$

and

$$\tan\phi=\frac{2\zeta\omega/\omega_0}{1-(\omega/\omega_0)^2}.$$

The curves below show the variation of X and ϕ with ω.

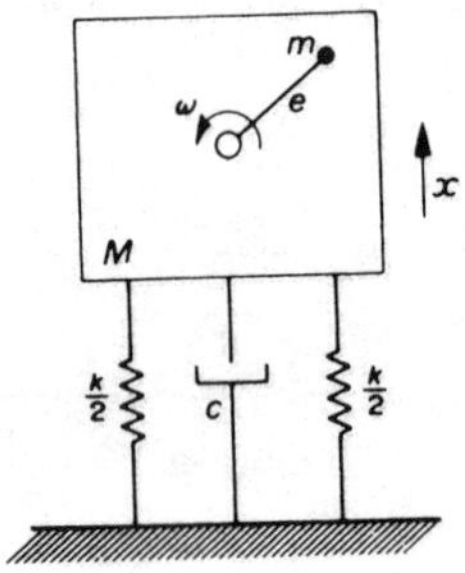

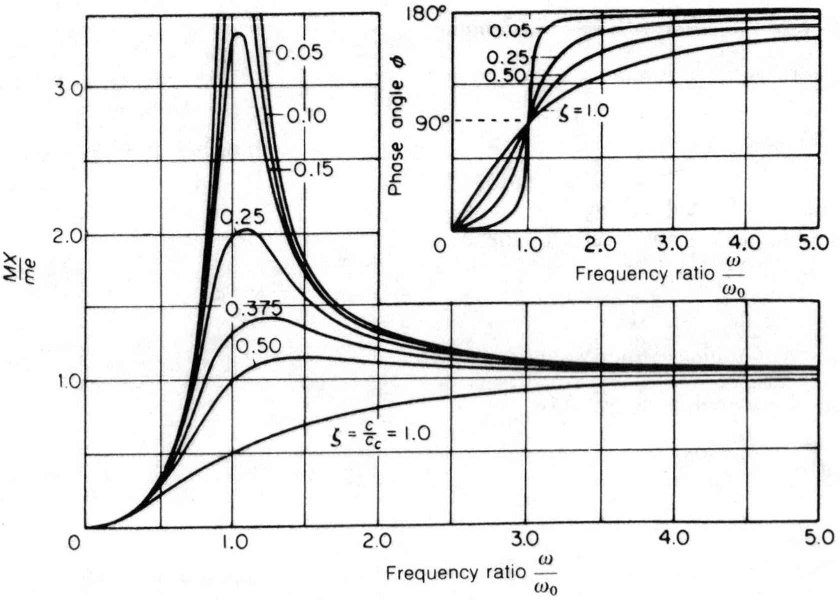

Displacement excitation

The governing equation is

$$\ddot{x} + 2\zeta\omega_0\dot{x} + \omega_0^2 x = 2\zeta\omega_0\dot{y} + \omega_0^2 y.$$

The ratio of the peak amplitudes of the mass and the base is

$$\frac{X}{Y} = \left[\frac{1 + (2\zeta\omega/\omega_0)^2}{\{1 - (\omega/\omega_0)^2\}^2 + \{2\zeta\omega/\omega_0\}^2}\right]^{1/2}$$

and their phase difference is given by

$$\tan\phi = \frac{2\zeta(\omega/\omega_0)^3}{1 - (\omega/\omega_0)^2 + (2\zeta\omega/\omega_0)^2}$$

These results are shown in the curves below. The amplitude ratio is also the ratio of the transmitted force to the exciting force in the vibration isolator with constants k and c separating mass m from ground.

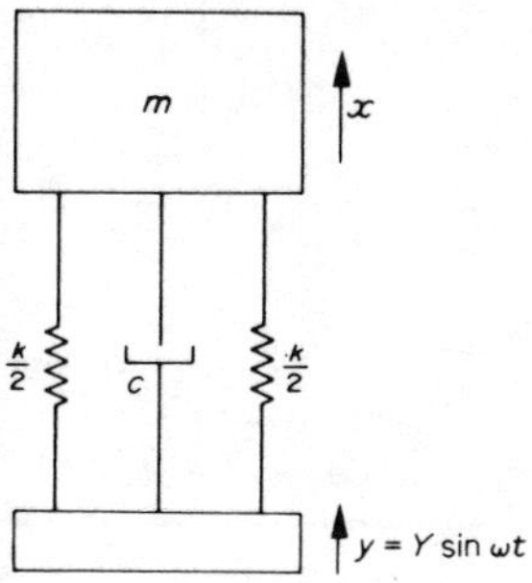

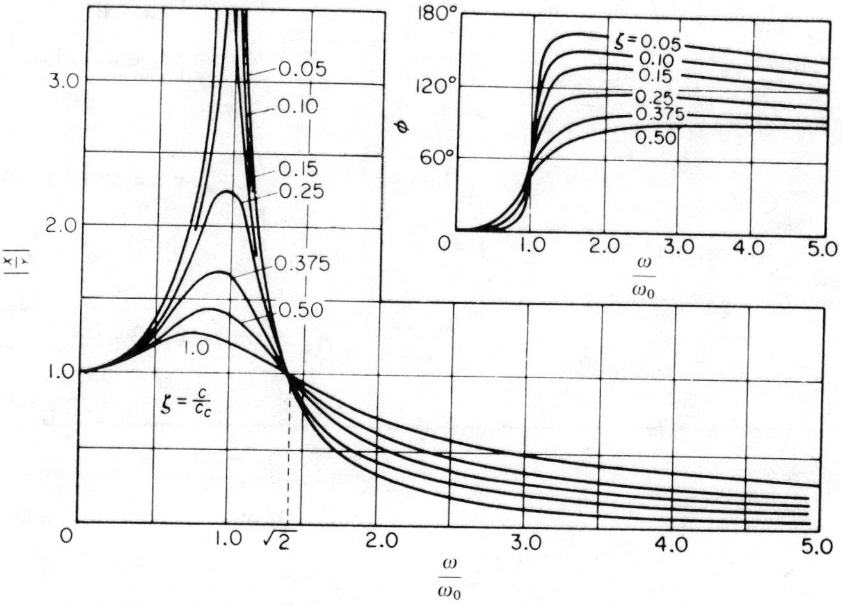

Vibration of beams of uniform section with uniformly distributed load

The natural frequencies of a beam of length L are given by

$$\omega_0 = k\sqrt{\frac{EI}{mL^4}}$$

where m is the mass per unit length.

The table below gives the value of k for the fundamental, 2nd and 3rd harmonics with different support conditions.

Support	1	2	3
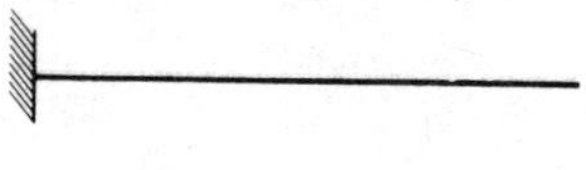	3·52	22·0	61·7
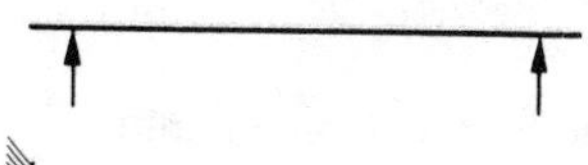	9·87	39·5	88·8
	15·4	50·0	104
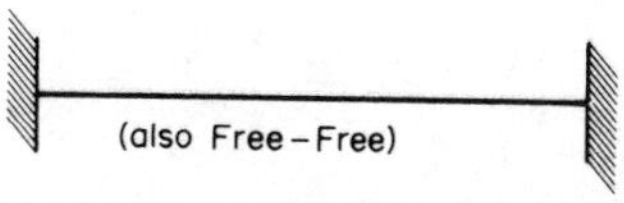 (also Free–Free)	22·4	61·7	121

Electricity

Electromagnetism

In the following, **E**, **D** are electric field strength and flux density; **H**, **B** are magnetic field strength and flux density; **J**, i are current density and current; q is electric charge; ρ, ρ_s are volume and surface charge density; **A**, ϕ are vector and scalar potentials; V, **S**, **l** and r are volume, area, length and distance; $\mathbf{a}_r$, **n** are radial and normal unit vectors; subscripts t and n denote tangential and normal; brackets represent retarded values; ϵ, μ and σ are permittivity, permeability and conductivity, and **v** is velocity.

Coulomb law

The force between point charges q_1 and q_2 is

$$\frac{q_1 q_2 \mathbf{a}_r}{4\pi\epsilon r^2}$$

Biot-Savart law: $d\mathbf{H} = \dfrac{i\,d\mathbf{l} \times \mathbf{a}_r}{4\pi r^2}$

Lorentz force

The force on a unit positive change is

$$\mathbf{E} + \mathbf{v} \times \mathbf{B}$$

Magnetic force

The force on a current element idl is

$$i d\mathbf{l} \times \mathbf{B}$$

and the force on a distributed current is

$$\mathbf{J} \times \mathbf{B}$$

per unit volume

Maxwell Equations

	Integral form	Differential form
Gauss law	$\oint_S \mathbf{D}.d\mathbf{S} = \int_V \rho\, dV$	$\operatorname{div} \mathbf{D} = \rho$
	$\oint_S \mathbf{B}.d\mathbf{S} = 0$	$\operatorname{div} \mathbf{B} = 0$
Faraday law	$\oint_C \mathbf{E}.d\mathbf{l} = -\int_S \dot{\mathbf{B}}.d\mathbf{S}$	$\operatorname{curl} \mathbf{E} = -\dot{\mathbf{B}}$
Ampere law / *Work law* / *Magnetic circuit law*	$\oint_C \mathbf{H}.d\mathbf{l} = \int_S (\mathbf{J} + \dot{\mathbf{D}}).d\mathbf{S}$	$\operatorname{curl} \mathbf{H} = \mathbf{J} + \dot{\mathbf{D}}$
Equation of Continuity	$\oint_S \mathbf{J}.d\mathbf{S} = -\int_V \dot{\rho}\, dV$	$\operatorname{div} \mathbf{J} = -\dot{\rho}$

Constitutive equations

$\mathbf{D} = \epsilon\mathbf{E}$; $\mathbf{B} = \mu\mathbf{H}$; $\mathbf{J} = \sigma\mathbf{E}$

Boundary conditions

$E_{t1} = E_{t2}$ $\quad$ $H_{t1} = H_{t2}$ (for no surface current)

$D_{n1} - D_{n2} = \rho_s$ $\quad$ $B_{n1} = B_{n2}$

At the surface of a perfect conductor: $\mathbf{D} = \rho_s \mathbf{n}$ and $\mathbf{n} \times \mathbf{H} = \mathbf{J}_s$

Potential functions

$$\phi = \frac{1}{4\pi\epsilon}\int_V \frac{[\rho]}{r} dV$$

$$\mathbf{A} = \frac{\mu}{4\pi}\int_V \frac{[\mathbf{J}]}{r} dV \quad \text{or} \quad \mathbf{A} = \frac{\mu}{4\pi}\int_C \frac{[i]}{r} d\mathbf{l}$$

$$\mathbf{B} = \nabla \times \mathbf{A}$$

$$\mathbf{E} = -\nabla\phi - \dot{\mathbf{A}}$$

if $\nabla \cdot \mathbf{A} = -\mu\epsilon\dot{\phi}$ then

$$\nabla^2\phi = \mu\epsilon\ddot{\phi} - \rho/\epsilon$$

$$\nabla^2\mathbf{A} = \mu\epsilon\ddot{\mathbf{A}} - \mu\mathbf{J}$$

Poisson's equation: $\nabla^2\phi = -\rho/\epsilon$

Laplace's equation: $\nabla^2\phi = 0$

Analysis of circuits

In the following Z is impedance and Y is admittance. Unless otherwise stated, V and I are r.m.s. (or d.c.) values of voltage and current.

Star-delta and delta-star transformation

Star to delta:

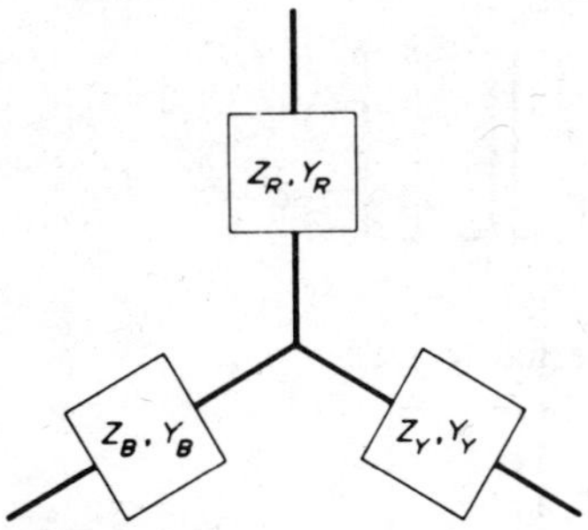

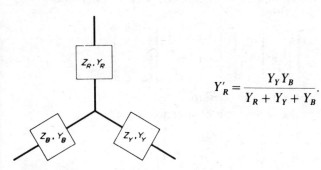

$$Y'_R = \frac{Y_Y Y_B}{Y_R + Y_Y + Y_B}.$$

Delta to star:

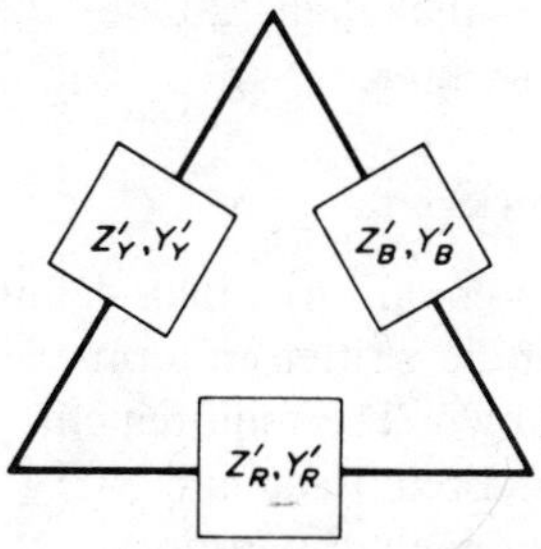

$$Z_R = \frac{Z'_Y Z'_B}{Z'_R + Z'_Y + Z'_B}.$$

Self-inductance of two coils

If the self-inductances of the individual coils are L_1 and L_2 and they are placed so that their mutual inductance is M, then the self-inductance of the combination, when connected in series, is $L_1 + L_2 \pm 2M$ dependent upon the relative directions of current flow. When connected in parallel the self-inductance of the combination is

$$(L_1 L_2 - M^2)/(L_1 + L_2 \mp 2M).$$

Reciprocity theorem

If a current is produced at point a in a linear passive network by a voltage source acting at a point b, then the same current would be produced at point b by the source acting at point a. The principle also holds for voltages produced by current sources.

Superposition principle

The response of a linear system to a number of simultaneously applied excitations is equal to the sum of the responses taken one at a time. When any one source is being considered all the others are de-activated; de-activation means that independent voltage sources are replaced by short circuits and independent current sources by open circuits.

Thévenin's theorem and equivalent circuit

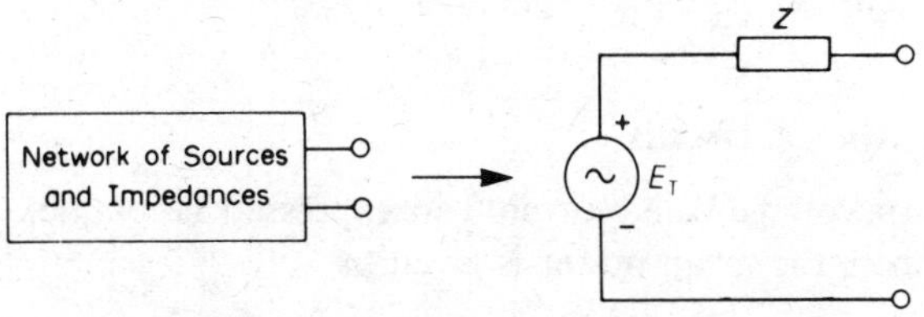

A two-terminal network containing sources and impedances can always be replaced, as far as any load is concerned, by a voltage source and impedance as shown. The value of E_T is the voltage which is measured at the terminals when open-circuited. The value of Z is the impedance presented at the open-circuited terminals when all independent sources are de-activated.

Norton's theorem and equivalent circuit

This is the equivalent to Thévenin's theorem in terms of a current source.

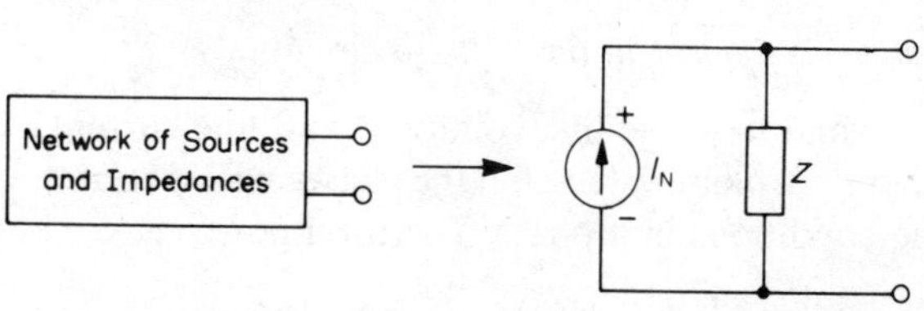

I_N is the current which flows in a short-circuit on the terminals and Z is as defined for the Thévenin equivalent circuit. It follows that the two equivalents are related by

$$E_T = I_N Z$$

Maximum power transfer from source to load

The power in the load is a maximum

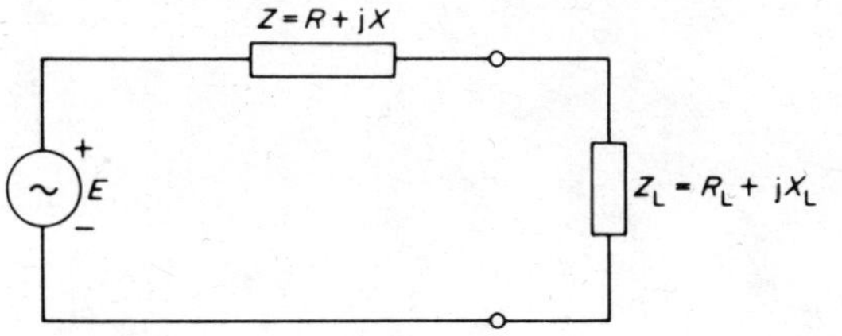

when the load is matched to the source, that is when

$$Z_L = Z^*$$

or

$$R_L = R, \qquad X_L = -X$$

and the load power is then $E^2/4R$ (The overall efficiency is not necessarily then 50% if Z is a Thévenin impedance.) If the load is restricted to resistance R_L only, then maximum power is obtained when

$$R_L = |Z|$$

Power in a.c. circuits

If r.m.s. voltage **V** and current **I** are expressed as complex numbers the mean power is given by

$$P = \mathrm{Re}\,\mathbf{VI}^* = \mathrm{Re}\,\mathbf{V}^*\mathbf{I} = |\mathbf{V}|\,|\mathbf{I}|\cos\phi$$

where ϕ is the angle of **I** relative to **V**; also the reactive component of volt-amperes is given by

$$Q = \mathrm{Im}\,\mathbf{V}^*\mathbf{I} = |\mathbf{V}|\,|\mathbf{I}|\sin\phi$$

if reactive VA is defined as + ve for leading current, or

$$Q = \mathrm{Im}\,\mathbf{VI}^* = -|\mathbf{V}|\,|\mathbf{I}|\sin\phi$$

if reactive VA is defined as + ve for lagging current. Reactive VA is sometimes referred to as reactive (or imaginary) power and its units designated VAR.

Power measurement in three-phase circuits

It is normal to quote line voltage V and line current I rather than phase values, but the phase angle ϕ relates to the conditions in a phase. The total power is

$$P = \sqrt{3}\,VI\cos\phi$$

for balanced star or delta connections.

In a system fed through n wires, it is necessary to take $n-1$ power readings for the sum to give the true total power. Therefore, for a delta-connected load and a 3-wire star-connected load, two wattmeter readings are sufficient. If the loads are balanced then

$$\tan\phi = \sqrt{3}\left(\frac{W_1 - W_2}{W_1 + W_2}\right)$$

Symmetrical components

The following matrix relation gives the various sequence components in terms of the unbalanced quantities.

$$\begin{bmatrix} V_1 \\ V_2 \\ V_0 \end{bmatrix} = \frac{1}{3}\begin{bmatrix} 1 & a & a^2 \\ 1 & a^2 & a \\ 1 & 1 & 1 \end{bmatrix}\begin{bmatrix} V_R \\ V_Y \\ V_B \end{bmatrix}$$

where a is the operator $\frac{1}{2}(-1 + j\sqrt{3})$

The matrix can be inverted to give

$$\begin{bmatrix} V_R \\ V_Y \\ V_B \end{bmatrix} = \begin{bmatrix} 1 & 1 & 1 \\ a^2 & a & 1 \\ a & a^2 & 1 \end{bmatrix}\begin{bmatrix} V_1 \\ V_2 \\ V_0 \end{bmatrix}$$

Other forms give 012 sequence rather than 120, or $1/\sqrt{3}$ multipliers in both equations.

Two-port or four-terminal networks

Each of the six possible pairs of equations which define the behaviour of a two-port can be written in terms of a matrix of parameters as shown. The requirement for reciprocity and the additional requirement for symmetry are also shown for each; a two-port containing only passive linear elements, and no independent or controlled source, satisfies the reciprocity condition.

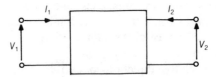

Parameter set	Equation	Reciprocity condition	Symmetry condition
Impedance, z	$\begin{Bmatrix} V_1 \\ V_2 \end{Bmatrix} = \begin{bmatrix} z_{11} & z_{12} \\ z_{21} & z_{22} \end{bmatrix} \begin{Bmatrix} I_1 \\ I_2 \end{Bmatrix}$	$z_{12} = z_{21}$	$z_{11} = z_{22}$
Admittance, y	$\begin{Bmatrix} I_1 \\ I_2 \end{Bmatrix} = \begin{bmatrix} y_{11} & y_{12} \\ y_{21} & y_{22} \end{bmatrix} \begin{Bmatrix} V_1 \\ V_2 \end{Bmatrix}$	$y_{12} = y_{21}$	$y_{11} = y_{22}$
Hybrid, h	$\begin{Bmatrix} V_1 \\ I_2 \end{Bmatrix} = \begin{bmatrix} h_{11} & h_{12} \\ h_{21} & h_{22} \end{bmatrix} \begin{Bmatrix} I_1 \\ V_2 \end{Bmatrix}$	$h_{12} = -h_{21}$	$\det H = 1$
Inverse hybrid, g	$\begin{Bmatrix} I_1 \\ V_2 \end{Bmatrix} = \begin{bmatrix} g_{11} & g_{12} \\ g_{21} & g_{22} \end{bmatrix} \begin{Bmatrix} V_1 \\ I_2 \end{Bmatrix}$	$g_{12} = -g_{21}$	$\det G = 1$
Transmission, a	$\begin{Bmatrix} V_1 \\ I_1 \end{Bmatrix} = \begin{bmatrix} a_{11} & a_{12} \\ a_{21} & a_{22} \end{bmatrix} \begin{Bmatrix} V_2 \\ I_2 \end{Bmatrix}$	$\det A = -1$	$a_{11} = -a_{22}$
Inverse transmission, b	$\begin{Bmatrix} V_2 \\ I_2 \end{Bmatrix} = \begin{bmatrix} b_{11} & b_{12} \\ b_{21} & b_{22} \end{bmatrix} \begin{Bmatrix} V_1 \\ I_1 \end{Bmatrix}$	$\det B = -1$	$b_{11} = -b_{22}$

Sometimes, especially in power systems, I_2 is defined in the opposite sense and two parameters in each set change sign accordingly. In that case it is common to use A, B, C, D for the transmission parameters, thus:

$$\begin{Bmatrix} V_1 \\ I_1 \end{Bmatrix} = \begin{bmatrix} A & B \\ C & D \end{bmatrix} \begin{Bmatrix} V_2 \\ I_2 \end{Bmatrix}$$

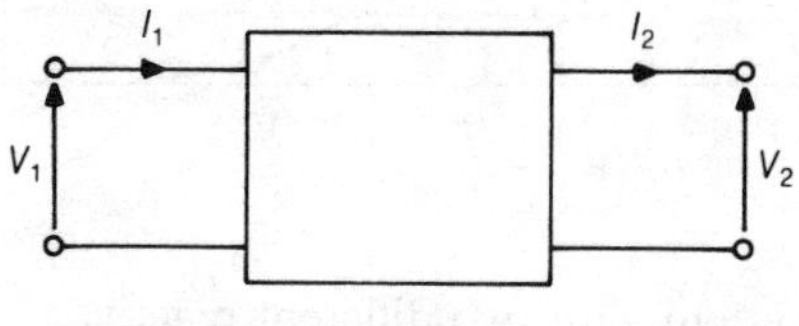

The *iterative impedance* Z_0 of a two-port is a function of its parameters such that when Z_0 is connected to the output terminals the impedance at the input terminals is also Z_0

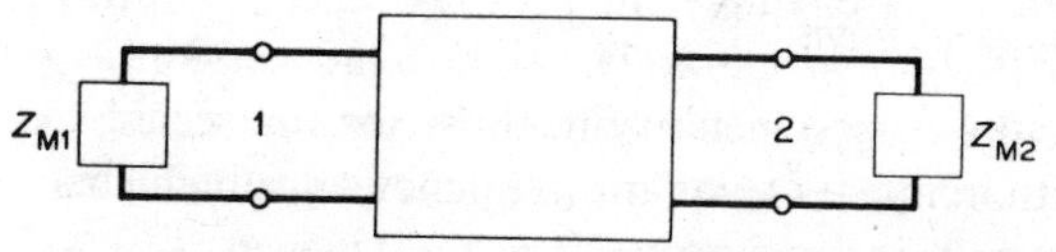

The *image impedances* Z_{M1} and Z_{M2} are functions of the parameters such that when connected as shown the impedance to the right at terminals 1 is Z_{M1} and the impedance to the left at terminals 2 is Z_{M2}; then

$$Z_{M1} = \sqrt{\frac{AB}{CD}} = \sqrt{\frac{a_{11}a_{12}}{a_{21}a_{22}}}; \qquad Z_{M2} = \sqrt{\frac{BD}{AC}} = \sqrt{\frac{a_{12}a_{22}}{a_{11}a_{21}}}$$

For a symmetrical circuit $Z_{M1} = Z_{M2} = Z_0$.

Resonance and response

Series resonant circuit

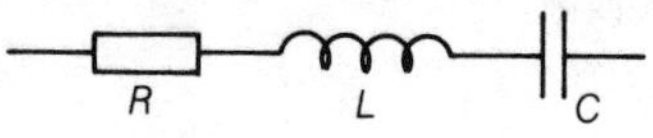

The *resonant frequency* (or undamped natural frequency) is that for which the impedance Z is real with the (minimum) value R, and is given by

$$\omega_0 = 1/\sqrt{LC}.$$

The *quality factor* or Q-factor is the voltage magnification at resonance, that is

$$Q = \frac{\text{voltage across } L \text{ or } C}{\text{voltage across whole circuit}} = \frac{\omega_0 L}{R} = \frac{1}{\omega_0 CR}$$

For any reactive component Q may be defined generally as

$$Q = \frac{2\pi\,(\text{maximum energy stored in } L \text{ or } C)}{\text{energy dissipated per cycle}}.$$

The impedance just off resonance, at a frequency

$$\omega = \omega_0 \pm \Delta\omega = \omega_0(1 \pm \epsilon)$$

where $\epsilon = \Delta\omega/\omega_0 \ll 1$, is given by

$$Z = R(1 \pm j2\epsilon Q).$$

If $2\epsilon Q = 1$, then $|Z| = \sqrt{2}R$ and, for constant voltage, the current taken is then a fraction $1/\sqrt{2}$ of its maximum value and the power is halved. Those frequencies at which $2\epsilon Q = 1$ are known as the half-power points and the interval between them, the half-power bandwidth, is

$$2\Delta\omega = \omega_0/Q$$

and at these points

$$2\Delta\omega/\omega_0 = 1/Q = 2\epsilon.$$

Parallel resonant circuit

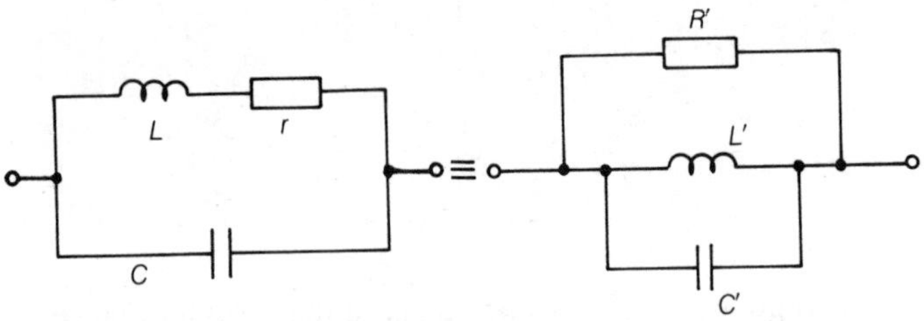

These circuits are equivalent with $L' = L, C' = C$ and $R' = Q^2 r$ if $Q = \omega_0 L/r \gg 1$. The right-hand version gives results complementary to those for the series resonant circuit. Its resonant frequency, at which the admittance Y is a minimum, is

$$\omega_0 = 1/\sqrt{L'C'}$$

and its quality factor

$$Q = \frac{R'}{\omega_0 L'} = \omega_0 C'R'.$$

The admittance just off resonance is

$$Y = \frac{1}{R'}(1 \pm j2\epsilon Q)$$

Bode diagrams

The complete diagram for any complex function of frequency $H(j\omega)$ has two parts, one showing magnitude and the other argument or phase, as functions of frequency. The frequency is plotted logarithmically and the magnitude is expressed in decibels, i.e. as $20\log_{10}|H|$ dB. In many cases the curves are closely approximated by their asymptotes.

A first-order system with one reactive element gives rise to functions with one time-constant T and a Bode diagram with a breakpoint or corner frequency at $\omega = 1/T$. The example below gives the diagram for the voltage transfer function of the circuit shown.

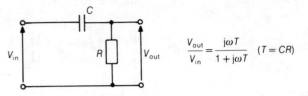

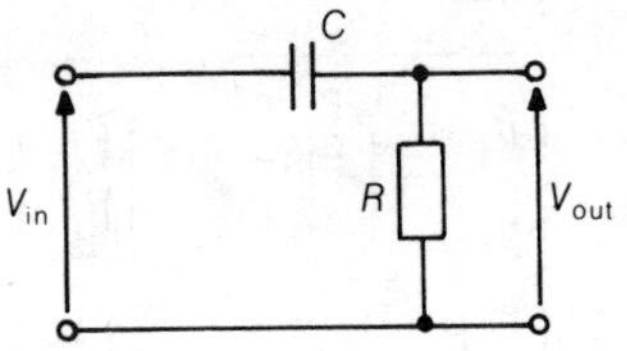

$$\frac{V_{out}}{V_{in}} = \frac{j\omega T}{1 + j\omega T} \quad (T = CR)$$

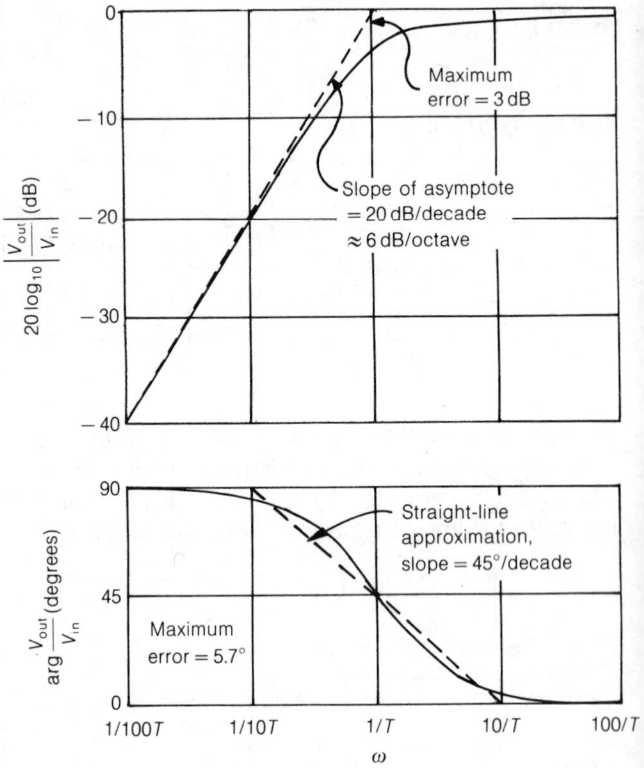

A second-order system with two different reactive (energy-storing) elements gives rise to functions containing a factor of the form

$$(j\omega)^2 + 2j\zeta\omega\omega_0 + \omega_0^2$$

or

$$(j\omega/\omega_0)^2 + 2j\zeta\omega/\omega_0 + 1$$

in which ω_0 is the resonant frequency and ζ the *damping ratio*, defined generally as

$$\zeta = \frac{\text{actual damping}}{\text{critical damping}}$$

(in reference to the transient response: see below) and related to the Q-factor by

$$\zeta = 1/2Q.$$

For small ζ the system shows resonant behaviour, as

in the example of the voltage transfer function and its Bode diagram given below. In this case the peak response occurs at the frequency

$$\omega_r = \omega_0\sqrt{1 - 2\zeta^2}$$

There is therefore no resonant peak when

$$\zeta > 1/\sqrt{2} \quad \text{or} \quad Q < 1/\sqrt{2},$$

and the maximum response then occurs for d.c., $\omega = 0$, and has magnitude 0 dB.

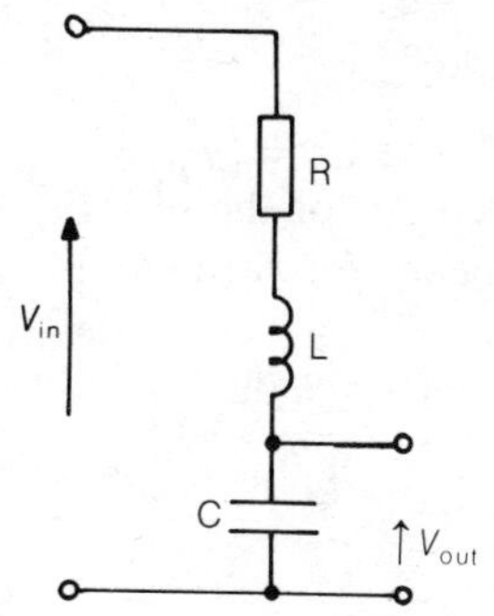

$$\frac{V_{out}}{V_{in}} = \frac{1}{(j\omega)^2 LC + j\omega RC + 1}$$

$$(\omega_0 = 1/\sqrt{LC},\ \zeta = \tfrac{1}{2}R\sqrt{C/L})$$

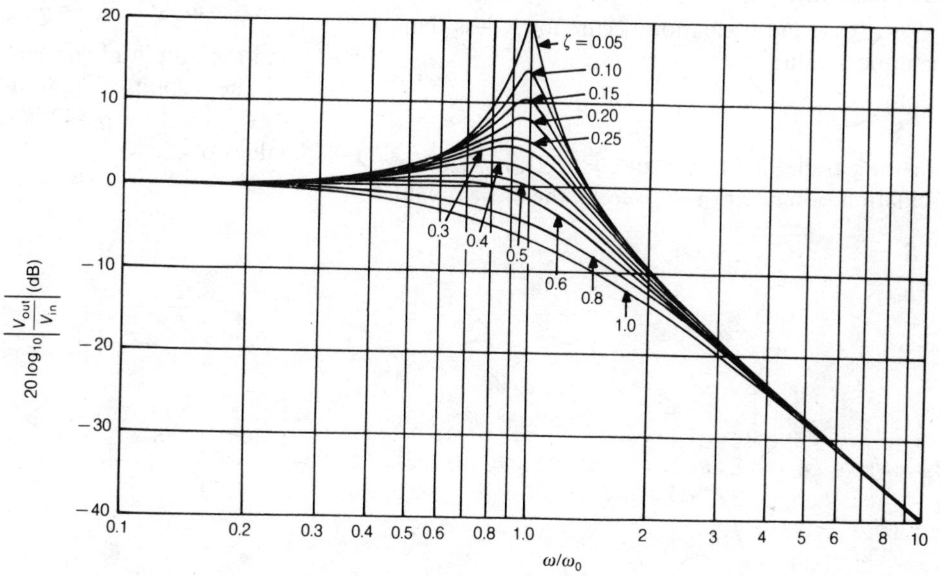

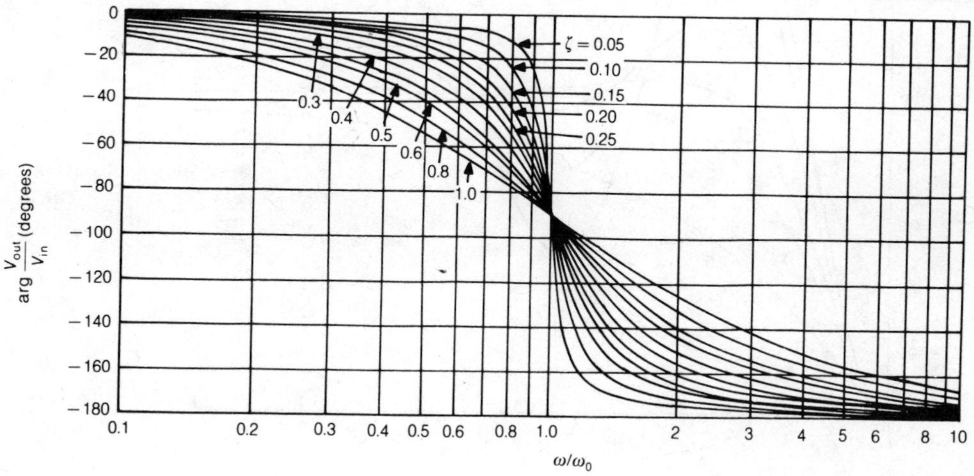

Transient response

The transient response of a first-order system shows an exponential approach to the steady-state output. For the RC circuit on the response V_{out} to an unit step V_{in} is as shown.

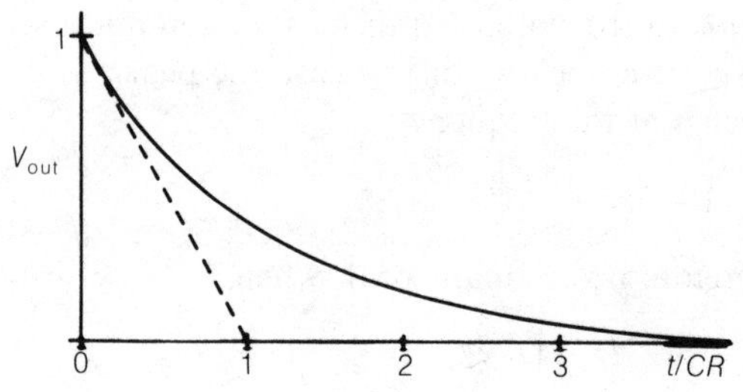

The transient response of a second order system is oscillatory if the damping is less than critical, i.e. for underdamping, $\zeta < 1$. For the series RLC circuit $\zeta = \frac{1}{2}R\sqrt{C/L}$ and oscillation occurs for R less than the critical value

$$R_c = 2\sqrt{L/C}.$$

For the parallel $R'L'C'$ circuit $\zeta = (1/2R')\sqrt{L'/C'}$ and oscillation occurs when R' exceeds the critical value

$$R'_c = \tfrac{1}{2}\sqrt{L'/C'}.$$

The frequency of oscillation is the *damped natural frequency** given by

$$\omega_n = \omega_0\sqrt{1-\zeta^2} = \omega_0\sqrt{1-(1/2Q)^2}$$

and the amplitude decays as $e^{-\alpha t}$ where $\alpha = \zeta\omega_0$.

The response V_{out} to a unit step V_{in} for the series RLC circuit on p. 149 is shown below for various values of ζ.

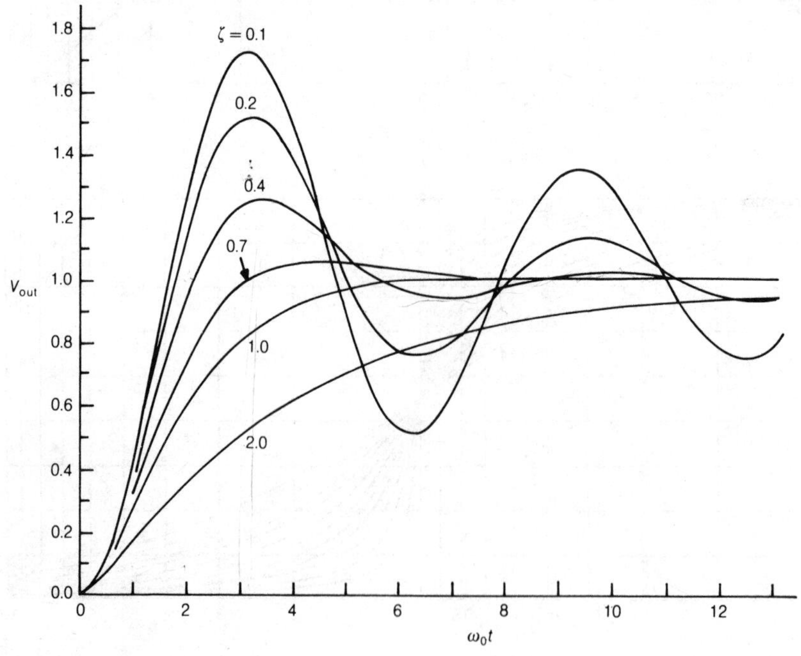

*The symbol ω_n is quite often used for the *undamped* natural frequency, here designated ω_0.

Poles and zeros

Both the response to a sinusoidal input and the transient response of a system can be deduced from the position of the poles and zeros of the appropriate function $H(s)$. The transfer function is written in the form

$$H(s) = \frac{K(s-z_1)(s-z_2)\cdots}{(s-p_1)(s-p_2)\cdots}$$

where $z_1, z_2 \cdots p_1, p_2 \cdots$ are its zeros and poles and can be plotted as points on an Argand diagram.

For the examples given earlier typical diagrams are shown below (pole ×, zero ○)

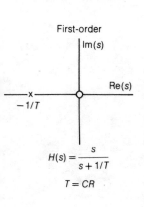

First-order

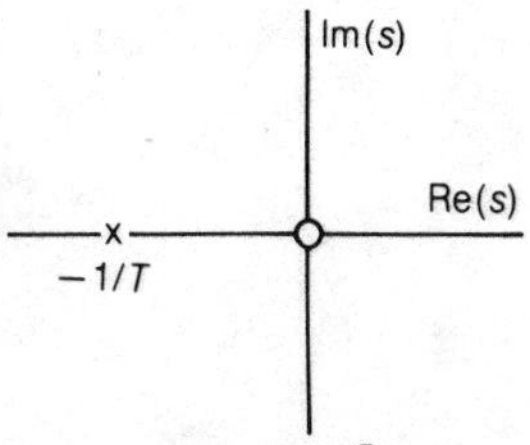

$$H(s) = \frac{s}{s + 1/T}$$

$T = CR$

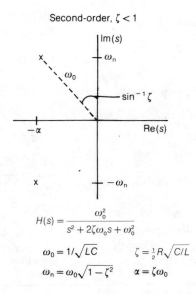

Second-order, $\zeta < 1$

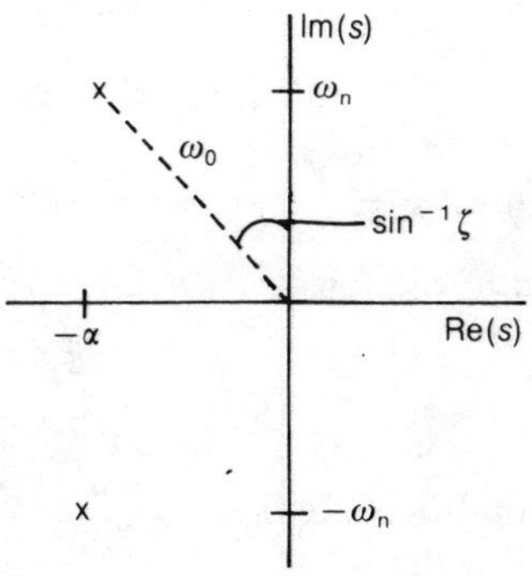

$$H(s) = \frac{\omega_0^2}{s^2 + 2\zeta\omega_0 s + \omega_0^2}$$

$\omega_0 = 1/\sqrt{LC}$ $\quad \zeta = \frac{1}{2}R\sqrt{C/L}$

$\omega_n = \omega_0\sqrt{1-\zeta^2}$ $\quad \alpha = \zeta\omega_0$

Transmission lines, microwaves and propagation

Transmission lines

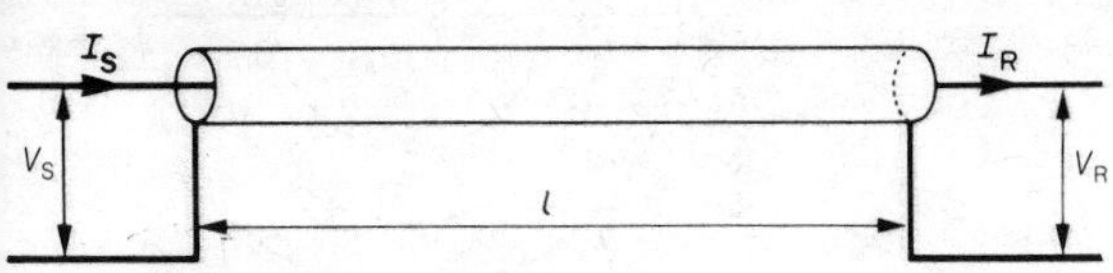

In the following, R, G, L, C are the resistance, conductance, inductance and capacitance per unit length.

The propagation constant, γ, is

$$\sqrt{\{(R + j\omega L)(G + j\omega C)\}} = \alpha + j\beta$$

The characteristic impedance, Z_0, is

$$\sqrt{\left(\frac{R + j\omega L}{G + j\omega C}\right)}$$

The voltage and current at a point distant x from the sending end are given by

$$\begin{bmatrix} V_x \\ I_x \end{bmatrix} = \begin{bmatrix} \cosh \gamma x & -Z_0 \sinh \gamma x \\ -\frac{1}{Z_0}\sinh \gamma x & \cosh \gamma x \end{bmatrix} \begin{bmatrix} V_S \\ I_S \end{bmatrix}$$

The matrix can be inverted with $x = l$ to yield

$$\begin{bmatrix} V_S \\ I_S \end{bmatrix} = \begin{bmatrix} \cosh \gamma l & Z_0 \sinh \gamma l \\ \frac{1}{Z_0}\sinh \gamma l & \cosh \gamma l \end{bmatrix} \begin{bmatrix} V_R \\ I_R \end{bmatrix}$$

If the line is terminated with an impedance Z_L the input impedance is

$$Z_S = Z_0 \frac{Z_L + Z_0 \tanh \gamma l}{Z_0 + Z_L \tanh \gamma l} = Z_0 \frac{1 + \rho \exp(-2\gamma l)}{1 - \rho \exp(-2\gamma l)}$$

where ρ is the reflection coefficient for voltage, defined as

$$\frac{Z_L - Z_0}{Z_L + Z_0} = \rho = \frac{\text{voltage of reflected wave}}{\text{voltage of incident wave}}$$

For a matched line ($Z_L = Z_0$)

$$\rho = 0 \quad \text{and} \quad Z_S = Z_0.$$

For a short-circuited line ($Z_L = 0$)

$$\rho = -1 \quad \text{and} \quad Z_S = Z_0 \tanh \gamma l.$$

For an open-circuited line ($Z_L \to \infty$)

$$\rho = 1 \quad \text{and} \quad Z_S = Z_0 \coth \gamma l.$$

The voltage standing-wave ratio is

$$\text{VSWR} = \frac{1 + |\rho|}{1 - |\rho|}$$

Attenuation, wavelength, phase and group velocity

The voltage at distance x can be written in the form

$$V = Ae^{+\gamma x} + Be^{-\gamma x}$$
$$= Ae^{\alpha x}e^{j\beta x} + Be^{-\alpha x}e^{-j\beta x}$$

where A, B are determined by the end conditions, and the two terms represent travelling waves in the negative and positive x-directions respectively; α represents an attenuation and β a change of phase with distance, and $\beta\lambda = 2\pi$ where λ is wavelength.

The phase velocity is $v_p = \omega/\beta$ and the group velocity is $v_g = d\omega/d\beta$.

On a dispersive line waves of different frequencies travel at different velocities and the relation between ω and β is known as the *dispersion relation.* In these cases the envelope of a wave containing different frequencies travels with the group velocity, v_g.

Instead of β a propagation constant or wave number k may be used; the attenuation constant α may be represented as the imaginary part of k. Introducing a complex k as $k' - jk''$ gives

$$j(k' - jk'') = k'' + jk' \equiv \alpha + j\beta$$

For a *lossless line*

$$R = G = 0; \qquad \alpha = 0;$$

$$\gamma = j\beta = j\omega\sqrt{LC}; \qquad Z_0 = \sqrt{\frac{L}{C}}; \qquad v_p = v_g = \frac{1}{\sqrt{LC}}.$$

For a *low-loss line*

$$R \ll \omega L; \qquad G \ll \omega C$$

$$\alpha = \frac{1}{2}\left(R\sqrt{\frac{C}{L}} + G\sqrt{\frac{L}{C}}\right); \qquad \beta \approx \omega\sqrt{LC}.$$

If $L/R = C/G$ the line is *distortionless* since both the attenuation and the velocity of propagation are then independent of frequency, and

$$\gamma = \alpha + j\beta = \sqrt{RG} + j\omega\sqrt{LC}$$

$$Z_0 = \sqrt{\frac{R}{G}} = \sqrt{\frac{L}{C}}$$

The *Smith Chart* provides a convenient graphical method of solving many transmission line problems. It consists of families of circles as shown below where

$$r = \frac{R}{Z_0} \quad \text{and} \quad x = \frac{X}{Z_0}$$

are the normalized components of the load impedance $Z = R + jX$.

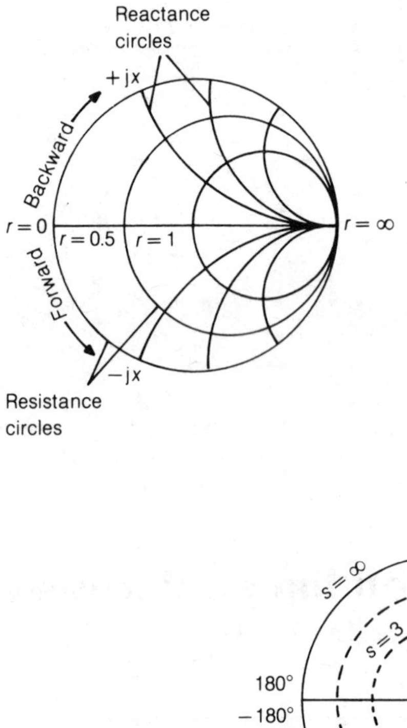

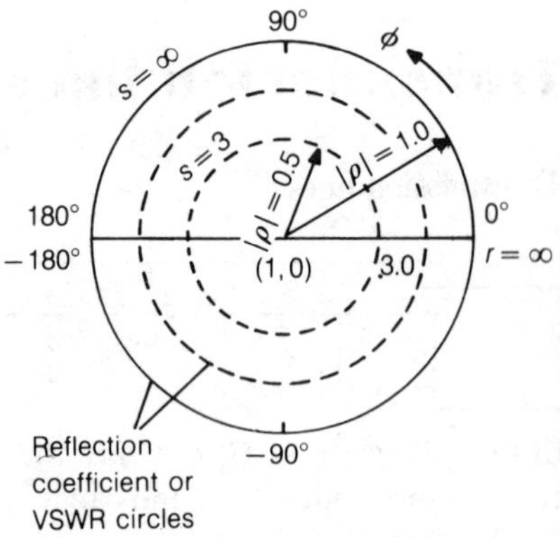

Distances measured from the centre give values of the reflection coefficients (ρ) and voltage standing wave ratio (VSWR, denoted by s):

$$\rho = \frac{Z/Z_0 - 1}{Z/Z_0 + 1} \qquad \text{VSWR} = \frac{1 + |\rho|}{1 - |\rho|}$$

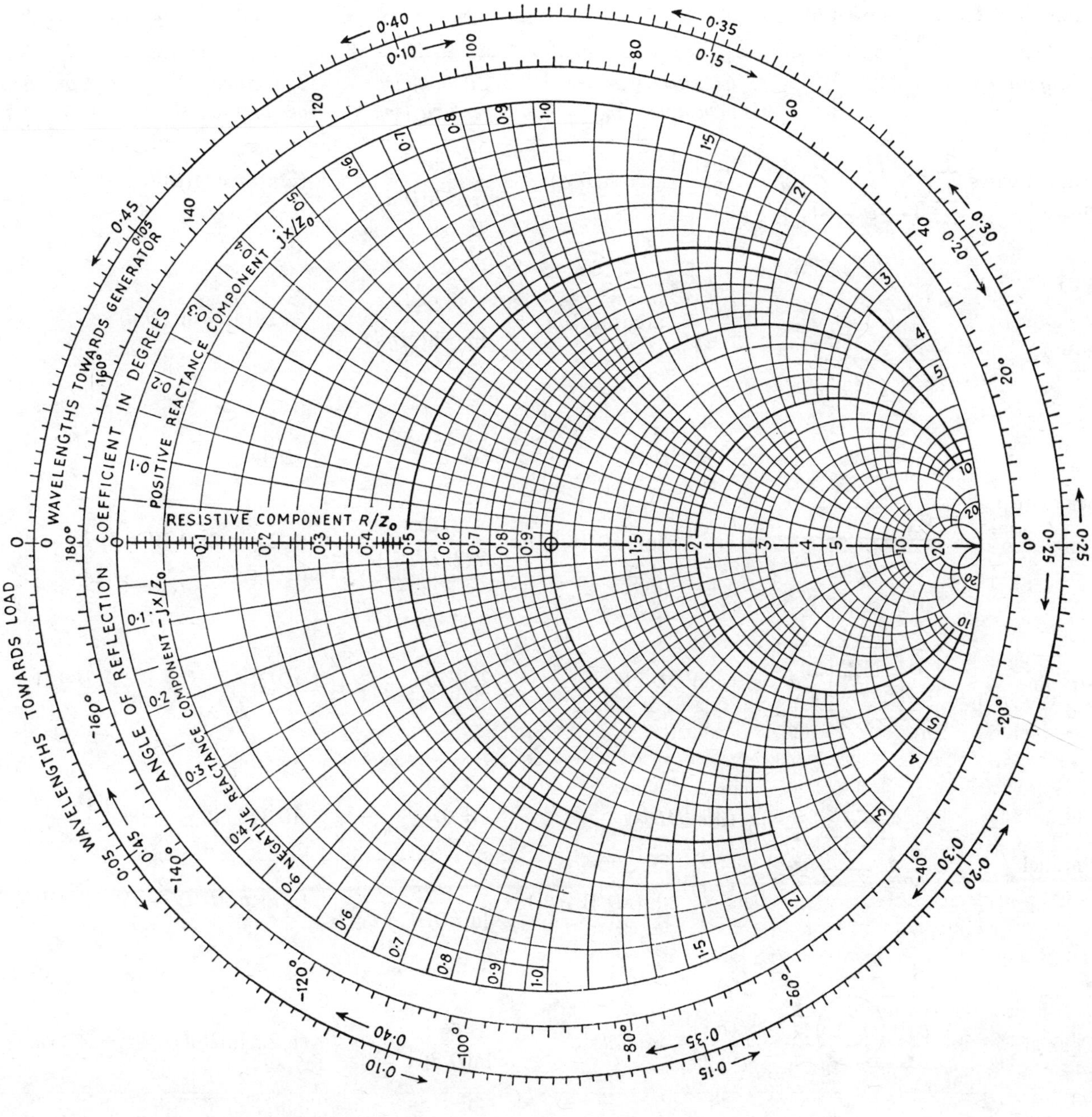
WAVELENGTHS TOWARDS GENERATOR
WAVELENGTHS TOWARDS LOAD
ANGLE OF REFLECTION COEFFICIENT IN DEGREES
POSITIVE REACTANCE COMPONENT jX/Z_0
NEGATIVE REACTANCE COMPONENT $-jX/Z_0$
RESISTIVE COMPONENT R/Z_0

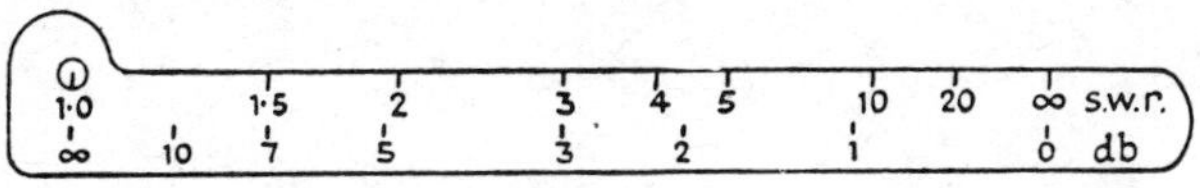
1·0 1·5 2 3 4 5 10 20 ∞ s.w.r.
∞ 10 7 5 3 2 1 0 db

Parameters for transmission lines

Configuration		*Characteristic impedance* (Z_0)	*Capacitance per unit length*	*Inductance per unit length*	*Conditions*
Parallel wires (air)	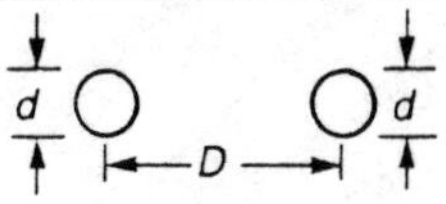	$\frac{(\mu/\epsilon)^{1/2}}{\pi}\operatorname{arcosh}(D/d)$	$\frac{\pi\epsilon}{\operatorname{arcosh}(D/d)}$	$(\mu/\pi)\operatorname{arcosh}(D/d)$	
Wire over ground	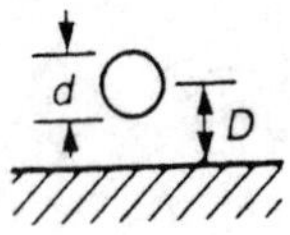	$\frac{(\mu/\epsilon)^{1/2}}{2\pi}\operatorname{arcosh}(2D/d)$	$\frac{2\pi\epsilon}{\operatorname{arcosh}(2D/d)}$	$(\mu/2\pi)\operatorname{arcosh}(2D/d)$	
Microstrip	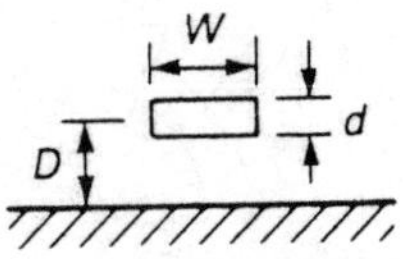	$(\mu/\epsilon)^{1/2}D/W$	$\epsilon W/D$	$\mu D/W$	$2D \ll W$ $d \ll 2D$
		$\frac{(\mu/\epsilon)^{1/2}}{2\pi}\ln(2\pi D/W + d)$	$\frac{2\pi\epsilon}{\ln(2\pi D/W + d)}$	$(\mu/2\pi)\ln(2\pi D/W + d)$	$2D \gg W$
Stripline	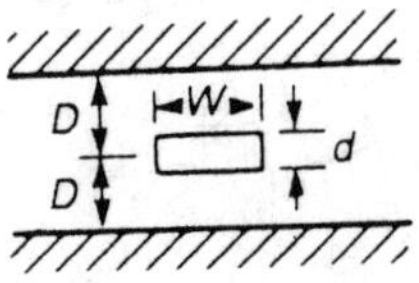	$(\mu/\epsilon)^{1/2}D/2W$	$2\epsilon W/D$	$\mu D/2W$	$D \ll W$ $d \ll D$
Parallel strip	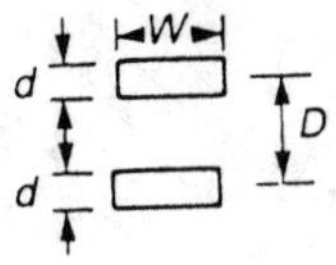	$(\mu/\epsilon)^{1/2}D/W$	$\epsilon W/D$	$\mu D/W$	$D \ll W$ $d \ll D$
		$\frac{(\mu/\epsilon)^{1/2}}{\pi}\ln(\pi D/W + d)$	$\frac{\pi\epsilon}{\ln(\pi D/W + d)}$	$(\mu/\pi)\ln(\pi D/W + d)$	$D \gg W$
Coaxial	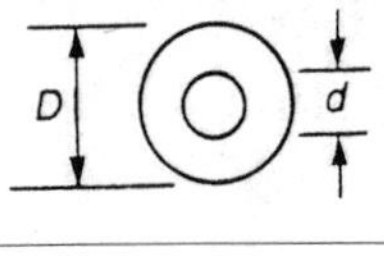	$\frac{(\mu/\epsilon)^{1/2}}{2\pi}\ln(D/d)$	$\frac{2\pi\epsilon}{\ln(D/d)}$	$(\mu/2\pi)\ln(D/d)$	$W \gg d$

Skin depth

The skin depth for plane conductors at frequency f (hertz) is given by

$$\delta = 1/\sqrt{\pi f \mu \sigma}$$

where μ is the permeability and σ the conductivity. The variation of δ with f for several metals is shown below.

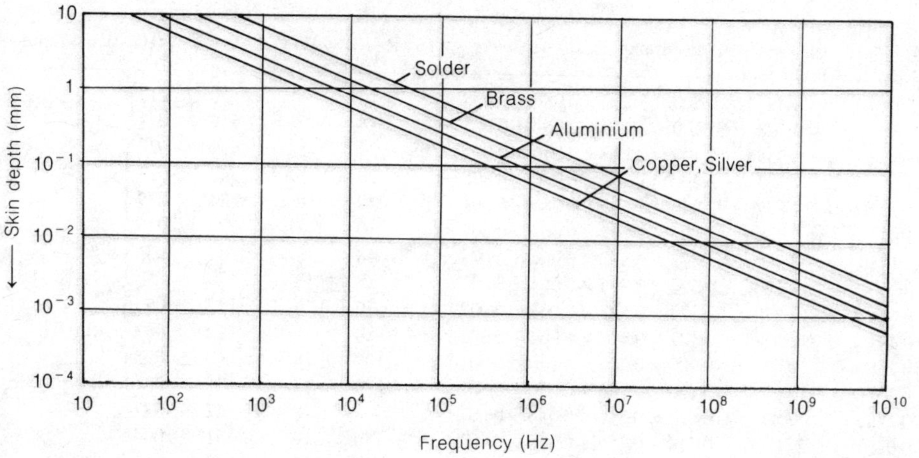

Rectangular waveguides

For any waveguide the phase-shift (or wave number) is

$$\beta = (\beta_0^2 - \beta_c^2)^{1/2}$$

where

$$\beta_0^2 = \omega^2 \epsilon \mu$$

and for a rectangular waveguide in a TE or TM mode

$$\beta_c^2 = \left(\frac{m\pi}{a}\right)^2 + \left(\frac{n\pi}{b}\right)^2$$

In this m and n denote the number of half-cycles along the x and y coordinates, for which the waveguide internal dimensions are a and b. At cut-off $\beta_0 = \beta_c$, and for evanescence $\beta_0 < \beta_c$. The waveguide wavelength λ_g is given by $2\pi/\beta$.

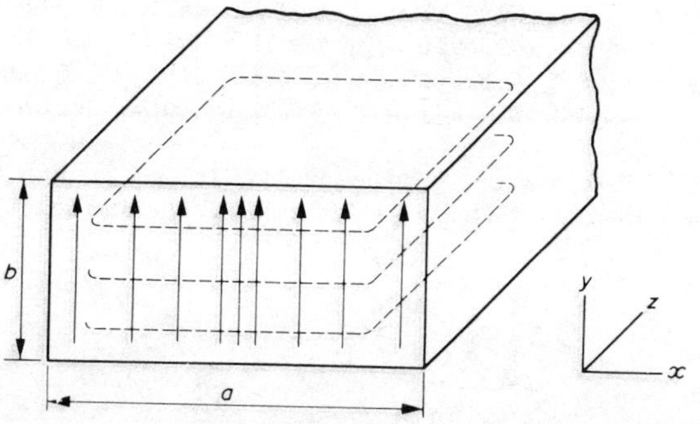

Electric (——) and magnetic (- - -) fields at a particular instant for a rectangular waveguide in the TE_{10} mode ($m = 1, n = 0$).

Data for standard rectangular waveguides, TE_{10} mode

f_c, λ	Frequency, free-space wavelength at cut-off (GHz, mm)
$\Delta f, W$	Recommended frequency range and power rating (GHz, MW)*
A	Theoretical attenuation per 100 m for pure copper at $1{\cdot}5 f_c$ (dB)†

The dimensions given are, respectively, *x*- and *y*-direction internal sizes, wall thickness and tolerance in *inches*. The designation letters prefix the numbers shown. The R designation is used in British, French and German civil standards; WG is a British military designation; WR is used by the (American) Electronic Industries Association; and MIL-W is an American military designation in which numbers indicate material as well as size: those listed are for copper.

						Designations				
f_c	λ	Δf	W	A	*Dimensions*	R	WG	WR	MIL-W	Band
0·908	330·2	1·14–1·73	15·1	0·497	6·500 × 3·250 × 0·080 ± 0·008	14	6	650	015	L
1·157	259·1	1·45–2·20	9·3	0·728	5·100 × 2·550 × 0·080 ± 0·008	18	7	510	021	
1·372	218·4	1·72–2·61	6·6	0·938	4·300 × 2·150 × 0·080 ± 0·006	22	8	430	027	
1·736	172·7	2·70–3·30	4·1	1·313	3·400 × 1·700 × 0·080 ± 0·005	26	9A	340	033	S
2·078	144·3	2·60–3·95	2·7	1·822	2·840 × 1·340 × 0·080 ± 0·004	32	10	284	039	
2·577	116·3	3·22–4·90	1·88	2·460	2.29 × 1·145 × 0·064 ± 0·003	40	11A	229	045	
3·152	95·10	3·94–5·99	1·17	3·430	1·872 × 0·872 × 0·064 ± 0·003	48	12	187	051	
3·711	80·78	4·64–7·05	0·90	4·120	1·590 × 0·795 × 0·064 ± 0·002	58	13	159	057	C
4·310	69·70	5·38–8·18	0·61	5·570	1·372 × 0·622 × 0·064 ± 0·002	70	14	137	063	
5·260	57·00	6·58–10·0	0·40	7·671	1·122 × 0·497 × 0·064 ± 0·002	84	15	112	069	
6·557	45·72	8·20–12·5	0·258	10·63	0·900 × 0·400 × 0·050 ± 0·001	100	16	90	075	X
7·869	38·10	9·54–15·0	0·201	12·90	0·750 × 0·375 × 0·050 ± 0·001	120	17	75	081	
9·488	31·60	11·9–18·0	0·138	17·09	0·622 × 0·311 × 0·040 ± 0·001	140	18	62	087	
11·571	25·91	14·5–22·0	0·093	23·00	0·510 × 0·255 × 0·040 ± 0·001	180	19	51	094	
14·051	21·34	17·6–26·7	0·051	35·76	0·420 × 0·170 × 0·040 ± 0·0008	220	20	42	100	
17·357	17·27	21·7–33·0	0·041	42·00	0·340 × 0·170 × 0·040 ± 0·0008	260	21	34	107	
21·077	14·22	26·4–40·1	0·028	56·70	0·280 × 0·140 × 0·040 ± 0·0008	320	22	28	007	Q
26·344	11·38	33·0–50·1	0·018	78·80	0·224 × 0·112 × 0·040 ± 0·0008	400	23	22	011	
31·391	9·550	39·3–59·7	0·0127	103·0	0·188 × 0·094 × 0·040 ± 0·0008	500	24	19	015	
39·877	7·518	49·9–75·8	0·0085	147·0	0·148 × 0·074 × 0·040 ± 0·0008	620	25	15	018	
48·369	6·198	60·5–92·0	0·0053	197·0	0·122 × 0·061 × 0·040 ± 0·0008	740	26	12	021	Millimetre
59·014	5·080	73·8–112	0·0036	266·0	0·100 × 0·050 × 0·040 ± 0·0008	900	27	10	024	
73·767	4·064	92·3–140	0·0023	370·0	0·080 × 0·040 × 0·030 ± 0·0008	1200	28	8	027	

*The rating W is based on a breakdown field of 3 kV/mm for air, a VSWR of 2 and a power safety factor of 2.
†For the attenuation at a frequency f other than $1{\cdot}5 f_c$, these figures should be multiplied by $0{\cdot}421\ \{(f/f_c)^2 + 1\}/\sqrt{[(f/f_c)\{(f/f_c)^2 - 1\}]}$.

Resonant cavities

In the table, λ is the resonant wavelength and δ the skin depth, given by $\sqrt{(2/\omega\mu\sigma)}$ for material of conductivity σ and permeability μ at angular frequency ω.

Resonator type		λ	Q
TE_{101}	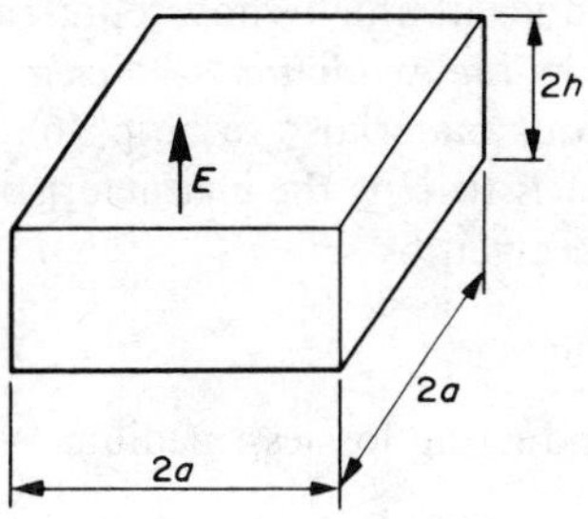	$2\sqrt{2}a$	$\frac{0{\cdot}353\lambda}{\delta}\cdot\frac{1}{1+0{\cdot}177\lambda/h}$
Circular cylinder TM_{010}	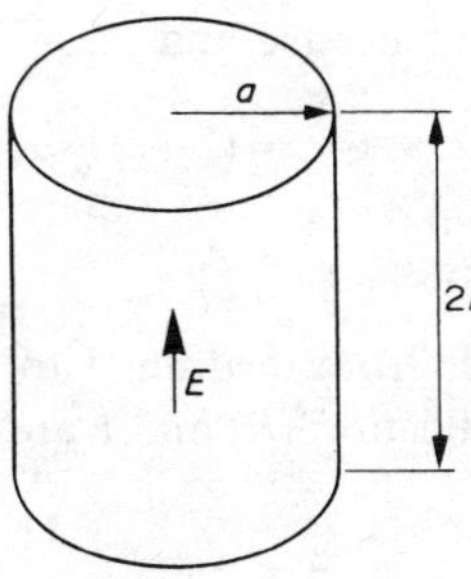	$2{\cdot}61a$	$\frac{0{\cdot}383\lambda}{\delta}\cdot\frac{1}{1+0{\cdot}192\lambda/h}$
Sphere	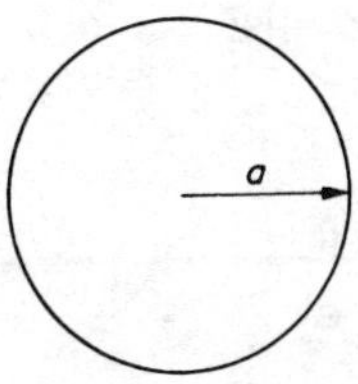	$2{\cdot}28a$	$0{\cdot}318\lambda/\delta$
Co-axial TEM	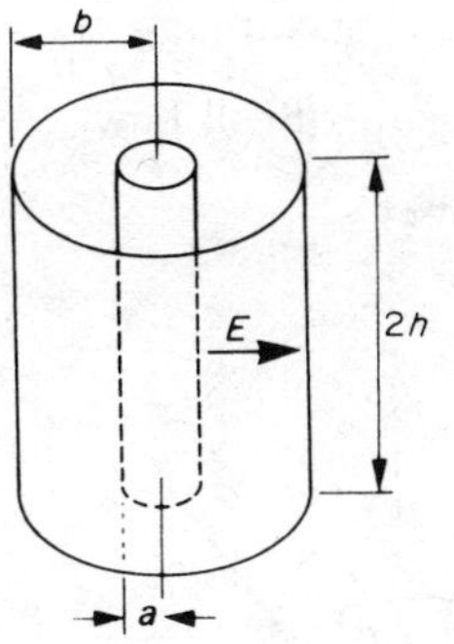	$4h$	$\frac{\lambda}{4\delta+7{\cdot}2h\delta/b}$ For optimum Q $b/a = 3{\cdot}6$ and $Z_0 = 77\ \Omega$

Radiation and aerials

The *Poynting vector* is

$$\mathbf{S} = \mathbf{E} \times \mathbf{H}$$

and for orthogonal fields in an isotropic non-conducting

medium of permeability μ and permittivity ϵ has the value

$$S = EH = E^2\sqrt{(\epsilon/\mu)}$$
$$= H^2\sqrt{(\mu/\epsilon)}$$

For sinusoidally time-varying fields the time-averaged Poynting vector is

$$\mathbf{S} = \text{Re}(\mathbf{E} \times \mathbf{H}^*)$$

where **E** and **H** are r.m.s. values. In a plane wave the average power density is $\frac{1}{2}EH$ where E and H are peak values.

An *isotropic radiator* emitting a mean power P produces a mean S of $P/4\pi r^2$ at distance r, and the r.m.s. electric field in free space is then

$$E = \sqrt{(30P)}/r$$

The *gain* of an aerial is the ratio of the power it emits per steradian in a given direction to the power per steradian emitted by a reference aerial of the same total power. Usually, the direction chosen is that of maximum power density and the reference aerial is an isotropic radiator.

The directivity may be measured either by the maximum gain or by the *beam width,* the angle contained between points at which the power density is half of the maximum.

The *radiation resistance* R_r of an aerial is such that the aerial radiates power $I^2 R_r$ when fed with r.m.s. current I. The *aperture* of a receiving aerial is the ratio of the power received to the power density of the incident field. The effective aperture of an aerial is greatest when it is matched; for a lossless aerial of gain G it is then given by $\lambda^2 G/4\pi$, where λ is the wavelength. The power received by a matched aerial is

$$P = V^2/4R_r$$

where V is the integral of the induced electric field along its length.

Non-isotropic radiators

	Current distribution	*Radiation resistance*	*Maximum gain*	*Beam width*	*Aperture*
Hertzian dipole	constant	$80\pi^2(l/\lambda)^2$	1·5	90°	$3\lambda^2/8\pi$
Half-wave dipole	half-cosine	73·1 Ω	1·64	78°	$30\lambda^2/73\pi$

(Here l is the total length of the aerial and λ the wavelength.)

Propagation

The propagation constant of a medium with permeability μ, permittivity ϵ and conductivity σ is

$$\gamma = \{j\omega\mu(\sigma + j\omega\epsilon)\}^{1/2} = j\omega\{\mu(\epsilon - j\sigma/\omega)\}^{1/2}$$

which may be written as $\alpha + j\beta$ or as $k'' + jk'$ (see page 152). If $\mu = \mu_0$ it may be written as

$$\gamma = j\frac{\omega}{c}\sqrt{\epsilon_r} = j\frac{\omega}{c}(n - jk)$$

where c is the velocity of light in vacuum, ϵ_r is the complex dielectric constant $(\epsilon - j\sigma/\omega)/\epsilon_0$ and $n - jk$ is the refractive index in which k now represents absorption and is known as *the extinction coefficient* (values of n and k for various materials are on p. 46).

The ratio of E to H is the characteristic impedance of the medium given by

$$Z = \sqrt{j\omega\mu/(\sigma + j\omega\epsilon)}$$

For a non-conducting lossless medium

$$\gamma = j\omega\sqrt{\mu\epsilon}, \qquad Z = \sqrt{\mu/\epsilon}$$

For a good conductor $\sigma \gg \omega\epsilon$ and

$$\gamma = (1 + j)\sqrt{\omega\mu\sigma/2}, \qquad Z = (1 + j)\sqrt{\omega\mu/2\sigma}$$

Reflection and transmission

For normal incidence from medium 1 to medium 2, the coefficients of reflection (R) and transmission (T) for the E-field are

$$R = \frac{Z_2 - Z_1}{Z_2 + Z_1} \qquad T = \frac{2Z_2}{Z_2 + Z_1}$$

where

$$Z_1 = \sqrt{\frac{\mu_1}{\epsilon_1}} \qquad Z_2 = \sqrt{\frac{\mu_2}{\epsilon_2}}$$

A wave incident at angle θ_i to the normal is refracted to an angle θ_t such that

$$(\sin\theta_i)/(\sin\theta_t) = n_2/n_1 \quad \text{(Snell Law)}$$

For the reflected wave

$$\theta_r = \theta_i$$

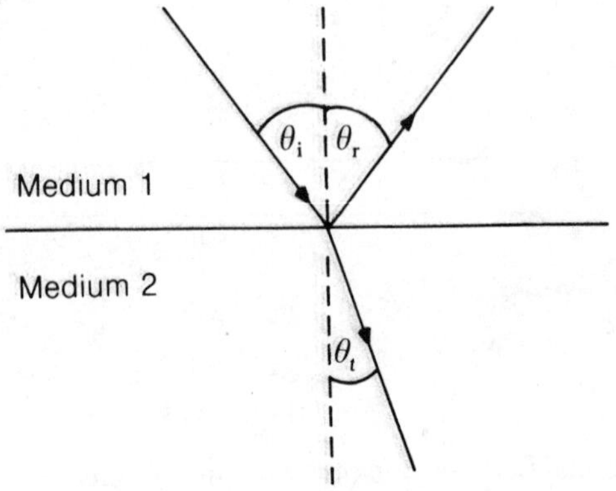

For a wave with its E-field parallel to the boundary (TE case) the coefficients of reflection and transmission are

$$R_{\text{TE}} = \frac{Z_2 \cos\theta_i - Z_1 \cos\theta_t}{Z_2 \cos\theta_i + Z_1 \cos\theta_t}$$

and

$$T_{\text{TE}} = \frac{2Z_2 \cos\theta_i}{Z_2 \cos\theta_i + Z_1 \cos\theta_t}$$

For a wave with its H-field parallel to the boundary (TM case)

$$R_{\text{TM}} = \frac{Z_2 \cos\theta_t - Z_1 \cos\theta_i}{Z_1 \cos\theta_i + Z_2 \cos\theta_t}$$

$$T_{\text{TM}} = \frac{2Z_2 \cos\theta_i}{Z_1 \cos\theta_i + Z_2 \cos\theta_t}$$

In general,

$$\text{Power reflectivity} = |R|^2$$

$$\text{Power transmissivity} = 1 - |R|^2$$

The *Brewster angle* θ_B is given by

$$\tan\theta_B = \sqrt{\frac{\epsilon_2}{\epsilon_1}} = \frac{n_2}{n_1}$$

and is the value of θ_i at which $R = 0$ for a TM wave.

When $n_1 > n_2$ the critical value of θ_i for total reflection is given by

$$\theta_c = \sin^{-1} n_2/n_1$$

Optical wave guides

For a symmetric-slab dielectric waveguide the guidance conditions are the following:

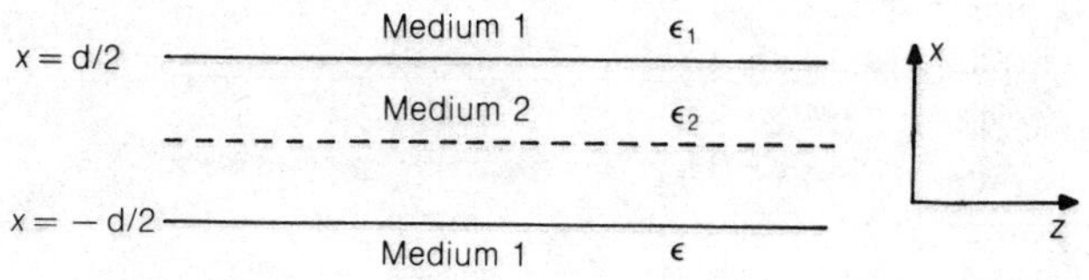

Even TE solutions

$$\tan(k_{2x}d/2) = \alpha_x/k_{2x}$$

where the dispersion relation gives

$$k_{2x} = \sqrt{\omega^2\mu\epsilon_2 - k_z^2}$$

= propagation constant in x direction in medium 2

and

$$\alpha_x = \sqrt{k_z^2 - \omega^2\mu\epsilon_1}$$

= propagation constant in x direction in medium 1

Odd TE solutions:

$$\cot(k_{2x}d/2) = -\alpha_x/k_{2x}$$

Components

Resistors

Colour Code	*Figure*	*Multiplier*	*Tolerance*	*Temperature coefficient (ppm/°C)*
Silver		10^{-2}		
Gold		10^{-1}		
Black	0	1		
Brown	1	10	±1%	100
Red	2	10^2	±2%	50
Orange	3	10^3	±5%	25
Yellow	4	10^4	±10%	15
Green	5	10^5		
Blue	6	10^6		10
Violet	7			5
Grey	8			
White	9			1

4 bands: 2 significant figures, multiplier, tolerance
5 bands: 3 significant figures, multiplier, tolerance
6 bands: 3 significant figures, multiplier, tolerance, temperature coefficient

If 4 or 5 bands, read from the band nearest to an end; if 6 bands, that indicating the temperature coefficient should be wider and indicates the right-hand end.

Letter code for multipliers for resistors:

$R = 1$, $K = 10^3$, $M = 10^6$, $G = 10^9$, $T = 10^{12}$

These are inserted in place of the decimal point.

Preferred values (BS 2488 E24 series):
10, 11, 12, 13, 15, 16, 18, 20, 22, 24, 27, 30, 33, 36, 39, 43, 47, 51, 56, 62, 68, 75, 82, 91.

Capacitors

A system similar to that used for resistors is used for polyester capacitors. Reading from the end remote from the connectors, the first two bands indicate significant figures; the third, fourth and fifth bands have the following significance.

Third band (multiplier)	Fourth band (tolerance)	Fifth band (working voltage)
Orange 0·001 μF	White ±10%	Red 250 V d.c.
Yellow 0·01 μF	Black ±20%	Yellow 400 V d.c.
Green 0·1 μF		

Letter code for multipliers for capacitors:
$p = 10^{-12}$, $n = 10^{-9}$, $\mu = 10^{-6}$, $m = 10^{-3}$

Semiconductor devices

Graphical symbols

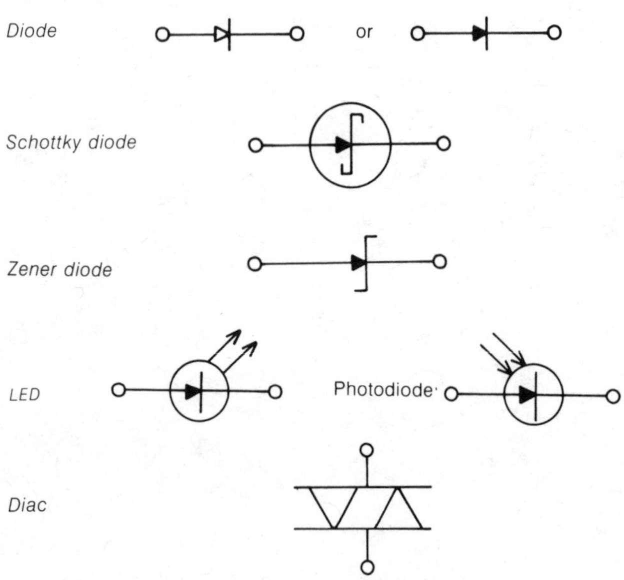

BJT (Bipolar Junction Transistor)

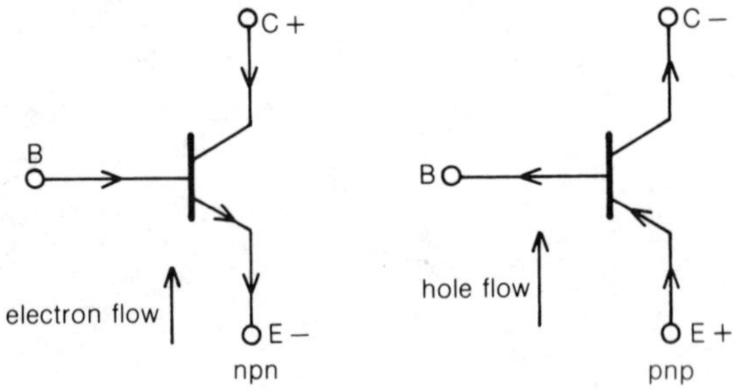

The polarities in each case are as shown.

JFET (Junction Field-Effect Transistor)

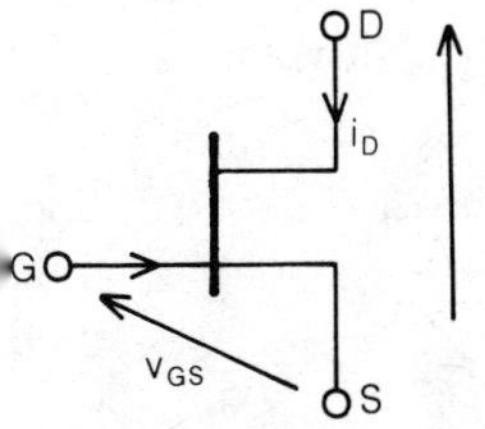

n-channel

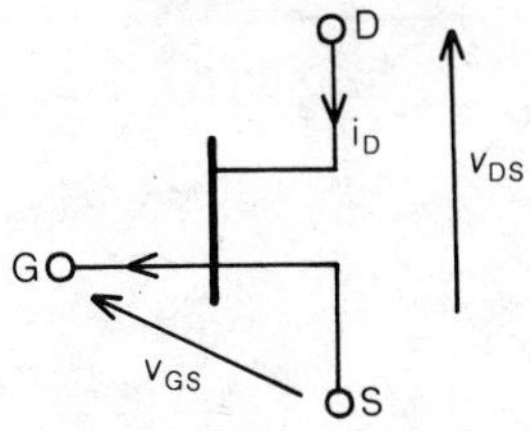

p-channel

The characteristics and polarities in each case are as shown below.

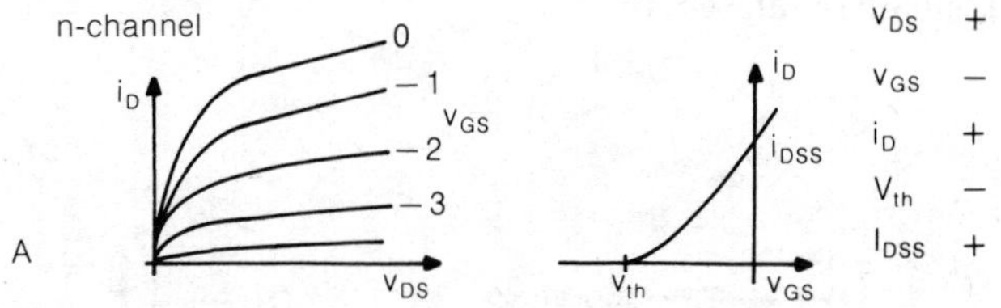

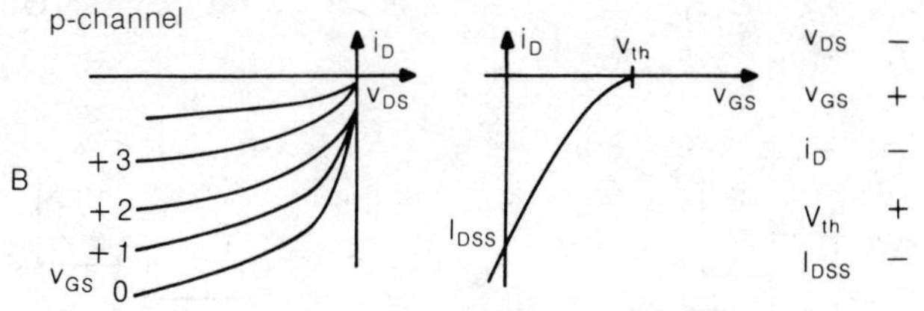

I_{DSS} = drain current at $v_{GS} = 0$
V_{th} = pinch-off voltage, i.e. v_{GS} for $i_D = 0$. The symbol V_p is frequently used.

MOSFET (Metal-Oxide-Silicon FET)
This device is also known as MOST or IGFET, Insulated-Gate FET

The characteristics and polarities in each case are as shown below.
The numbers are typical values of v_{GS} in volts.

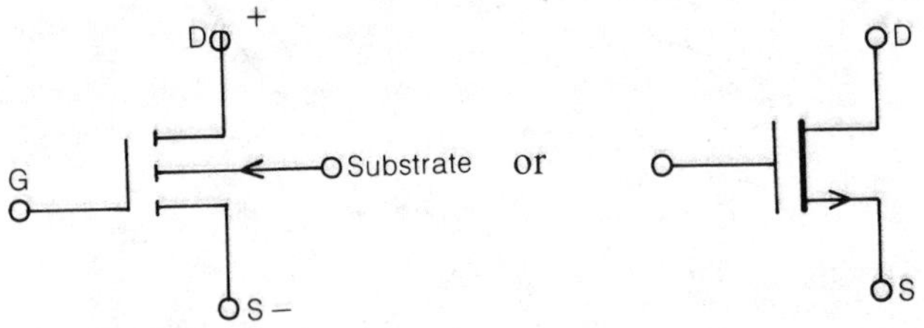

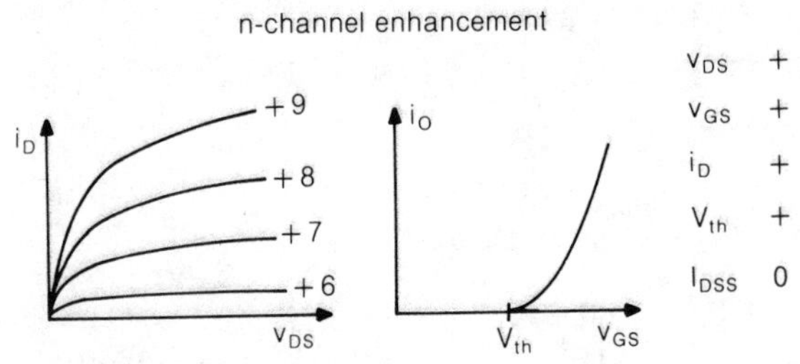

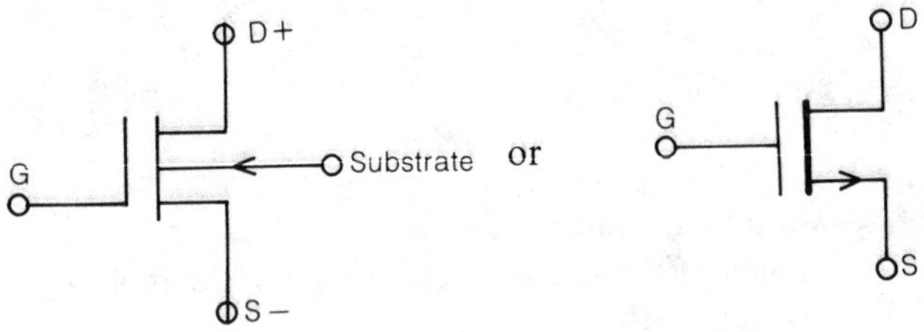

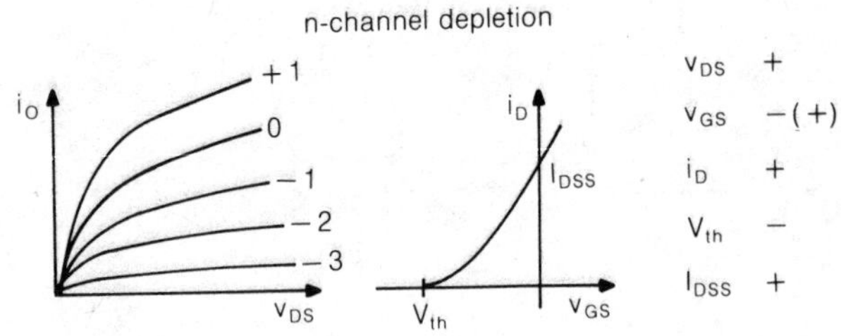

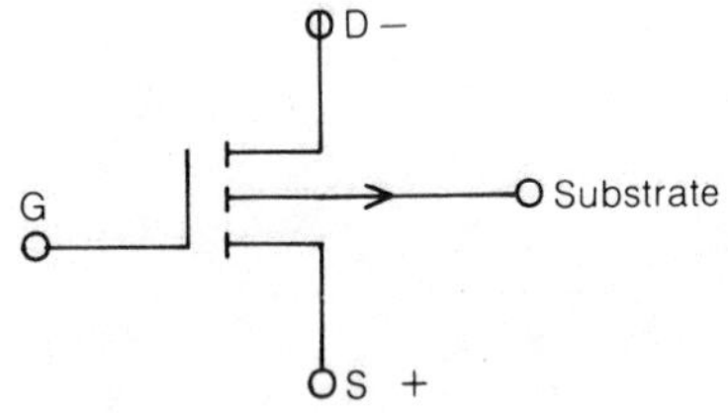

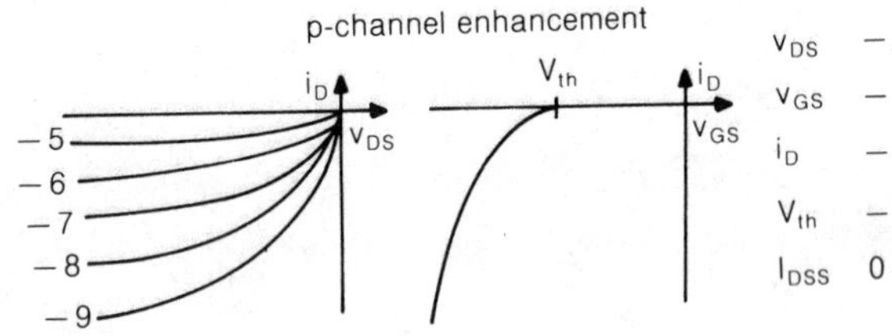

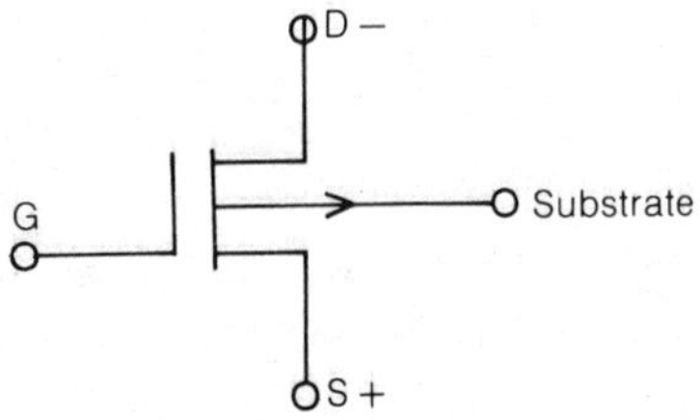

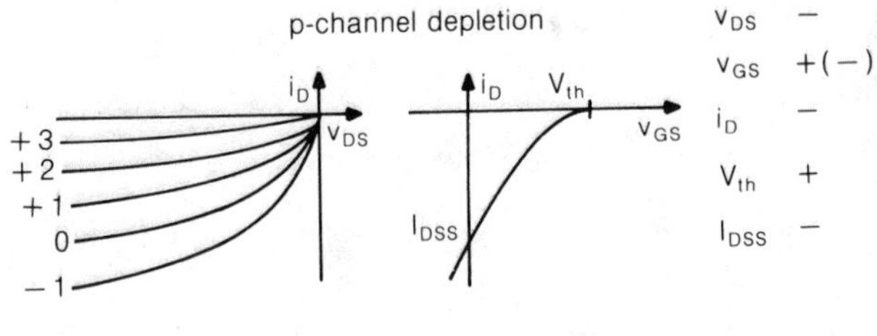

Thyristors

Initials	*Names and typical ratings*	*Symbol*	*Static characteristics*	*Initial*	*Names and typical ratings*	*Symbol*	*Static characteristics*
SCR	Silicon controlled rectifier 5 kV, 5 kA		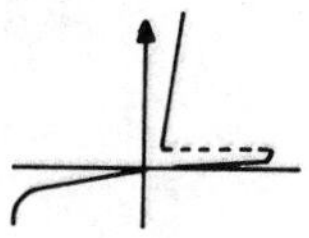	PUT	Programmable unijunction transistor 40 V, 15 A		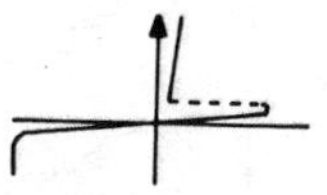
ASCR	Asymmetrical silicon controlled rectifier 2 kV, 2 kA		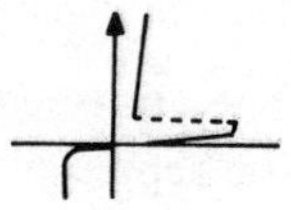	SUS	Silicon unilateral switch 30 V, 1 A	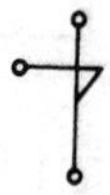	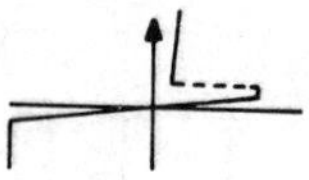
Triac	Triode a.c. switch 800 V, 50 A		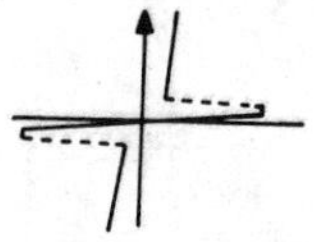	SBS	Silicon bilateral switch 30 V, 1 A	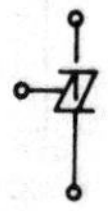	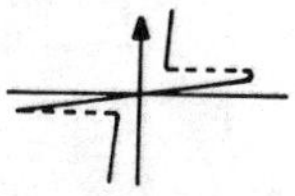
GTO	Gate turn-off thyristor 4 kV, 4 kA		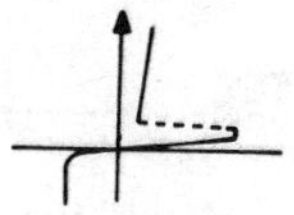	SCS	Silicon controlled switch 100 V, 1 A	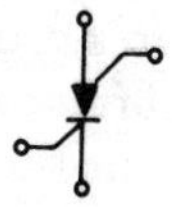	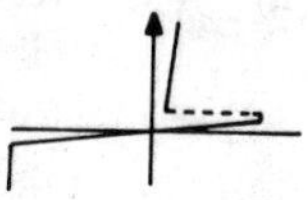
SIDAC	Switching diode SIDAC Breakover diode 400 V, 1 A	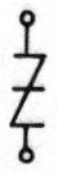	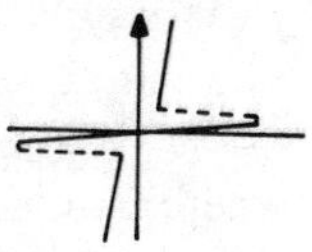				

Small signal models

Notation

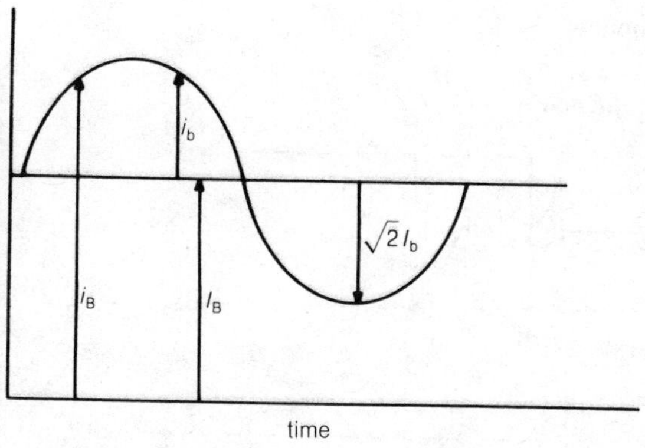

When considering time-varying quantities upper-case symbols (V, I) with appropriate subscripts are used for d.c., average, or root-mean-square values. Lower-case symbols are used for instantaneous values and for a.c. components. Thus, in the diagram,

I_B = average (d.c.) value.

I_b = r.m.s. value of a.c. component

i_B = total instantaneous value

i_b = instantaneous value of a.c. component.

For circuit parameters, lower-case subscripts refer to small-signal values, e.g.

$$h_f = \frac{\partial I_2}{\partial I_1}$$

Upper-case subscripts refer to large-signal (or d.c.) values, e.g.

$$h_F = \frac{I_2}{I_1}$$

A second subscript letter b, B, e, E, c or C can be added to the first to indicate which terminal of the transistor is common to input and output.

BJT models

h-parameters

(For definitions see p. 147)

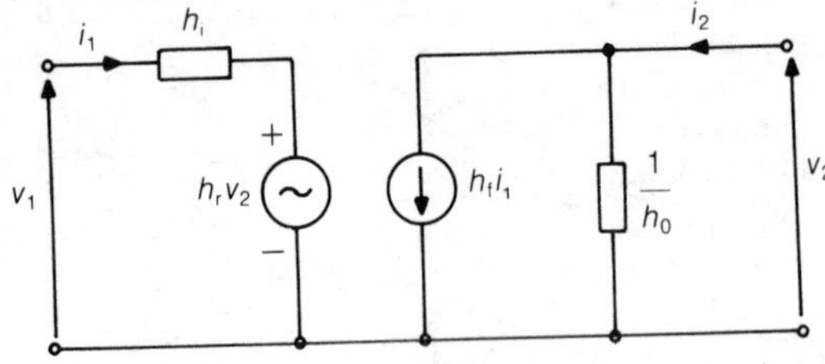

$$\begin{bmatrix} v_1 \\ \\ i_2 \end{bmatrix} = \begin{bmatrix} h_{11} & h_{12} \\ h_i & h_r \\ & \\ h_f & h_o \\ h_{21} & h_{22} \end{bmatrix} \begin{bmatrix} i_1 \\ \\ v_2 \end{bmatrix}$$

$h_{11} = h_i =$ input impedance with output short-circuited to a.c.

$h_{12} = h_r =$ reverse voltage transfer with input open-circuited to a.c.

$h_{21} = h_f =$ forward current ratio with output short-circuited to a.c. (current gain)

$h_{22} = h_o =$ output admittance with input open-circuited to a.c.

The magnitudes of the current gains h_{fb} and h_{fe} for common base and common emitter circuits, respectively, are sometimes denoted by α and β and satisfy the relationships

$$\alpha = \frac{\beta}{1+\beta}; \qquad \beta = \frac{\alpha}{1-\alpha}; \qquad 1-\alpha = (1+\beta)^{-1}$$

The three h-parameter sets are then related as follows:

Common base	*Common emitter*	*Common collector*
$h_{ib} = h_{ie}(1-\alpha)$	$h_{ie} = h_{ib}(1+\beta)$	$h_{ic} = h_{ie}$
$h_{rb} = h_{ie}h_{oc}(1-\alpha) - h_{re}$	$h_{re} = h_{ib}(1+\beta) - h_{rb}$	$h_{rc} = 1$
$h_{fb} = -\alpha$	$h_{fe} = \beta$	$h_{fc} = -(1+\beta)$
$h_{ob} = h_{oe}(1-\alpha)$	$h_{oe} = h_{ob}(1+\beta)$	$h_{oc} = h_{oe}$

r-parameters and T equivalent circuit

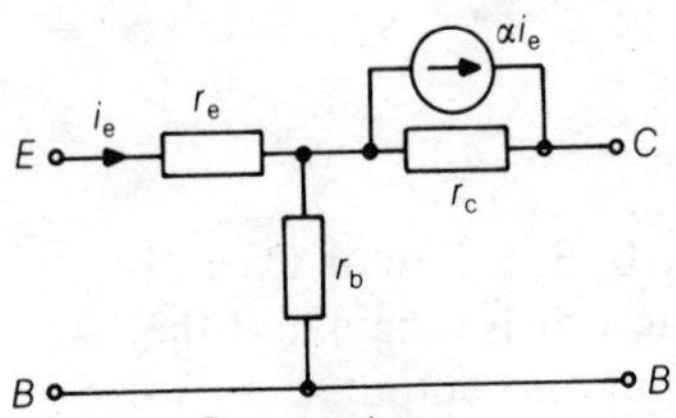

Common base

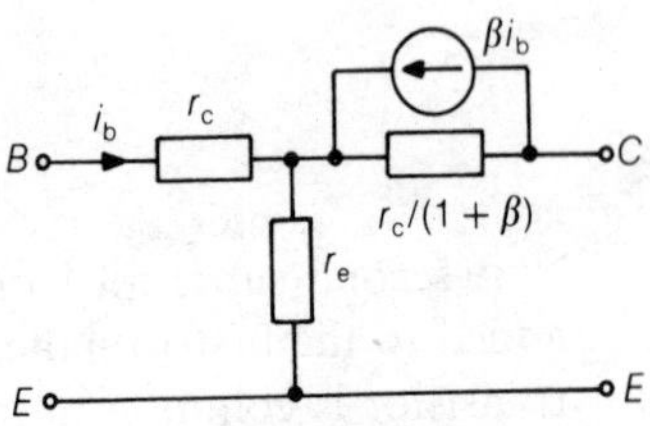

Common emitter

$$r_e = \frac{h_{re}}{h_{oe}} = h_{ib} - \frac{h_{rb}}{h_{ob}}(1+h_{fb})$$

$$r_b = h_{ie} - \frac{h_{re}}{h_{oe}}(1+h_{fe}) = \frac{h_{rb}}{h_{ob}}$$

$$r_c = \frac{1+h_{fe}}{h_{oe}} = \frac{1-h_{rb}}{h_{ob}}$$

Hybrid π parameters

The above equivalent circuits contain no provision for the representation of high-frequency effects. The hybrid π equivalent circuit includes capacitors and is appropriate for BJTs at high frequencies.

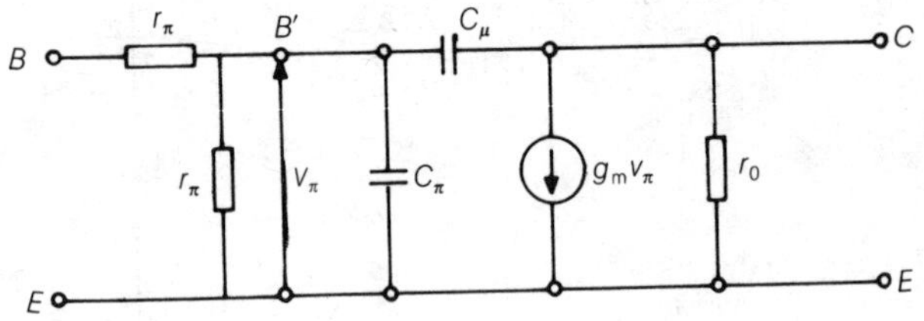

g_m is the transconductance $\partial i_C/\partial v_\pi$

At high frequency

$$h_{fe}(\text{or } \beta) = h_{fe0}/(1 + j\omega C_\pi r_\pi)$$

where h_{fe0} is the d.c. value of h_{fe} and C_μ is neglected because it is very much less than C_π. The value of h_{fe} will fall to $(1/\sqrt{2})h_{fe0}$ when $\omega C_\pi r_\pi = 1$. The cut-off frequency is defined as

$$\omega_c = \frac{1}{C_\pi r_\pi} \quad \text{or} \quad f_c = \frac{1}{2\pi c_\pi r_\pi}.$$

FET models

Low frequencies

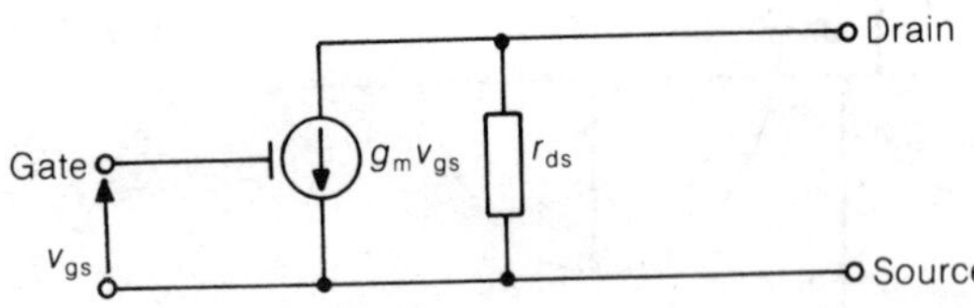

Higher frequencies

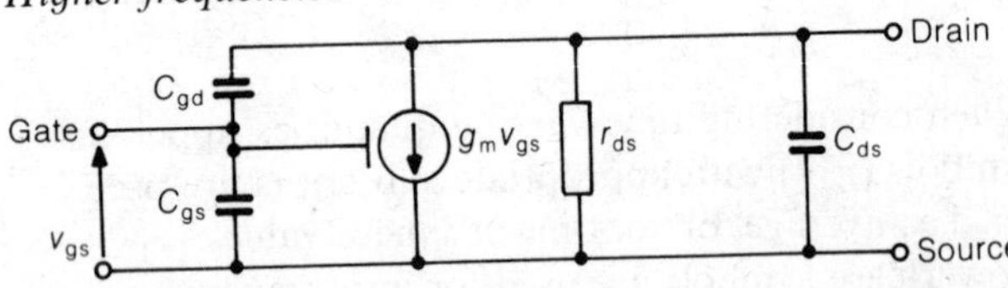

In these g_m is the transconductance $\partial i_D/\partial v_{GS}$ and r_{ds} the drain resistance $\partial v_{DS}/\partial i_D$.

The Ebers-Moll large-signal model

The generalized model for a n–p–n BJT is:

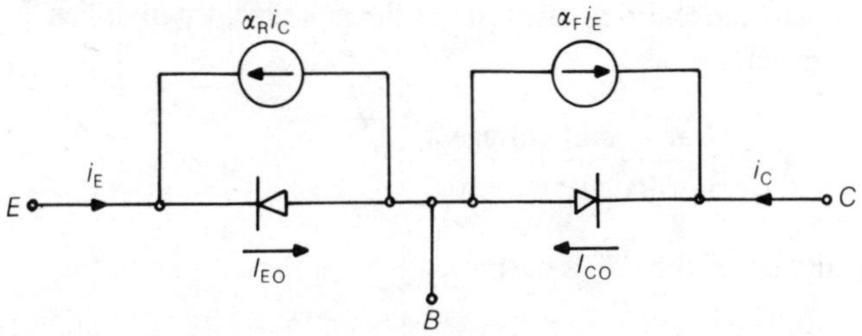

For the current directions shown, the Ebers-Moll equations are

$$i_C = -\alpha_F i_E - I_{CO}(e^{v_{BC}/V_T} - 1)$$
$$i_E = -\alpha_R i_C - I_{EO}(e^{v_{BE}/V_T} - 1).$$

In these α_F and α_R are the forward and reverse common-base current ratios, I_{EO} and I_{CO} are the reverse leakage currents of the diodes, and V_T is the voltage kT/e (approximately 25 mV at room temperature).

For a BJT operating normally in the active region, BE is forward biased and BC reverse biased. The above circuit then reduces to the following alternative forms:

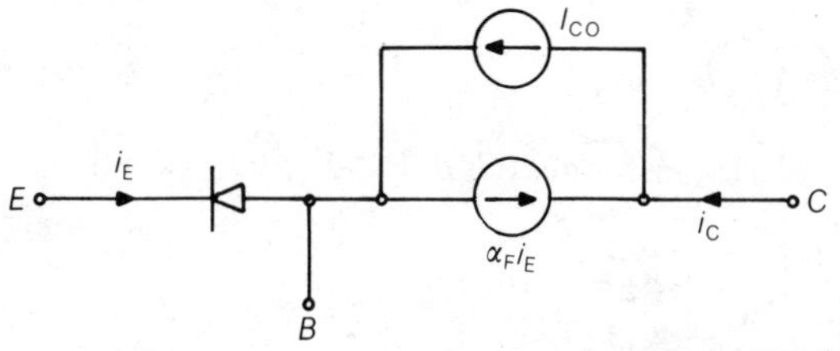

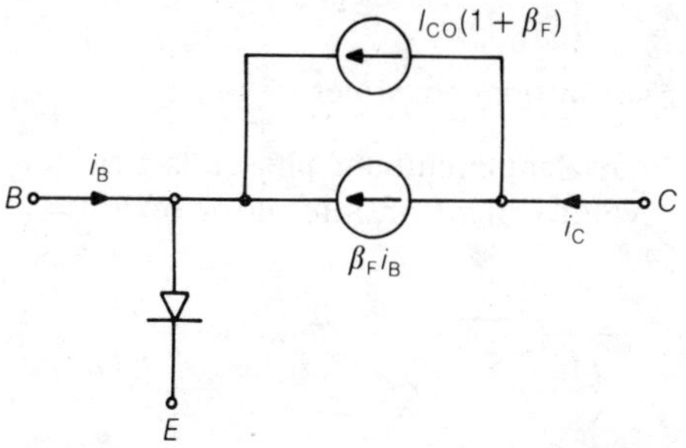

Here β_F is the forward common-emitter current gain. The leakage currents I_{CO} and $I_{CO}(1+\beta_F)$ are often neglected.

Electrical machines

In the following Φ is flux per pole, P number of poles, n number of turns in series per phase, N speed (rev/s), ω speed (rad/s) and f frequency (Hz).

D.C. machines

$$\text{E.M.F.} = \frac{\Phi ZNP}{A}$$

where Z is the total number of conductors and A the number of parallel paths in which they are arranged.

$$\text{Torque} = \frac{\Phi ZPI}{2\pi A}$$

where I is the total armature current.

For ordinary windings $A = P$ if lap-wound, and $A = 2$ if wave-wound (whatever the value of P).

If flux is constant $\Phi ZP/2\pi A$ is the e.m.f. or torque constant K of the machine such that

$$\text{E.M.F.} = K\omega \quad \text{and} \quad \text{Torque} = KI$$

A.C. machines

$$\text{E.M.F.} = 4{\cdot}44\ \Phi nf \text{ (r.m.s.)}$$

$$\text{Synchronous speed} = 120 f/P \text{ rev/min} = 60 f/p \text{ rev/min} = 2\pi f/p \text{ rad/s}$$

where p is the number of pole-pairs.

$$\text{Regulation} = \frac{\text{change in voltage (or speed) between no load and full load}}{\text{rated voltage (or speed)}}$$

Transformers

Equivalent circuits:

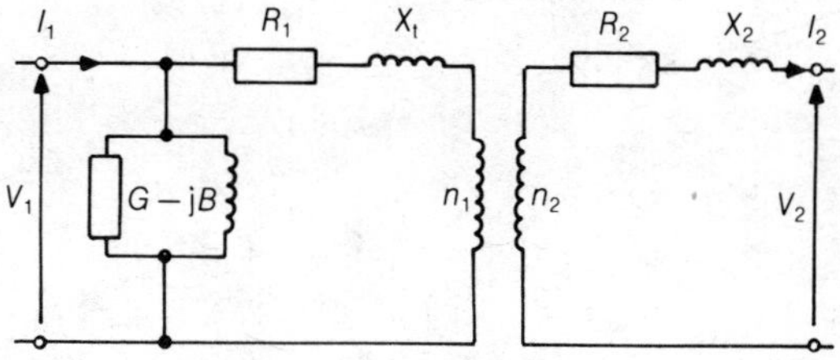

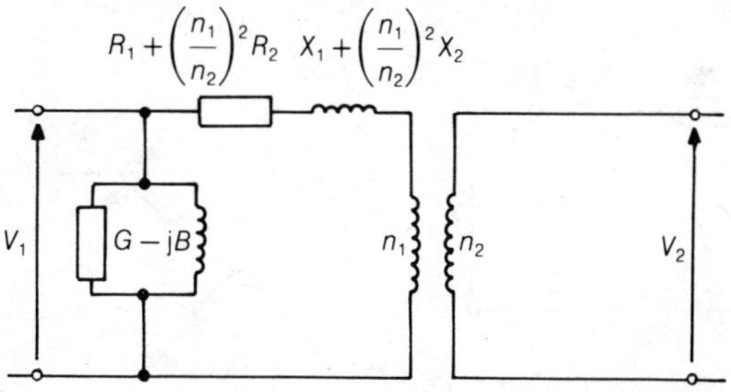

An impedance Z_2 in the secondary may be considered to be replaced by an equivalent impedance Z_2' in the primary, where

$$Z_2' = Z_2 \left(\frac{n_1}{n_2}\right)^2$$

For an ideal transformer ($G, B, R_1, X_1, R_2, X_2 = 0$)

$$\frac{V_1}{V_2} = \frac{n_1}{n_2}; \qquad I_2 n_2 = -I_1 n_1$$

Synchronous machines

Equivalent circuit and phasor diagram for one phase (cylindrical rotor, alternator operation):

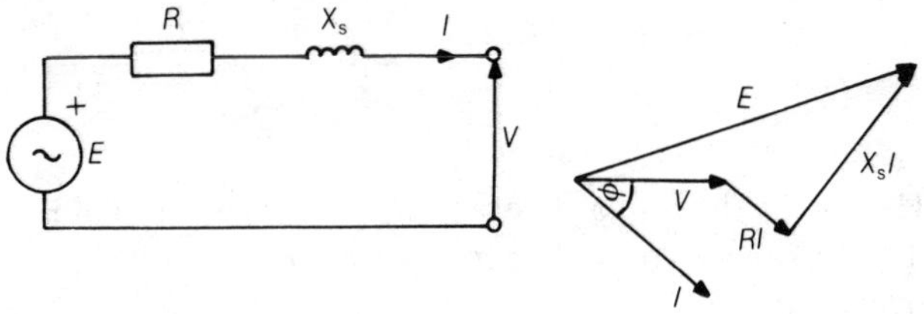

$$V = E - (R + jX_s)I$$

where R is armature resistance and X_s is synchronous reactance (including leakage and armature reaction). Usually $R \ll X_s$.

For operation as a motor the same equivalent circuit and sign convention for I can be used; the phasor diagram is then:

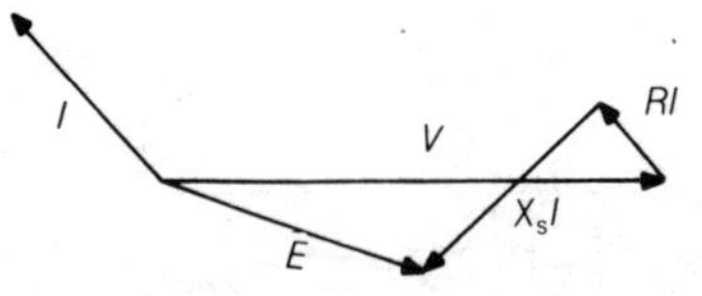

Alternatively the positive direction of I can be redefined for I', to give:

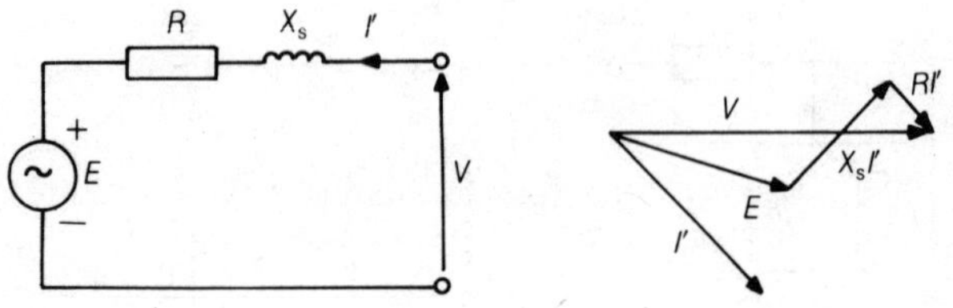

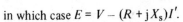

in which case $E = V - (R + jX_s)I'$.

The power per phase it losses are neglected is $VE \sin \delta / X_s$.

The synchronous reactance X_s can be determined from open- and short-circuit tests of the machine, when driven as a generator, as

$$X_s = \frac{\text{open-circuit voltage}}{\text{short-circuit current}}$$

at a given excitation current.

Asynchronous or induction motor

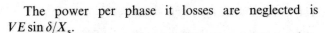

Equivalent circuit:

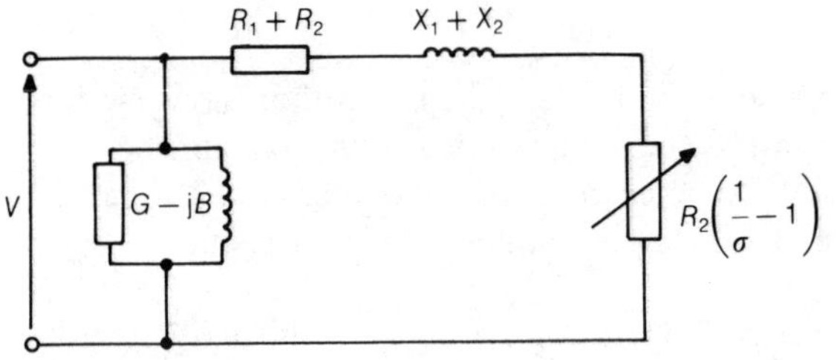

R_2 and X_2 are values referred to the stator as determined by a locked-rotor test. The slip is defined as

$$\sigma = \frac{\omega_s - \omega}{\omega_s}$$

where ω_s and ω are the synchronous and actual speeds, respectively.

The slip is also given by

$$\sigma = \frac{\text{rotor copper loss}}{\text{gross power input to rotor}}$$

The torque per phase is

$$T = \frac{V^2 R_2 \sigma / \omega_s}{(R_1 \sigma + R_2)^2 + \sigma^2 X^2}$$

where $X = X_1 + X_2$, and has a maximum

$$T_{max} = \frac{V^2}{2X\omega_s}$$

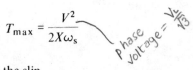

phase voltage = $V_L/\sqrt{3}$

at the slip

$$\sigma = R_2 / X$$

provided that $R_1 \ll X$.

Electrical properties of materials

In this section unless otherwise stated k denotes a propagation constant or wave number and k_B the Boltzmann constant; e is the magnitude of the electronic charge.

Free charges

The mobility of charge carriers of mass m and charge e in a medium is

$$\mu = \frac{e}{m}\tau$$

where τ is their average collision time or relaxation time for momentum transfer. If their number density is w, they produce a conductivity

$$\sigma = ne\mu = \frac{ne^2\tau}{m}$$

at frequencies ω such that $\omega\tau \ll 1$. If $\omega\tau \gtrsim 1$ the conductivity becomes $\sigma/(1+\mathrm{j}\omega\tau)$.

If the medium has permittivity ϵ, its propagation constant for a wave at frequency ω is given by

$$k^2 = \omega^2\mu\epsilon\left\{1 + \frac{\sigma}{\mathrm{j}\omega\epsilon(1+\mathrm{j}\omega\tau)}\right\}$$

in which μ now signifies the magnetic permeability. For $\omega\tau \gg 1$ this becomes

$$k^2 = \mu\epsilon(\omega^2 - \omega_p^2)$$

where ω_p is the *plasma frequency* given by

$$\omega_p^2 = \frac{\sigma}{\epsilon\tau} = \frac{ne^2}{m\epsilon}.$$

The wave is evanescent, and the medium reflecting, for $\omega < \omega_p$.

In the presence of an applied magnetic field **B** the propagation constant is given by

$$k^2 = \omega^2\mu\epsilon\left[1 + \frac{\sigma}{\mathrm{j}\omega\epsilon\{1+\mathrm{j}(\omega \pm \omega_c)\tau\}}\right]$$

where ω_c is the *cyclotron frequency* eB/m.

Free electrons confined by an infinite potential barrier have density of energy states per unit volume in the energy interval E to $E + \mathrm{d}E$ given by

$$Z(E)\mathrm{d}E = 4\pi\left(\frac{2m}{h^2}\right)^{3/2} E^{1/2}\mathrm{d}E$$

in which h is Planck's constant. The probability of an electron occupying an available state of energy E is given by the Fermi–Dirac function

$$F(E) = \{\mathrm{e}^{(E-E_F)/k_BT} + 1\}^{-1}.$$

In this T is absolute temperature and E_F the Fermi energy, defined in general as that value of E for which $F = \frac{1}{2}$ (or, if $T = 0$, above which $F = 0$). For free electrons the Fermi energy is

$$E_F = \frac{h^2}{m}\left(\frac{3n}{8\pi}\right)^{2/3}.$$

In the classical (Maxwell–Boltzmann) approximation the probability of a particle having energy E is proportional to the Boltzmann factor $\exp(-E/k_BT)$.

Thermionic emission

The Richardson–Dushman equation for the current density due to thermionic emission of electrons from a surface of work function ϕ at absolute temperature T is

$$J = AT^2\mathrm{e}^{-\phi/k_BT}$$

in which A is a constant of theoretical value $4\pi emk_B^2/h^3$, or $1.2\ \mathrm{M\,A\,m^{-2}\,K^{-2}}$.

Semiconductors

In the following m^* denotes effective mass, μ mobility and D diffusion coefficient; subscripts e and h denote electrons and holes respectively.

If the $E - k$ relation is assumed to be independent of direction, that is if the surfaces of constant energy are spherical in k-space, then the effective mass of an electron within an energy band is the scalar quantity

$$m_e^* = \frac{h^2}{4\pi^2\partial^2E/\partial k^2}$$

and the same expression gives the effective mass of a hole m_h^*. The density of states for electrons at the bottom of the conduction band is then

$$Z_e(E) = 4\pi\left(\frac{2m_e^*}{h^2}\right)^{3/2}(E - E_g)^{1/2}$$

and that for holes at the top of the valence band is

$$Z_h(E) = 4\pi\left(\frac{2m_h^*}{h^2}\right)^{3/2}(-E)^{1/2};$$

in these E is measured from the top of the valence band and E_g is the energy gap. The approximate effective number density of electrons in the conduction band is

$$n = 2\left(\frac{2\pi m_e^* k_BT}{h^2}\right)^{3/2}\mathrm{e}^{-(E_g-E_F)/k_BT}$$

and the corresponding density of holes in the valence band is

$$p = 2\left(\frac{2\pi m_h^* k_BT}{h^2}\right)^{3/2}\mathrm{e}^{-E_F/k_BT}.$$

An intrinsic semiconductor has $n = p$ and its Fermi energy is

$$E_F = \tfrac{1}{2}E_g + \tfrac{3}{4}k_B T \ln \frac{m_h^*}{m_e^*}.$$

An extrinsic semiconductor has

$$np = n_i^2$$

for any impurity level, where n_i is the value of n and p in the intrinsic material at the same temperature.

The conductivity of a semiconductor is the sum of electron and hole contributions, that is

$$\sigma = ne\mu_e + pe\mu_h.$$

In a region where n and p vary with distance x a diffusion current is superimposed on that due to conductivity. The total current density in an electric field E is then

$$J = J_e + J_h = (ne\mu_e + pe\mu_h)E + eD_e \frac{\partial n}{\partial x} - eD_h \frac{\partial p}{\partial x}.$$

The diffusion coefficient for a charge carrier is related to its mobility by the *Einstein relation*

$$\frac{D}{\mu} = \frac{k_B T}{e}.$$

The *continuity equation* for electrons is

$$\frac{\partial n}{\partial t} = -r_e + \frac{\nabla \cdot \mathbf{J}_e}{e}$$

where r_e is the net loss rate per unit volume by recombination; for holes

$$\frac{\partial p}{\partial t} = -r_h - \frac{\nabla \cdot \mathbf{J}_h}{e}.$$

The net recombination rates can be expressed as

$$r_e = \frac{\Delta n}{\tau_e}; \qquad r_h = \frac{\Delta p}{\tau_p}$$

where τ_e, τ_h are the lifetimes and $\Delta n, \Delta p$ the excess populations, over the equilibrium values, of electrons and holes respectively.

The *diffusion length* for a species with lifetime τ is

$$L = \sqrt{D\tau}$$

The *p–n* junction

The *rectifier equation* for the forward current across a junction with forward bias voltage V is

$$I = I_0(e^{eV/k_B T} - 1)$$

where I_0 is the reverse saturation current.

The contact potential of an abrupt junction has the magnitude

$$V_0 = \frac{k_B T}{e} \ln \frac{N_A N_D}{n_i^2}$$

where N_A, N_D are the densities of ionized acceptor and donor atoms on the p and n sides respectively. If the densities of mobile carriers are assumed to be negligible, and N_A, N_D uniform, for distances x_p and x_n on the respective sides, then the width of the depletion layer for a reverse bias voltage V is

$$w = x_p + x_n = \left\{\frac{2\epsilon(V_0 + V)}{e(N_A + N_D)}\right\}^{1/2} \left\{\sqrt{\frac{N_D}{N_A}} + \sqrt{\frac{N_A}{N_D}}\right\}$$

and the junction capacitance per unit area is

$$C = \left\{\frac{e\epsilon N_A N_D}{2(V_0 + V)(N_A + N_D)}\right\}^{1/2}$$

The Hall effect

The *Hall coefficient* R_H of a material is the ratio of the transverse electric field induced to the product JB, where J is the current density flowing through a magnetic flux density B perpendicular to it. When the charge carriers are electrons of density n

$$R_H = \frac{1}{ne}$$

and if holes of density p

$$R_H = -\frac{1}{pe},$$

the direction of $\mathbf{J} \times \mathbf{B}$ being taken as positive. When both electrons and holes are present, with mobilities μ_e and μ_h,

$$R_H = \frac{n\mu_e^2 - p\mu_h^2}{e(n\mu_e + p\mu_h)^2}.$$

Dielectrics

The *Clausius-Mossotti* equation for the polarizability of non-polar molecules of number density N in a dielectrically isotropic material is

$$\alpha = \frac{\epsilon_r - 1}{\epsilon_r + 2} \frac{3\epsilon_0}{N}$$

where ϵ_r is the dielectric constant of the material. The polarization, or total dipole moment per unit volume, in an applied field E is

$$P = (\epsilon_r - 1)\epsilon_0 E = \chi\epsilon_0 E$$

where χ is the *electrical susceptibility*.

Orientational polarization

In an applied field E molecules having a permanent dipole moment μ_d each contribute to the resultant total moment a fraction of μ_d given by the *Langevin function*

$$L(a) = \coth a - \frac{1}{a}$$

when a is $\mu_d E/k_B T$. For small a, $L \approx a/3$.

If the dipoles have a relaxation time τ for alignment in an a.c. field, then the effect of frequency ω on orientational polarization is given by the *Debye equations* for the dielectric constant ϵ_r. If ϵ_r is written $\epsilon' - j\epsilon''$, then these are:

$$\epsilon' = \epsilon_H + \frac{\epsilon_L - \epsilon_H}{1 + \omega^2\tau^2}$$

$$\epsilon'' = \frac{\omega\tau}{1 + \omega^2\tau^2}(\epsilon_L - \epsilon_H)$$

in which ϵ_L is the value of ϵ_r at frequencies for which $\omega\tau \ll 1$, and ϵ_H its value when $\omega\tau \gg 1$.

Resonant polarization

If bound charges resonate at an undamped frequency ω_0 with a damping ratio ζ, the effect of frequency ω on their induced polarization is given by the following equations for their contribution to ϵ_r:

$$\epsilon' = 1 + (\epsilon_L - 1)\left\{\frac{1 - r^2}{(1 - r^2)^2 + 4\zeta^2 r^2}\right\}$$

$$\epsilon'' = (\epsilon_L - 1)\left\{\frac{2\zeta r}{(1 - r^2)^2 + 4\zeta^2 r^2}\right\}$$

In these r is the frequency ratio ω/ω_0 and ϵ_L the value of ϵ' for $r \ll 1$.

Dielectric loss

In an electric field of r.m.s. value E the loss per unit volume due to conductivity σ is σE^2. In terms of a complex dielectric constant this becomes

$$\mathrm{w} = \omega\epsilon''\epsilon_0 E^2 = \omega\epsilon E^2 \tan\delta$$

where ϵ is the permittivity $\epsilon'\epsilon_0$ and $\tan\delta$, the *loss tangent* is ϵ''/ϵ'.

If $\sigma \ll \omega\epsilon$, or $\tan\delta \ll 1$, the attenuation of a wave is given by the attenuation constant

$$\alpha = \tfrac{1}{2}\sigma\sqrt{\frac{\mu}{\epsilon}}$$

due to conductivity σ, where μ is the permeability; or, if $\mu = \mu_0$,

$$\alpha = \frac{1}{2}\frac{\omega}{c}\sqrt{\epsilon'}\tan\delta = \frac{\omega k}{c}$$

where c is the speed of light in vacuum and k or $\tfrac{1}{2}\sqrt{\epsilon'}\tan\delta$ is the extinction coefficient (the imaginary part of the refractive index). The attenuation in decibels per kilometre is then

$$20\log_{10} e^{1000\alpha} = 8686\alpha = 1{\cdot}820 fk \times 10^{-4}$$

where f is the frequency in hertz.

Logic

Boolean algebra

$A.0 = 0$

$A.1 = A$

$A.A = A$

$A.\overline{A} = 0$

$A + 0 = A$

$A + 1 = 1$

$A + A = A$

$A + \overline{A} = 1$

$\overline{\overline{A}} = A$

$A + B = B + A$, $AB = BA$ (Commutative law)

$A(BC) = (AB)C$, $A + (B + C) = (A + B) + C$ (Associative law)

$A(B + C) = AB + AC$, $A + BC = (A + B)(A + C)$ (Distributive law)

$A + AB = A$, $A(A + B) = A$ (Absorption law)

$\overline{AB} = \overline{A} + \overline{B}$, $\overline{A + B} = \overline{A}.\overline{B}$ (De Morgan's theorems)

Alternative symbols

OR: $+$ $\vee$

AND: $\cdot$ $\wedge$

The symbol for exclusive OR is $\oplus$

Powers of two

Power of 2	*Decimal*	*Power of 2*	*Decimal*
2^0	1	2^8	256
2^1	2	2^9	512
2^2	4	2^{10}	1 024
2^3	8	2^{11}	2 048
2^4	16	2^{12}	4 096
2^5	32	2^{13}	8 192
2^6	64	2^{14}	16 384
2^7	128	2^{15}	32 768

Number systems

Decimal, hexadecimal, octal and binary number equivalents

Decimal	*Hex*	*Octal*	*Binary*	*Decimal*	*Hex*	*Octal*	*Binary*
0	0	0	0	8	8	10	1000
1	1	1	1	9	9	11	1001
2	2	2	10	10	A	12	1010
3	3	3	11	11	B	13	1011
4	4	4	100	12	C	14	1100
5	5	5	101	13	D	15	1101
6	6	6	110	14	E	16	1110
7	7	7	111	15	F	17	1111

Logic symbols

The symbols on the left conform to BS 3939: Section 21 (IEC 117–15). Those on the right are the earlier American Military Specification symbols, now ANSI Y32-14-1973.

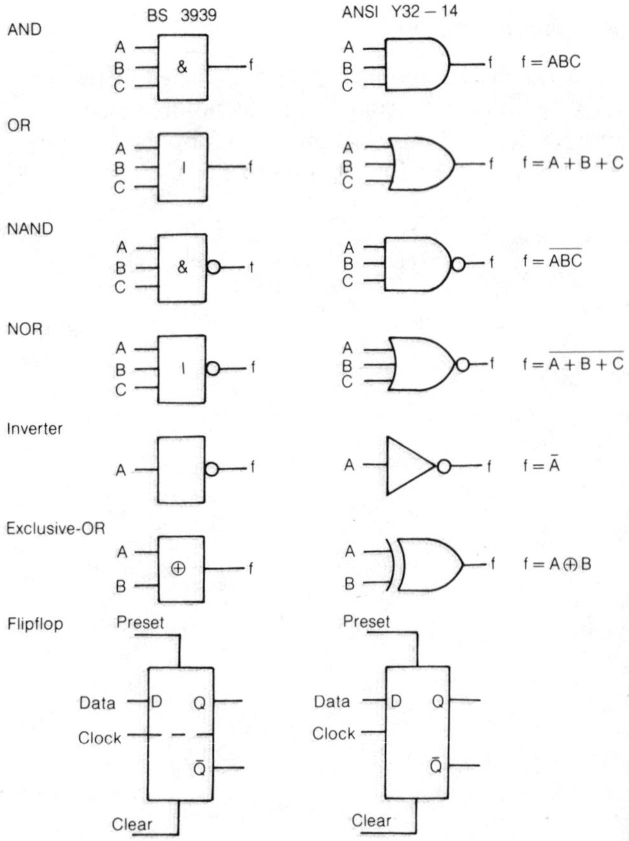

ASCII (American Standard Code for Information Interchange)

or ISO code, also known as CCITT Alphabet No 5

6	0	0	0	0	1	1	1	1
Bits 5	0	0	1	1	0	0	1	1
4	0	1	0	1	0	1	0	1
0 0 0 0	NULL	DLE	SPACE	0		P	@	p
0 0 0 1	SOH	DC1	!	1	A	Q	a	q
0 0 1 0	STX	DC2	”	2	B	R	b	r
0 0 1 1	ETX	DC3	#	3	C	S	c	s
0 1 0 0	EOT	DC4	$	4	D	T	d	t
0 1 0 1	ENQ	NACK	%	5	E	U	e	u
0 1 1 0	ACK	SYNC	&	6	F	V	f	v
0 1 1 1	BELL	ETB	’	7	G	W	g	w
1 0 0 0	BS	CNCL	(	8	H	X	h	x
1 0 0 1	HT	EM	)	9	I	Y	i	y
1 0 1 0	LF	SS	*	:	J	Z	j	z
1 0 1 1	VT	ESC	+	;	K	[	k	{
1 1 0 0	FF	FSR	,	<	L	£	l	|
1 1 0 1	CR	GSR	–	=	M	]	m	}
1 1 1 0	SO	RSR	.	>	N	∧	n	¬
1 1 1 1	SI	USR	/	?	O	—•—	o	DELETE
3 2 1 0 Bits								

An eighth bit is available for parity purposes.

Miscellaneous

Gauges for wire and sheet metal

The sizes given refer to the thickness of sheet or the diameter of wire.

I.S.O. (metric) sizes

The following sizes in millimetres form the R40 series; of these the first and alternate sizes are the R20 series, of which the first and alternate sizes are the R10 series. The order of preference is R10, R20, R40.

R10, R20	0·020	0·040	0·080	0·160	0·315	0·63	1·25	2·5	5·0	10·0	20·0
	0·021	0·042	0·085	0·170	0·335	0·67	1·32	2·65	5·3	10·6	21·2
R20	0·022	0·045	0·090	0·180	0·355	0·71	1·40	2·8	5·6	11·2	22·4
	0·024	0·048	0·095	0·190	0·375	0·75	1·50	3·0	6·0	11·8	23·6
R10, R20	0·025	0·050	0·100	0·200	0·40	0·80	1·60	3·15	6·3	12·5	25·0
	0·026	0·053	0·106	0·212	0·425	0·85	1·70	3·35	6·7	13·2	
R20	0·028	0·056	0·112	0·224	0·45	0·90	1·80	3·55	7·1	14·0	
	0·030	0·060	0·118	0·236	0·475	0·95	1·90	3·75	7·5	15·0	
R10, R20	0·032	0·063	0·125	0·250	0·50	1·00	2·00	4·0	8·0	16·0	
	0·034	0·067	0·132	0·265	0·53	1·06	2·12	4·25	8·5	17·0	
R20	0·036	0·071	0·140	0·28	0·56	1·12	2·24	4·5	9·0	18·0	
	0·038	0·075	0·150	0·30	0·60	1·18	2·36	4·75	9·5	19·0	

Standard wire gauge

The following gauge numbers correspond to the sizes given in thousandths of an inch; equivalent sizes in millimetres are also given to two or three significant figures. The gauge is now obsolescent.

SWG	in x 10^3	mm	SWG	in x 10^3	mm	SWG	in x 10^3	mm	SWG	in x 10^3	mm
7/0	500	12·7	9	144	3·66	24	22	0·559	39	5·2	0·132
6/0	464	11·8	10	128	3·25	25	20	0·508	40	4·8	0·122
5/0	432	11·0	11	116	2·95	26	18	0·457	41	4·4	0·112
4/0	400	10·2	12	104	2·64	27	16·4	0·417	42	4·0	0·102
3/0	372	9·45	13	92	2·34	28	14·8	0·376	43	3·6	0·091
00	348	8·84	14	80	2·03	29	13·6	0·345	44	3·2	0·081
0	324	8·23	15	72	1·83	30	12·4	0·315	45	2·8	0·071
1	300	7·62	16	64	1·63	31	11·6	0·295	46	2·4	0·061
2	276	7·01	17	56	1·42	32	10·8	0·274	47	2·0	0·051
3	252	6·40	18	48	1·22	33	10·0	0·254	48	1·6	0·041
4	232	5·89	19	40	1·02	34	9·2	0·234	49	1·2	0·030
5	212	5·38	20	36	0·914	35	8·4	0·213	50	1·0	0·025
6	192	4·88	21	32	0·813	36	7·6	0·193			
7	176	4·47	22	28	0·711	37	6·8	0·173			
8	160	4·06	23	24	0·610	38	6·0	0·152			

Standard screw threads

In the following, the diameter given is the nominal diameter which is the basic value of the major diameter (over the crests of an external thread).

I.S.O. metric

The table below gives only the first-choice threads in the coarse and fine series. (The constant-pitch series includes threads up to 300 mm diameter with pitches of 0·35, 0·5, 0·75, 1, 1·25, 1·5, 2, 3, 4 and 6 mm.) The dimensions are in millimetres. The theoretical thread depths are approximately 0·613 and 0·541 of the pitch for external and internal threads respectively.

	Pitch			Pitch	
Diameter	Coarse	Fine	Diameter	Coarse	Fine
1·6	0·35	–	16	2	1·5
2	0·4	–	20	2·5	1·5
2·5	0·45	0·35*	24	3	2
3	0·5	0·35*	30	3·5	2
4	0·7	0·5*	36	4	3
5	0·8	0·5*	42	4·5	3
6	1	0·75*	48	5	3
8	1·25	1	56	5·5	4
10	1·5	1·25	64	6	4
12	1·75	1·25	†		

* These threads from part of the constant-pitch series.
† Threads of diameters greater than 70 mm in the coarse and fine series are taken from the constant-pitch series, with pitches of 6 mm (coarse) or 4 mm (fine) and first-choice diameters 72, 80, 90, 100, 110, 125, 140, 160, 180, 200, 220, 250 and 280 mm.

Unified

The diameters are in inches. There is also an extra-fine series, and a constant-pitch series with threads up to 6 in diameter and pitches corresponding to 32, 28, 20, 16, 12, 8, 6 and 4 threads per inch. The threads are identical in form to the I.S.O. metric threads.

	Threads per inch			Threads per inch	
Diameter	Coarse	Fine	Diameter	Coarse	Fine
$\frac{1}{4}$	20	28	$1\frac{3}{8}$	6	12
$\frac{5}{16}$	18	24	$1\frac{1}{2}$	6	12
$\frac{3}{8}$	16	24	$1\frac{3}{4}$	5	
$\frac{7}{16}$	14	20			
$\frac{1}{2}$	13	20	2	$4\frac{1}{2}$	
			$2\frac{1}{4}$	$4\frac{1}{2}$	
$\frac{9}{16}$	12	18	$2\frac{1}{2}$	4	
$\frac{5}{8}$	11	18	$2\frac{3}{4}$	4	
$\frac{3}{4}$	10	16	3	4	
$\frac{7}{8}$	9	14			
1	8	12	$3\frac{1}{4}$	4	
			$3\frac{1}{2}$	4	
$1\frac{1}{8}$	7	12	$3\frac{3}{4}$	4	
$1\frac{1}{4}$	7	12	4	4	

Whitworth

The diameters are in inches; the two series are British Standard Whitworth (BSW) and British Standard Fine (BSF). Sizes which are not recommended are excluded here. The theoretical thread depth is approximately 0·640 of the pitch. Whitworth threads are now obsolescent.

	Threads per inch			Threads per inch	
Diameter	BSW	BSF	Diameter	BSW	BSF
$\frac{3}{16}$	24		2	$4\frac{1}{2}$	7
$\frac{1}{4}$	20	26	$2\frac{1}{4}$	4	6
$\frac{5}{16}$	18	22	$2\frac{1}{2}$	4	6
$\frac{3}{8}$	16	20	$2\frac{3}{4}$	$3\frac{1}{2}$	6
$\frac{7}{16}$	14	18	3	$3\frac{1}{2}$	5
$\frac{1}{2}$	12	16	$3\frac{1}{4}$		5
$\frac{9}{16}$		16	$3\frac{1}{2}$	$3\frac{1}{4}$	$4\frac{1}{2}$
$\frac{5}{8}$	11	14	$3\frac{3}{4}$		$4\frac{1}{2}$
$\frac{3}{4}$	10	12	4	3	$4\frac{1}{2}$
$\frac{7}{8}$	9	11	$4\frac{1}{4}$		4
1	8	10	$4\frac{1}{2}$	$2\frac{7}{8}$	4
$1\frac{1}{8}$	7	9	5	$2\frac{3}{4}$	4
$1\frac{1}{4}$	7	9	$5\frac{1}{2}$	$2\frac{5}{8}$	4
$1\frac{1}{2}$	6	8	6	$2\frac{1}{2}$	4
$1\frac{3}{4}$	5	7			

British Association (B.A.)

The dimensions are in millimetres. No. 0 B.A. is not recommended and only even numbers are first-choice. The theoretical thread depth is 0·6 of the pitch. B.A. threads are now obsolescent.

No.	Diameter	Pitch	No.	Diameter	Pitch
0	6·0	1·00	13	1·2	0·25
1	5·3	0·90	14	1·0	0·23
2	4·7	0·81	15	0·90	0·21
3	4·1	0·73	16	0·79	0·19
4	3·6	0·66	17	0·70	0·17
5	3·2	0·59	18	0·62	0·15
6	2·8	0·53	19	0·54	0·14
7	2·5	0·48	20	0·48	0·12
8	2·2	0·43	21	0·42	0·11
9	1·9	0·39	22	0·37	0·10
10	1·7	0·35	23	0·33	0·09
11	1·5	0·31	24	0·29	0·08
12	1·3	0·28	25	0·25	0·07

References

The following is a selection of sources from which further information can be found.

Mathematics

1. Abramowitz, M. and Stegun, L. A. *Handbook of Mathematical Functions*, 3rd edn New York, Dover, 1965.
2. British Association, continued by Royal Society, *Mathematical Tables*. Cambridge, C.U.P., various
3. Bronshtein, I. N. and Semendyayev, K. A. *A Guide Book to Mathematics*. New York, Springer Verlag, 1973.

Materials

4. Brandes, E. A. (ed.) *Smithells' Metals Reference Book*, 6th edn London, Butterworth, 1983.
5. *Metals Handbook*, 9th edn Metals Park, Ohio, ASM International, 1978–87.
6. British Steel, p.l.c. *Iron and Steel Specifications*, 7th edn 1989.
7. Brandrup, J. and Immergut, E. H. (eds) *Polymer Handbook*, 3rd edn New York, Wiley, 1989.
8. Bever, M. B. (ed.) *Encyclopedia of Materials Science and Engineering*. Oxford, Pergamon Press, 1986.
9. Tottle, C. R. (ed.) *Encylopaedia of Metallurgy and Materials*. Plymouth, Macdonald and Evans, 1984.
10. Willardson, R. K. and Beer, A. C. (eds) *Semiconductors and Semimetals*. New York, Academic Press, 1966–88.
11. Touloukian, Y. S. *et al. Thermophysical Properties of Matter*. New York, Plenum, 1970–79.

Thermodynamic tables

12. *U.K. Steam Tables*. London, Arnold, 1970.
13. Haar, L. *et al. NBS–NRC Steam Tables*. London, Hemisphere, 1984.
14. Irvine, T. F. and Liley, P. E. *Steam and Gas Tables*. London, Academic Press, 1984.
15. Ražnjević, K. *Handbook of Thermodynamic Tables and Charts*. Washington, Hemisphere, 1976.
16. Predvoditelev, A. S. *et al. Tables of Thermodynamic Functions of Air*. Trans. Assoc. Tech. Services, New Jersey, 1962.
17. Stewart, R. B. *et al. Thermodynamic Properties of Refrigerants*. Atlanta, Amer. Soc. of Heating, Refrigerating and Air-conditioning Engineers, 1986.
18. Watson, J. T. R. (ed.) *Viscosity of Gases*. Edinburgh, HMSO, 1972.

Civil and mechanical engineering

19. Young, W. C. *Roark's Formulas for Stress and Strain*, 6th edn New York, McGraw-Hill, 1989.
20. Blake, L. S. (ed.) *Civil Engineer's Reference Book*, 4th edn London, Butterworth, 1989.
21. Merritt, F. S. (ed.) *Standard Handbook for Civil Engineers*, 3rd edn New York, McGraw-Hill, 1983.
22. Parrish, A. (ed.) *Mechanical Engineer's Reference Book*, 11th edn London, Butterworth, 1973.
23. *Handbook on Structural Steelwork*. London, British Constructional Steelwork Association and others, 1990.
24. Kutz, M. (ed.) *Mechanical Engineers' Handbook*. New York, Wiley, 1989.

Electrical and control engineering

25. Mazda, F. F. (ed.) *Electronic Engineer's Reference Book*, 6th edn London, Butterworth, 1989.
26. Fink, D. G. and Christiansen, D. (eds) *Electronic Engineers' Handbook*, 3rd edn New York, McGraw-Hill, 1989.
27. Grayson, M. (ed.) *Encyclopedia of Semiconductor Technology*. New York, Wiley, 1984.
28. Markus, J. and Weston, C. *Essential Circuits Reference Guide*. New York, McGraw-Hill, 1988.
29. Laughton, M. A. and Say, M. G. (eds) *Electrical Engineer's Reference Book*, 14th edn London, Butterworth, 1985.
30. Fink, D. G. and Beaty, H. W. (eds) *Standard Handbook for Electrical Engineers*, 12th edn New York, McGraw-Hill, 1987.
31. Freeman, R. L. (ed.) *Reference Manual for Telecommunications Engineering*. New York, Wiley, 1985.
32. Singh, M. G. (ed.) *Systems and Control Encyclopedia*. Oxford, Pergamon Press, 1987.

General and miscellaneous

33. Kaye, G. W. C. and Laby, T. H. *Physical and Chemical Constants*, 15th edn London, Longman, 1986.
34. *Handbook of Chemistry and Physics*. Baco Raton, Florida, CRC Press, yearly.
35. *Kempe's Engineers Year Book*. London, Morgan-Grampian, yearly.
36. *British Standards Catalogue*. London, BSI, yearly.

Index